Case Studies in Star Formation

Case Studies in Star Formation offers an overview of our current observational and theoretical understanding in the molecular astronomy of star formation. The book is divided into six sections: the first introduces an overview of star formation and the essential language, concepts, and tools specific to molecular astronomy studies. Each subsequent section focuses on individual sources, beginning with a description of large-scale surveys. The volume covers low- and high-mass star formation, ionisation and photodissociation regions, and concludes with the extragalactic perspective. Conventional textbooks begin with principles, ending with a few convenient examples. Through copious examples, *Case Studies* reflects the reality of research, which requires the creative matching of ongoing observations to theory and vice versa, often raising as many questions as answers. This supplementary study guide enables graduate students and early career researchers to bridge the gap between textbooks and the wealth of research literature.

DUNCAN MACKAY is an honorary Senior Research Fellow at the University of Kent, Canterbury. He has four decades of teaching and research experience in astrophysics and pedagogical practice and has published and lectured professionally on cross-disciplinary issues for many years.

MARK THOMPSON is head of the School of Physics & Astronomy at the University of Leeds. He is an expert observational astronomer in the far-infrared to radio wavelength regime with more than 20 years of experience in observing Galactic star formation and international survey projects.

JAMES URQUHART is a Lecturer in Physics and Astrophysics and head of the Centre of Astronomy and Planetary Sciences at the University of Kent, Canterbury. He has contributed to almost 200 scientific publications in the areas of star formation and Galactic structure.

CAMBRIDGE OBSERVING HANDBOOKS FOR RESEARCH ASTRONOMERS

Today's professional astronomers must be able to adapt to use telescopes and interpret data at all wavelengths. This series is designed to provide them with a collection of concise, self-contained handbooks, which cover the basic principles peculiar to observing in a particular spectral region, or to using a special technique or type of instrument. The books can be used as an introduction to the subject and as a handy reference for use at the telescope or in the office.

Books currently available in this series:

1. *Handbook of Infrared Astronomy*
   I. S. Glass
4. *Handbook of Pulsar Astronomy*
   D. R. Lorimer and M. Kramer
5. *Handbook of CCD Astronomy*, Second Edition
   Steve B. Howell
6. *Introduction to Astronomical Photometry*, Second Edition
   Edwin Budding and Osman Demircan
7. *Handbook of X-ray Astronomy*
   Edited by Keith Arnaud, Randall Smith, and Aneta Siemiginowska
8. *Practical Statistics for Astronomers*, Second Edition
   J. V. Wall and C. R. Jenkins
9. *Introduction to Astronomical Spectroscopy*
   Immo Appenzeller
10. *Observational Molecular Astronomy*
   David A. Williams and Serena Viti
11. *Practical Optical Interferometry*
   David F. Buscher

Case Studies in Star Formation

A Molecular Astronomy Perspective

DUNCAN MACKAY
University of Kent, Canterbury

MARK THOMPSON
University of Leeds

JAMES URQUHART
University of Kent, Canterbury

Shaftesbury Road, Cambridge CB2 8EA, United Kingdom

One Liberty Plaza, 20th Floor, New York, NY 10006, USA

477 Williamstown Road, Port Melbourne, VIC 3207, Australia

314–321, 3rd Floor, Plot 3, Splendor Forum, Jasola District Centre, New Delhi – 110025, India

103 Penang Road, #05–06/07, Visioncrest Commercial, Singapore 238467

Cambridge University Press is part of Cambridge University Press & Assessment, a department of the University of Cambridge.

We share the University's mission to contribute to society through the pursuit of education, learning and research at the highest international levels of excellence.

www.cambridge.org
Information on this title: www.cambridge.org/9781009277440
DOI: 10.1017/9781009277433

First published 2023

Printed in the United Kingdom by TJ Books Limited, Padstow Cornwall

A catalogue record for this publication is available from the British Library.

ISBN 978-1-009-27744-0 Hardback

in memory
Geoffrey Hill Macdonald
1943–2020

Contents

Preface

Science is not reliable because it provides certainty. It is reliable because it provides us with the best answers we have at present.

– Carlo Rovelli†

To place *Case Studies* in its pedagogical context the reader could do worse than look to a statement by João Alves, on his appointment in October 2020 as the new editor of *Star Formation News* (*SFN*), a broadsheet that has served the astronomy community for over 30 years (www.starformation.news). Alves wrote:

An essential goal of the SFN is to welcome young researchers into the field, help them navigate the community, and inspire them to make their mark. A paradox in today's incredibly easy access to large amounts of information is PhD students' hyper-specialization. In part, this is a structural problem arising from what is expected from young researchers to succeed. The SFN web will minimize this drawback and avoid dividing the SFN into sub-fields to expose the reader to a broader view, following Bo [Reipurth]'s original design. Given the current information growth, this will be a difficult path to tread, but the spirit will be kept. (SFN #334, 2020)

In the same spirit, for graduate students and early researchers, *Case Studies* introduces an overview of our current understanding of star formation from a molecular astronomy perspective, making no initial assumptions beyond that which we might expect of an undergraduate knowledge base in the more familiar facets of astronomy, physics, and chemistry. The Introduction identifies the key stages in the formation process of both low- and high-mass stars, linearly tracked from their diffuse interstellar raw material origins through to protostellar nuclear ignition and the early impact of that on the progenitor molecular envelope. This chapter also introduces some of the essential language and ideas specific to molecular astrophysics and astrochemistry, plus several observational and theoretical tools to be applied in subsequent chapters. The Introduction concludes by referencing some standard textbooks that offer greater detail on the subjects raised.

For the bulk of the book, each of its five parts begins with examples of recent large-scale surveys, before focusing on individual representative sources. Of the five, the first focuses on low-mass star formation, the second on high-mass star formation, the third on ionised regions and their interactions with hot molecular cores, the fourth on photodissociation regions, and the fifth takes us to star formation in galaxies beyond our own. Each chapter concludes with a brief summary of contents and links to the ongoing pedagogical purpose. Since *Case Studies* is targeted principally at postgraduate students and early researchers, unlike a standard textbook our motivation is to reflect some of the research realities in which a creative pursuit of understanding seeks to match unfolding observations to the theory that underpins them. This contrasts with the conventional teaching route more familiar at the undergraduate level in which we would begin with first principles and end with convenient examples.

Real life is not necessarily convenient, and typically questions more often than answers arise at the postgraduate level. In offering more detailed exemplars than are customarily found in textbooks, *Case Studies* presents a sufficient descriptive introduction to each individual source to give the uninitiated reader some sense of the wider Galactic and extragalactic environment. In addition, copious references are offered to the research literature readily available through the online Astrophysical Data System (ADS) operated by the Smithsonian Astrophysical Observatory (SAO) under a grant from NASA, known as 'adsabs' (https://ui.adsabs.harvard.edu). At various points throughout the text, the principal observational techniques of radio, millimetre, and submillimetre astronomy are placed in their practical research context, as are references to the major single-dish and interferometric instruments – past, present, and those approaching commission.

We hope the text may also be as useful to the teacher as to the student of a taught course in bringing together much disparate information, as well as to the self-studies of early researchers. Molecular astronomy specifically informs our understanding of the stages of star formation through the chemical tracers of physical conditions and dynamics. Both large-scale surveys and individual source studies reinforce or undermine what it is we think we know about the particular molecular species that we are using as an astronomical probe. This close interchange between observation, theoretical modelling, and laboratory experiment is the route through which molecular astronomers are continually contributing to the development of knowledge of star formation both within our own Galaxy and increasingly to that occurring far beyond.

† Carlo Rovelli, *Reality Is Not What It Seems: The Journey to Quantum Gravity*, trans. Simon Carnell and Erica Segre. London: Penguin Books (2017).

Acronyms

ALMA	Atacama Large Millimetre Array
APEX	Atacama Pathfinder Experiment
ATLASGAL	APEX Telescope Large Area Survey of the Galaxy
BIMA	Berkeley Illinois Maryland Association
CSO	Caltech Submillimetre Observatory
ESA	European Space Agency
ESO	European Space Observatory
FUSE	Far Ultraviolet Spectroscopic Explorer
GLIMPSE	Galactic Legacy Infrared Mid-Plane Survey
IRAC	Infrared Array Camera
IRAM	Institut de Radioastronomie Millimetrique
IRAS	Infrared Astronomical Satellite
JCMT	James Clerk Maxwell Telescope
JVLA	Jansky Very Large Array
JWST	James Webb Space Telescope
e-MERLIN	enhanced Multi-Element Remote-Linked Interferometer Network
MIPS	Multiband Imaging Photometer
NASA	National Aeronautics and Space Administration
NOEMA	Northern Extended Millimetre Array
NRAO	National Radio Astronomy Observatory
PdBI	Plateau de Bure Interferometer
PILS	Protostellar Interferometric Line Survey
SCUBA	Submillimetre Common User Bolometer Array
SMA	Submillimetre Array
VLA	Very Large Array
WISE	Wide Field Infrared Explorer

PART I

Introduction

1
An Overview of Star Formation

1.1 Introduction

Within our current understanding of star formation as observed through the lens of molecular astronomy, we can identify the following as the most significant areas of contemporary research: prestellar cores, hot cores and hot corinos, accretion and protoplanetary disks, photodissociation regions, stellar jets, disk winds, outflows, and masers. Of course, these sit within the wider considerations of dense molecular clouds on many scales, from the giant molecular clouds (GMCs) to fragments, filaments, and clumps. In turn, our understanding of molecular clouds depends on our understanding of molecular excitation, energy balance, gas and grain surface reaction kinetics, cosmic ray ionisation, and photochemistries. Some of these facets are more appropriate to low-mass star formation regions (LMSFRs), some to their high-mass equivalents (HMSFRs). While theoretical understanding of low-mass star formation (up to 2 $M_{\odot}$.) is currently quite well constrained, that of high-mass stars (>10 $M_{\odot}$) is only now emerging with the help of the latest high-resolution telescopes that offer acutely focused observations as well as deep field surveys of multiple sources at great distances.

All the steps along the way to star formation sit in an evolutionary schema that starts in molecular clouds concentrated particularly in the spiral arms of the disks of galaxies such as our own. However, rates of star formation even in the Milky Way differ in different regions. For example, rates are slower in the Taurus clouds, much more rapid in Orion or W3 or W49. Observations show even more vigorous star formation rates in some external galaxies, such as the 30 Doradus region in our close neighbour the Large Magellanic Cloud (LMC), and in more distant 'starburst galaxies' such as NGC 253 or Arp 220.

With few exceptions stars form in clusters, in anything from ten to a million or more in relatively close proximity. A cluster of high-mass stars (an OB association) may occupy a volume of space defined by just a few tens of parsecs early on in its main sequence life, but it will progressively spread to cover a region likely to be hundreds of parsecs wide. OB stars burn out in a few tens of million years, having driven apart their clustered near neighbours through radiation pressure, mechanical turbulence, gas dispersion, and rapid reductions in gravitational stability. They also scatter many lower-mass stars whose formation they will have triggered, although these smaller stars linger to evolve much more slowly. We see such distributions in Orion, Upper Scorpius , and Upper Centaurus-Lupus. These local clusters that are still embedded in their parent molecular cloud cores are obviously at an early stage of evolution, shrouded in dust and often observable only at infrared wavelengths. We will look at surveys of such clusters, as well as the examples of Sgr B2, W43, and Orion BN/KL in the High-Mass Star Formation (HMSF) section. Cluster subgroups such as Orion 1a, 1b, and 1c are clear evidence of the recursive nature of sequential, self-propagating star formation in which the impacts of high-mass stars, expanding HII regions, stellar winds, and occasional supernova explosions lead to widespread gas compression and the triggering of fresh star formation. The Trapezium stars are another familiar example in the visible, along with the Pleiades, which mark the current end of a subgroup sequence of formation and destruction.

At its simplest, the evolutionary sequence for low-mass, solar-type stars proceeds from the diffuse interstellar medium (ISM) to dense cloud to collapsing prestellar core to protostellar envelope and accretion disk, before nuclear ignition and a main sequence lifetime ending in a planetary nebula re-seeding the ISM. In contrast, recent statistical analysis of multiple high-mass star formation regions shows quite clearly how the absence of low- and intermediate-mass stars associated with hot massive proto-cluster cores strongly suggests that high-mass star formation kick-starts the formation of lower mass stars which emerge subsequently. It seems likely that high-mass clusters are triggering the collapse of cold dense prestellar cores, which themselves fragment and engender the binary and multiple stars we observe in two-thirds of solar mass main sequence systems. Let us add a little more detail to that overview, starting for simplicity, if not stellar formation logic, with the low-mass case.

Low-Mass Star Formation (LMSF)

1.2 Diffuse Clouds

The average particle density in the interstellar medium (ISM) is about 1 cm^{-3}. Where the diffuse ISM is in temperature and pressure equilibrium with its surroundings, this neutral hydrogen (HI) gas forms diffuse clouds under gravity in which gas particle densities are low (a few tens per cubic centimetre) and temperatures moderate (80–100 K). We can also extend the diffuse cloud classification from the purely atomic to the diffuse, and the translucent, depending on overall column densities, molecular hydrogen to atomic hydrogen ratios, or even some CO (carbon monoxide) abundance. Size and scale vary considerably, but typically these clouds are ~0.5 pc in diameter and ~3 $M_\odot$ in total. They are optically thin to the interstellar radiation field, with column densities correlating with visual extinction A_v ~0.05 magnitude, and it is the penetration of Far Ultraviolet (FUV) radiation that determines their chemistry. The tenuous molecular composition of diffuse clouds is studied at submillimetre wavelengths through absorption (typically ground state lines) against strong background sources, often extragalactic. Even the highest column density clouds show principally diatomic molecules, and just a few triatomics, both neutral and ionised, dominated by the elements of hydrogen, carbon, and oxygen. The precise abundance of the overwhelmingly most common molecule, H_2, is determined by the balance between its accretion, diffusion, and reaction rates on the surfaces of dust grains in competition with the rate of impinging photo-dissociating UV. Given the effectiveness of photodesorption processes, dust surfaces in diffuse clouds are regarded as essentially clean of molecular species for most of the cloud's diffuse lifetime.

1.3 Molecular Clouds

About 80 per cent of the molecular hydrogen in our Galaxy exists, however, in the much denser conditions of GMCs, each with masses commonly over 10^6 $M_\odot$. These enormous self-gravitating reservoirs of molecular gas and dust are where stars in the Milky Way exclusively form, although at a gas conversion efficiency rate of only ~5 per cent due to the multiple complex internal shock, turbulence, and radiation dynamics arising from star formation (particularly HMSF) that disrupt the localised continuity of gravitational collapse and gas coalescence. In classifying what we will find in our case studies to be far from homogenous molecular cloud conditions, we can generalise to the extent of noting that GMC column densities do average over 10^{22} H-nuclei cm^{-2} and show visual extinctions up to A_v ~8 magnitude. Remember, these are broad averages, and we will be identifying much higher-density locations within GMCs in some of the case studies that follow.

As in the diffuse cloud case, the thermal balance within a GMC determines its stability, and we will look at this in a little more detail in Chapter 6. Photoelectric effects dominate in the outer regions, while cosmic ray heating and local turbulence are most significant in the inner regions . However, dense molecular clouds are typically cold (~10 K). This is because the gas is predominantly molecular and, with the exception of H_2 (which has no dipole moment), molecules typically have many available rotational levels through which collisional energies are radiated. The kinetic energy of molecular gas is therefore efficiently reduced in comparison with an atom-dominated cloud. Balancing abundance against cooling efficiency, ^{12}CO offers the greatest cooling contribution in lower-density molecular conditions ($<10^3$ cm^{-3}). At higher densities, the ^{12}CO transitions become optically thick and the cooling efficiency of the less common isotopologue, ^{13}CO, can compete, as can that of certain molecular ions and neutral hydrides, including H_2O.

Where FUV photons penetrate molecular clouds, they ionise, dissociate, and heat the gas. Atomic H ionising photons are absorbed in a thin transition zone (column density ~10^{19} cm^{-2}) in which almost fully ionised atomic gas becomes almost fully neutral. FUV photons with energies <13.6 eV dissociate H_2 but ionise C, so an HI/CII region forms the next 'layer'. In dense photodissociation regions (PDRs), dust at visual extinction A_v ~4 magnitudes marks the H_2 front, and beyond that come C/CO zones. This is a simplification of reality, as the case studies in Chapters 14 and 15 will show. However, where dust grains within GMCs are protected from externally impinging FUV, here the gas and dust temperatures drop, and gas-phase chemistry is now driven, as we noted, by cosmic ray ionisation that initiates complex ion–molecule reaction networks. Within such a dark cloud, the dust-grain surfaces become the sites of migration and reaction between atoms and radicals, with gas–grain exchange through accretion and desorption. With progressive cooling during cloud collapse, ice layers accumulate as storehouses of the gas–grain molecular products, many of which are saturated and/or complex, only to be released if and when the gas and dust are subsequently warmed by star formation activity.

1.4 Dense Prestellar Cores

Against the gravitational forces within a cooling dense molecular cloud, turbulent motions from a variety of processes both disrupt and amplify existing inhomogeneities in gas and dust distribution, resulting in the formation of extra-dense filaments. The densest of these (on a scale ~0.1 pc) may collapse under their own gravity, forming a central nascent protostar with an accretion disk (radius ~ ten thousand AU). The disk actually expands over time, with magnetic braking working against the conservation of angular momentum while continuing accretion onto the central protostar is enabled through an energy balance maintained by jets and outflows (disk winds) emerging from the central object perpendicular to the disk plane.

Prior to any appearance of a nascent protostar within a central density peak, the initial dense , cold core has a typical gas-phase particle density of 10^6–10^7 cm^{-3} and temperature ~10 K. In Chapter 3 we will meet the many cores of the Ophiuchus molecular cloud with characteristic sizes of several thousand AU. At this point, on a scale at which the current generation of interferometers has really come into its own, we might ask what is it that is so valuable that is gleaned from the higher resolution of the molecular composition of accretion disks and their associated winds? After all, could any of these molecular species survive the disruption, dispersal, and material aggregations that follow stellar ignition? We must look to the outer reaches of protoplanetary disks and later in their evolutionary development for circumstances favourable to molecular formation chemistries that might produce the molecular species we currently observe in the fleeting passage of outer Solar system objects close to Earth. However, in the early stages of star formation, in trying to understand the multiple steps associated with that formation process, molecular emission tells us much about the physical conditions of each stage, all of which will become clearer through the case study examples to follow.

Along with the many smaller molecular species observed in the gas phase of cold, dense prestellar cores, there are complex carbon molecules offering direct evidence for active microscopic dust grain surface chemistries. Also evident is deuterium fractionation , and these observations collectively give distinctive clues as to core conditions of temperature, density, and radiation flux. Ices are also widely observed in absorption towards many reddened background stars, having accumulated on the surface of dust grains. H_2O ices are known to begin accumulating at densities ~10^3 cm^{-3} and temperatures ~15 K. Other ices, such as CH_4, NH_3, and some CO_2, also start to freeze out under these conditions with rapid increases as the core collapses and particle densities increase. Alternative routes for these simple saturated molecules to form in the gas in the quantities observed are woefully inefficient. Rather, the molecules are undoubtedly forming on grain surfaces through elemental reactions, including ubiquitous successive hydrogenation.

By the time the collapsing core reaches a gas-phase density ~10^5 cm^{-3}, the dominant volatile carbon species, CO, is undergoing rapid depletion onto the grains. This abrupt large-scale freeze-out of CO produces a distinct H_2O-poor (apolar) ice phase which has been observed directly as well as re-created in laboratory analogues. This CO-rich ice reacts with, among other species, atomic H to form H_2CO and CH_3OH – molecules we shall meet repeatedly in the case studies to follow, along with even more carbon-based complex organic molecules (COMs). We can reiterate that these species are certainly not made efficiently through collisional gas-phase reactions alone. However, this is cold carbon-chain chemistry, not to be confused with a warm carbon-chain chemistry (WCCC) that we will also find associated with low-mass protostars at a later stage of their evolution. For theorists and modellers, a major puzzle of recent years has been by what process or processes the COMs observed in the gas phase are actually desorbed from grain ices. The dust temperatures are deemed too low for thermal desorption, yet non-thermal processes such as photodesorption, cosmic-ray-induced spot heating, or localised exothermic reaction desorption give unsatisfactory results in any but species-specific cases. This is where current laboratory research into ice analogues will prove of crucial interest.

To pick up on the deuterium fractionation evidence, it is characteristic of cold prestellar cores that they have a high abundance of deuterated molecules such as DCO^+, DCN, and HDCO, with ratios to their undeuterated counterparts at least three orders of magnitude higher than the overall [D]/[H] ratio (which is $\sim 2 \times 10^{-5}$ in the ISM). Even doubly and triply deuterated species have been observed. The reasons for their abundance are that their formation reactions are exothermic and therefore efficient in such cold conditions, while the dominant destruction pathway for H_3^+ and H_2D^+, the principal initiators of fractionation, involves CO which by this stage, as we said, is quickly freezing out. Chapter 3 will reconsider deuteration in more detail. Chemical modelling suggests a lifetime for the cold, dense prestellar freeze-out phase of about 10^5 years before we next identify distinctive protostellar characteristics. For those unfamiliar with the many molecular species introduced throughout the following chapters, Appendix A directs the reader to some conveniently tabulated data.

1.5 Cold Protostellar Envelopes

Prestellar cores emit at infrared and longer wavelengths. Collapse of the core progresses, with the gravitational free-fall time extended perhaps by ambipolar diffusion and turbulence from within, irrespective of possible external influences such as shocks from neighbouring stellar activity. Once a central hydrostatic object coalesces within the core, we can say we have a protostar, and it will grow with accretion of surrounding envelope material from the engendered circumstellar disk. We will see in the LMSF chapters that protostellar envelopes are detected through millimetre-wave dust continuum emission, luminosities being roughly proportional to the total mass of dust in the envelope. Centimetre-wave radio continuum detections arise from accretion shocks at the protostellar surface, this luminosity being proportional to the total luminosity of the star, and envelope and protostellar masses can then be deduced from these luminosities. High ratios of millimetre-wave luminosity to total luminosity also show just how much the envelope mass may exceed that of the central protostar, through an accretion phase we expect to last up to 10^5 years. Infall motions themselves can be traced through molecular and molecular ion rotational lines, and both red and blue shifts are observed along the line of sight resulting from the large-scale rotation of the envelope gas around the protostar.

Within an opaque protostellar envelope, very close to a low-mass protostar, small (<100 AU), warm, dense regions have been discovered in recent years, designated 'hot corinos', the prototypical example being that associated with IRAS 16293-2422 which we will examine in Chapter 3. This gas is rich in complex organic molecules (COMs) – those with at least six atoms – with line profiles suggesting the emission arises from the inner surface of the accretion disk. However, while the disks themselves can have lifetimes of several million years, hot corino observations seem limited to the earliest stages of accretion disk evolution, and it seems likely that the complex organics are quite quickly degraded to CO in the warm gas as the protostar evolves.

1.6 Jets and Disk Winds

Along with the accretion disk itself, protostellar systems typically show outflow activity in which some of the accretion energy converts to kinetic energy in a collimated bipolar jet or disk wind. The nearby Perseus molecular cloud, for example, which we will meet in Chapter 4, shows hundreds of jets associated with embedded protostars, which we can trace in rotationally excited H_2 as well as emission lines from SiO, SO, and CO. These outflows generate turbulence in the surrounding envelope, driving their dispersal, and effectively limiting the available accretion mass. Equally, outflows can drive shocks that may well trigger star formation in neighbouring clouds and, especially in the high-mass case, account for the tendency for stars to form in clusters. Wherever the velocity of ejected material exceeds the sound speed, the surrounding gas and dust cannot respond dynamically until the material arrives. When that happens, the shocked gas is compressed, heated, and accelerated before subsequently cooling through line emission, and these changes in conditions are reflected in observable changes directly to the gas phase and to its composition resulting from the sputtering of ices. While H_2 and CO, as so often, are the dominant coolants, the common shock indicators in molecular clouds close to emerging stars are H_2O, SiO, SO, and SO_2, which we will meet particularly in the HMSF sources.

1.7 Protoplanetary Disks

As we will see when we look at specific examples, the impinging FUV (Far Ultraviolet), EUV (Extreme Ultraviolet), and X-ray stellar radiation emerging from the protostar creates dense photodissociation conditions at a disk's inner surface. Deeper into the disk, the magnetised gas is turbulent and grain collisions promote growth into larger aggregates, which is how planet formation begins. Accretion continues, jet and disk outflows drive strong shock waves into the disk and envelope, small shock-heated knots called Herbig–Haro (HH) objects appear where jets impact surroundings, and broad, diffuse lobes of high-velocity ambient gas are swept up and accelerated to high velocities. With the dispersal of the envelope cloud, the protostar and disk become visible, and we have what are called T Tauri stars ($<2\ M_\odot$) or Herbig AeBe stars if larger (2–8 $M_\odot$). Total accretion timescales are ~2 Myr, while disk dissipation takes ~3 Myr. With a variety of evolutionary stages, the timescale for low- and intermediate-mass stars to reach the main sequence is of order 10–100 Myr respectively.

Observations of the transition from accretion disk to protoplanetary disk have begun to be made possible by the latest high-resolution instrumentation. Close to a star, densities and temperatures are high and chemical equilibrium conditions rapidly attained. Further out from the star, differing conditions engender zones of particular molecular mixes before freeze-out onto cold grains at the outer margins. While gas–grain interactions dominate in the icy mid-plane, there is turbulent diffusion mixing material vertically and radial transport moving material laterally in the longer term. Dust grains will collide, stick, and potentially grow into aggregates. In several case studies we will look at key micro-chemical gas–grain processes occurring in the various molecular zones that undoubtedly do precede any aggregation of dust that might initiate planetesimal formation. Any subsequent evolution towards larger aggregations having a gravitational impact sufficient to engender, say, kilometre-size planetesimals that precede the formation of those planets that we now see associated with at least five thousand stars in our own Galaxy is a question for study elsewhere. The schematic of Figure 1.1 summarises the stages of low-mass star formation, giving spatial and temporal estimates.

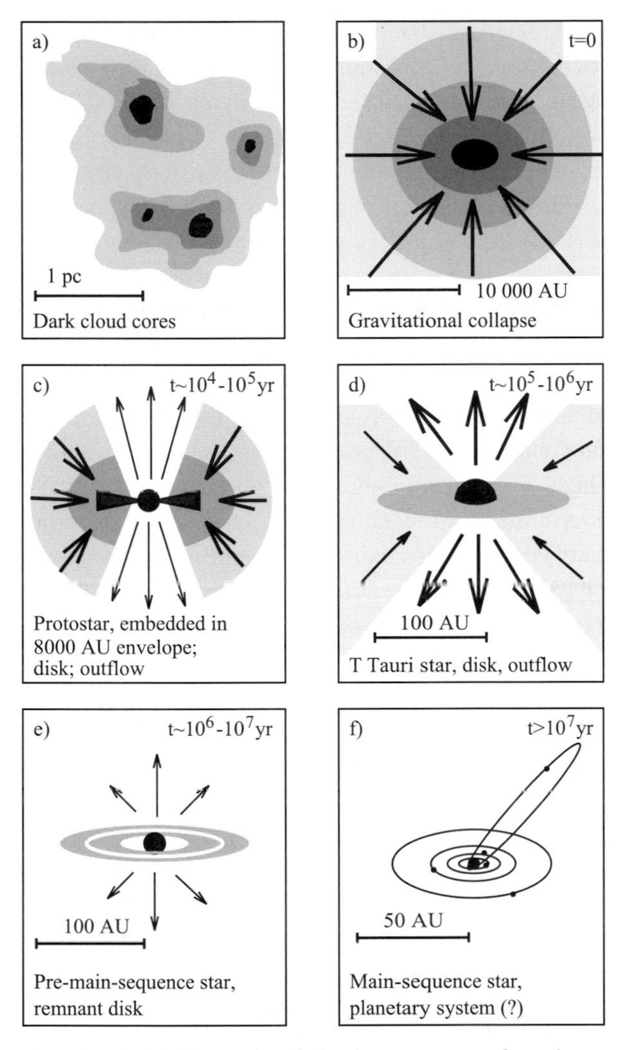

Figure 1.1 Schematic of the low-mass star formation process, with order-of-magnitude spatial and temporal scales (Hogerheijde 1998, based on Shu et al., 1987).

High-Mass Star Formation

1.8 Dark Cloud to Main Sequence

To reiterate, low-mass stars are of solar mass or less and not more than 2 $M_\odot$; intermediates are classified as between 2 $M_\odot$ and 8 $M_\odot$; and high-mass stars anything between 10 $M_\odot$ and 150 $M_\odot$ (at the theoretical extreme). Although the combined luminosities of high-mass stars in any galaxy dominate the total luminosity, their lifetimes are short in astronomical terms. Given the Galactic initial mass function (IMF), there are intrinsically fewer of them and, from the observer's perspective, they are generally further away (typically of order several kiloparsecs) than the average sample of low-mass stars. However, the development of ever higher-resolution instrumentation has increasingly allowed investigation of HMSFRs within, and even outside, our own Galaxy in ever greater detail. The James Webb Space Telescope launched in the last week of 2021 is just the latest in that line of development spanning the past fifty years.

With the observational constraints to date, there is as yet no entirely agreed-upon scenario for the formation of high-mass stars with quite the assurance we have just described for low-mass stars. The observations we do have and the theory both suggest that high-mass star formation takes place over a few 10^5 yr. High-mass protostars once formed also reach the main sequence over a comparable timescale while still embedded within their parent molecular cloud. Since they also invariably form in clusters, study of an individual high-mass star in relative isolation is an additional difficulty, so the overall challenge of HMSF studies is appreciable.

The physical complexity of an evolving HMSFR cluster is matched by the chemical complexity observed at millimetre and submillimetre wavelengths, so for the observer we can choose to adopt a four-stage categorisation of detectable evolutionary development. First, a gravitationally collapsing molecular cloud will fragment, with protostars forming in the cold, dense clumps. With typical temperatures ~10–20 K, these are only seen in extinction in the near- and mid-IR but seen in emission at far-IR and submillimetre wavelengths. Many of them are classified as Infrared Dark Clouds (IRDCs), sitting in front of diffuse background emission from the Galactic disk. Secondly, with collapse continuing, accretion disks and bipolar outflow systems form, and protostar heating raises the temperature of the envelope cloud. Continuing accretion boosts both protostellar mass and luminosity, bringing the protostar into visibility at infrared wavelengths. At this point, we have what is called a high-mass protostellar object (HMPO), characterised by high bolometric luminosity ($>100\ L_\odot$), strong thermal dust continuum, but only weak centimetre emission.

Thirdly, as the envelope temperature increases and molecular ice species on the dust grain surfaces evaporate, they initiate a rich chemical gas-phase mix in a so-called hot molecular core (HMC). While HMPO environments may show some similarities with conditions associated with HMCs, the chemical composition of the gas around HMCs is richer due to higher temperatures.

Why? Because, by definition, HMPOs are protostellar: they never reach a nuclear fusion stage. While not only warming their surroundings less than is evident in a YSO case, HMPOs never evolve sufficiently to emit ionising radiation. In contrast, as a massive object which has reached the YSO stage continues to accumulate matter, a fourth developmental stage after the HMC is observable, following from the increasingly strong radiation that dissociates and ionises the surrounding envelope gas. This is the evidence of ionised atomic hydrogen (HII), seen first in hypercompact (HCHII) regions which increase in size to ultra-compact (UCHII) region scale, emitting strong free–free emission at centimetre wavelengths. With the massive YSO remaining embedded within its natal cloud, the four stages are snapshots of a continuous evolution. Also, given that high-mass star formation typically occurs in clusters, it is not surprising that HMCs are commonly observed adjacent to UCHII regions, where one star in the cluster is likely to be more evolved than its immediate neighbour, detailed examples of which will be described in Chapters 11 and 12.

Among possible HMSF scenarios, it may be that the reality is closely similar to that of its low-mass stellar counterpart, with high-mass equivalents of dense cloud cores simply on a larger scale . In such a monolithic collapse scenario, the transition would be from dense molecular cloud core to protostellar envelope . However, since high-mass stars reach the main sequence in just a few hundred thousand years, one consequence of an accumulation of extremely large quantities of matter through rapidly accelerating accretion could be that the resulting high luminosity stops or even reverses the process through radiation pressure. That assumes a practically all-encompassing envelope, but if accretion were exclusively via an equatorial planar disk the majority of luminosity would not be trapped, and so the scale of resulting star not limited. This kind of competitive accretion seems the most likely scenario. One earlier suggestion, that of the collisional coalescence of stars already forming within a dense cloud, now seems very unlikely given a lack of sufficient stellar density within observed clusters.

1.9 Hot Cores

Closely associated with luminous high-mass stars are HMCs, compact (<0.1 pc) regions of warm (>100 K), dense (>10^7 cm^{-3}) gas. However, most are not yet well resolved, and there are others clearly energised in other ways. The archetypal Hot Core and its Compact Ridge companion in Orion, first studied in detail 30 years ago, is a case in point, being associated rather with the explosive decay of a multiple-star system, presenting features that are externally heated clumps, as we shall see in Chapter 9. The hot core in G34.26, described in Chapter 12, is equally intriguing for different reasons. Among the general discoveries of hot core chemistry, we can observe that nitrogen-bearing molecular species (such as HCN and CH_3CN), oxygen-bearing species (such as H_2CO), and sulphur-bearing species (such as SO and SO_2) are ubiquitous. HMCs offer the richest and most complex chemistries observed at millimetre wavelengths, with distinctions between oxygen-bearing and nitrogen-bearing species observed in a variety of hot core structures.

For the molecular astronomer, what makes hot cores particularly fascinating is the presence of high abundances of saturated (fully hydrogenated) species in the gas phase, including many COMs that are undoubtedly formed through grain-surface reaction networks as well as gas–grain interchange. The prototypical hot core G29.96-0.02 is also one that shows clear evidence of bulk rotation and nearby bipolar outflows from its associated star which have a likely influence, examined in Chapter 8. One of the COMs species that we will look at in detail in Chapter 12, and meet repeatedly elsewhere, is methyl cyanide (CH_3CN), which is a particularly useful tracer in delineating kinematical structure as well as higher temperatures in HMC objects. The chemistry of COMs, their origin and fate in warmed but cooling gas–grain circumstances, puts valuable constraints on modelling and observational interpretations. The high-resolution instruments now available, and those coming on stream in the near future, will undoubtedly reveal that the spatial structures within hot cores are likely to be a complex mix of components contributing to emission.

1.10 Compact HII Regions

High-mass stars may or may not begin hydrogen burning before accretion is complete, but surface temperatures, determined by accretion magnitudes and rates, can be sufficient to initiate ionising radiation. The flux of hydrogen-ionising photons may therefore generate an ionised hydrogen (HII) region while infall is still occurring. It seems likely that infalling material could compress the emergent HII region close to the stellar surface, forming a hypercompact (HCHII) region, until the balance between infall decrease and ionising photon flux increase leads to expansion of the ionised gas volume and formation of an ultracompact (UCHII) region. In simple circumstances the expansion is only halted when pressure equilibrium is reached between the ionised gas and the surrounding ISM; however, the reality of high-mass star formation clusters is that the mechanical energy of stellar winds outweighs that of thermal pressure, so HII regions are rarely neatly spherical structures. While the stellar UV is ionising hydrogen, the separated protons and electrons are subsequently free to recombine as neutral atomic hydrogen. Since only about one in three recombinations goes directly into the ground state, there is a subsequent cascade of radiative de-excitation detectable as line emission across a wide frequency range. These radio recombination lines (RRLs) are particularly useful to observers because they are largely unaffected by dust obscuration and provide kinematical, temperature, and density information for the ionised gas.

Given the complexity of forces behind the variety of compact HII structures, from the first IRAS space telescope observations in the infrared nearly four decades ago, it was clear that HII regions are not a homogeneous class of objects. Differing originating processes were characterised for the more evolved cases. Many have a 'cometary' appearance, meaning a dense leading edge and a more diffuse tail, as if generated by a star moving supersonically through its molecular cloud. G34.26+0.15 in Chapter 12 shows an archetypal cometary morphology. Alternatively, we might envisage the UCHII expansion into a region of rapidly reduced density, for example at the edge of a dense cloud, with an explosive 'champagne' escape flow resulting. A third possibility is that a massive star perhaps interacts with a clumpy molecular cloud and the ionised stellar wind becomes mass loaded by ablation and photoionisation of the clumps it passes through, terminating at a recombination (neutralising) front. Each schema has its theoretical justifications, and most HII regions are sustainable objects with masses $\sim$1 $M_\odot$ over timescales $\sim 10^5$ yr. These regions typically have densities $\sim 10^4$ cm^{-3} on scales $\sim$0.05 pc and, since they are deeply embedded in their parent envelope, no stellar luminosity directly escapes. We can observe them through its conversion into infrared radiation following heating and re-radiation from dust.

1.11 Photodissociation Regions (PDRs)

A variety of circumstances may result in the emergence of a photodissociation layer (sometimes also referred to as a photon-dominated region), but the most obvious is at the edge of a dense envelope that is subjected to an impinging stellar flux. This could be a reflection nebula subject to a relatively low radiation field (as Chapter 15 will show for the Horsehead Nebula in Orion), or it could be the edge of an HII region subject to a more powerful stellar field (as is the case for the Orion Bar in Chapter 14). While many of the physical and chemical processes within a PDR are closely similar to those in the atom-dominated ISM of diffuse and translucent clouds equally subject to photo-dissociating radiation, the PDR designation is usually applied specifically to the dense molecular cloud regions closest to bright, hot, high-mass stars. The PDR then lies between the ionised gas of the HII region and the neutral gas in which it is embedded, with a thin ionisation front between the two defining the region in which ionisation reduces from 100 per cent to zero, and in which dust dominates extinction.

Multiple indicators of the degree of ionisation within this front include bright atomic cooling lines (such as CI, CII, OI, and SII), fluorescent and thermal emission of rovibrational and pure rotational lines of H_2, polycyclic aromatic hydrocarbons (PAHs), dust mid- and far-IR continuum, and molecular rotational lines of CO, CN, HCN, and HCO^+, among others. Chapters 14 and 15 will look at those PDR molecular emissions in the high-flux and low-flux cases in more detail. For the moment, we can recognise that photon-induced chemistries are controlled by dust extinction as a function of depth in the cloud, with both distinct and mixed zones traced by different molecular species.

1.12 Masers

With clustered HMSF typically occurring at less than 1 pc separation, the gas kinematics can be thoroughly mixed up and complex to analyse. Being closely linked to a young, massive star's precise evolutionary stage, rotating accretion disks, bipolar outflows, expanding ionised hydrogen bubbles, and masers all contribute to our understanding of that stellar development process. Maser emission lines, being bright and highly beamed, are often used to trace the quite precise distance and proper motion measurements of these high-energy conditions. Masers are typically classified according to the conditions of their excitation. Those that highlight conditions where the radiant temperature is less than the kinetic temperature are likely to be collisionally pumped. Those that characterise conditions associated with proximity to UCHII regions or outflows, on the other hand, experience radiant temperature as the dominant cause of pumping. Engendered by the population inversion of already excited gas-phase molecules, common molecular masers include OH and SiO (radiatively pumped), H_2O (collisionally pumped), and CH_3OH (responsive to both pumping mechanisms and categorised as Class I and II, respectively). Factors such as velocity gradient as well as radiation field are intimately linked in maser formation, and we will consider those conditions of maser emission through many of the examples to come in the case studies.

Molecular Astrophysics

1.13 Molecular Excitation

For a detailed account of molecular emission and absorption, the reader is directed to the Further Reading titles at the end of this chapter (Tielens, 2021, being particularly recommended). For the basics, suffice it to say that molecular astronomical observations in the gas phase depend upon the emission and absorptions arising from rovibrational transitions of interstellar and circumstellar molecules. Detailed theoretical knowledge of level populations, critical densities, molecular excitation temperatures, collisional de-excitation rates, and the optical depth dependence of each underpins all observational studies and is the prerequisite for determining the physical conditions associated with the star formation medium. The conditions in which emission enables astronomers to deduce likely chemical reaction networks are governed by temperature, density, ultraviolet photon field, and energetic ion (cosmic ray) flux. It is the relationship between these physical parameters and the molecular line emission to which they give rise that furthers our understanding of the interstellar medium generally and the star formation process in particular.

1.14 Level Populations

Interaction with photons or cosmic rays, and collisions with other molecules, atoms, or electrons, both induce molecular transitions. The simplest molecular cloud collisions are likely to be with ubiquitous H_2 or He. The resulting excitations involve jumps from lower- to higher-energy rotational quantised states, and the subsequent de-excitation and release of photons of a particular energy and wavelength present as line emission. Such emission occurs at a rate determined either by the Einstein *A*-coefficient, A_{ji}, for radiative de-excitation, or spontaneous radiative decay. The average time for spontaneous decay is $1/A_{ji}$. The equivalent coefficient for absorption between the same two levels is B_{ij}. Collisional de-excitation, with an energy transfer exclusively between collision partners, also occurs, in which case, again, no radiation is emitted.

1.15 Critical Densities and Excitation Temperatures

Where local thermodynamic equilibrium (LTE) prevails, the level populations are described by a Boltzmann distribution and the relative populations of the two levels are a simple function of temperature. Typically, molecular gas in the interstellar and circumstellar medium is not in LTE, and determining level populations requires we solve statistical equilibrium equations. In these calculations, above a critical particle density, collisions will dominate the de-excitation (and therefore observable emission) process. At densities below the critical density for each particular transition, each exciting collision is followed by a radiative de-excitation, so an upper-level population 'collapse' follows. If we define γ_{ji} as the rate coefficient for collisional de-excitation in the thermalised gas above critical density, then we can define that critical density for each specific transition as $n_{\mathrm{cr}} \sim A_{ji} / \gamma_{ji}$. Maximum emission efficiency typically occurs at gas densities close to critical, so even tracing non-uniform molecular cloud density is possible using different molecular species and different transitions. The excitation temperature, T_{ex}, of a given molecular transition is dependent upon the critical density; hence, where emission lines become visible, the critical density becomes an approximate particle density diagnostic, just as the excitation temperature serves as an approximation to the gas kinetic temperature.

Astrochemical Basics

1.16 Gas-Phase and Grain-Surface Reactions

Given the generally non-LTE conditions in interstellar and circumstellar molecular clouds, in which low-temperature and low-density gas is subject to high UV photon or energetic ion fluxes, the gas-phase chemistry is dominated by kinetic rather than thermodynamic factors. Reaction types and their reaction rates, and the factors that differentiate these, therefore largely determine the gas composition within a molecular cloud, with dust-surface reactions and gas–dust exchanges significantly supplementing the purely gas-phase chemical network under specific circumstances, such as close to newly forming stars. Most interstellar gas-phase reactions are bimolecular, with rates of formation and destruction determined by collision cross section as a function of velocity averaged over the velocity distribution of reaction partners. These bimolecular reactions include neutral–neutral, ion-neutral, charge transfer, and radiative association, while unimolecular reactions include photodissociation, and associative detachment. Appendix D gives some tabulated generic examples of rates for the variety of reaction types.

The factors determining gas–grain interactions and grain-surface processes involve accretion efficiency, sticking probability, binding energy, and surface mobility. Thereafter most reactions occur between radicals (species with unpaired electrons) depending upon activation energy barriers and tunneling probability factors. A variety of desorption mechanisms then arise in differing circumstances, driven by heat (thermal desorption), UV photons (photodesorption), local 'hot spots' arising from exothermic reactions (chemical desorption), shocks (sputtering), and heavy atom impacts (cosmic ray–driven desorption). Large, interstellar, carbon-based molecules composed of tens of atoms have their own specific chemistries. Polycyclic aromatic hydrocarbons (PAHs) are planar, fullerenes are spherical (e.g. the C_{60} 'soccer ball') or ellipsoid. When a PAH absorbs a UV photon, it acquires a high excitation temperature (~1,000 K) for a very short time (~1 s), emitting then in the IR region over an extended period (a day, month, or year, depending on how close to the emitting star) until it goes back down to ~10 K. While in an excited state, additional reactions can include ionisation or fragmentation into smaller hydrocarbons, plus reaction with other gas-phase species, particularly atomic H and electrons. For a standard text, Tielens (2021), listed under Further Reading at the end of this chapter, gives greater detail on these topics and is recommended.

1.17 Chemical Modelling

In attempting to match observation with theory, while there are variations in types of computational modelling of reaction networks in astrophysical environments – from steady-state single-point, through time-dependent single-point, to time-dependent depth-dependent, each exclusively gas-phase, grain-surface, or incorporating gas–grain exchange – all of them involve calculation of a set of rate equations as functions of particle abundance and time. These typically comprise systems of ordinary differential equations representing formation and loss of each species. Depending on the chosen application, inputs are likely to include initial elemental abundances, gas densities, gas temperatures, cosmic ray ionisation rates, radiation field strengths, dust extinctions, and perhaps freeze-out and desorption factors. All utilise a reaction database with its rate coefficients derived from both theory and experiment, with computations involving networks of thousands of atomic and molecular species. Appendix A directs the reader to examples of such databases. The output is usually a set of gas-phase or grain-surface fractional abundances for each species relative to hydrogen over a given time period, which for star formation studies is typically of most interest between 10^3 and 10^5 years.

Observational Basics

1.18 Antenna Temperature and Optical Depth

Submillimetre and radio telescopes provide us with antenna temperatures, T_a, which we relate to fundamental molecular constants and astronomical parameters such as column densities, number densities, and gas temperatures. Even in the simple case of LTE , where level populations are dominated by collisions in gas at uniform temperature, the derivations for linear and non-linear molecules are different and best studied in detail in a standard text (that of Williams & Viti, 2013, referenced in Further Reading, being an excellent example). The relationship between T_a at frequency ν and optical depth τ can be given by the following equation:

$$\tau = h/\Delta\nu\left\{N_u B_{ul}(e^{h\nu/kT} - 1)\right\},$$

where N_u is the column density of the upper state and $\Delta\nu$ is the full width at half-maximum line width in units of velocity. The optical depth is computed by considering the probability that the emitted photon may escape from the locality, allowing for an absorption at a rate $B_{ul}\rho$, where B_{ul} is the Einstein coefficient for radiatively induced de-excitation and ρ is the energy density. Using standard equations (for which again derivations can be found in standard texts), given that the ratio of Einstein coefficients is expressible as

$$A_{ul} / B_{ul} = 8\pi h\nu^3 / c^3$$

and the relationship between antenna temperature and optical depth is expressible as

$$T_a = hc^3 N_u A_{ul} / 8\pi k\nu^2 \Delta\nu(\Delta\Omega_s / \Delta\Omega_a)(1 - e^{-\tau}/\tau)\tau,$$

we can also find

$$N_u = 8\pi k\nu^2\Omega / hc^3 A_{ul}(\Delta\Omega_\alpha / \Delta\Omega_\sigma)(\tau / 1 - e^{-\tau}),$$

in which the omega terms are source and antenna solid angles.

In summary, optically thin transitions will give telescope antenna temperatures proportional to the column density in the upper level of the emission transition observed. If we assume all the transitions are thermalised and that we know the kinetic temperature, the column density derived from the single transition can be equated with the total column density for the species. If the emission is optically thick, then the opacity results in an underestimate of the upper-level column density as well as the molecular rotational temperature. In fact, degrees of optical depth can be deduced from a rotation diagram analysis for both the LTE case and, with corrections, the non-LTE case. Before briefly looking at this, we should be sure of the following observational basics.

1.19 Velocity Distribution

Velocity maps are widely used to separate out different gas components from one another. Gas clouds are invariably moving, and in particularly complex ways close to newly forming stars. We look for measurable shifts in the observed emission or absorption lines relative to our local standard of rest. The mean speed of Galactic rotation in our stellar neighbourhood is ~250 km s^{-1}. There is also a peculiar Solar motion with respect to that ~16.5 km s^{-1}. Earth is locked into that rest frame, and relative to it, along a line of sight we may observe red- or blue-shifted lines. A numerically positive velocity difference indicates gas receding from us, and a negative shift indicates gas approaching. However, unless the gas movement is directly towards or away from us precisely aligned along the line of sight, these are not 'actual' relative velocities. Most movement will be somewhere between line of sight and plane of sky projection. However, most important is the fact that emissions from different species that show the same velocity shift are likely to be in the same spatial location.

1.20 Column Density, Beam Dilution, and Relative Abundance

Three simple constraints fundamental to observational astronomy are column density, beam dilution, and relative abundance. Imagine we look down a 100 cm-length tube with a 1 cm-diameter circular cross section and see 100 particles. Assuming an even distribution of particles inside the tube, from the formula for the volume of a cylinder ($h\pi r^2$) we deduce an average of 1 particle in every 0.78 cm^3 of space, which is the 'volume density'. Take that volume density and multiply it by the tube length and we have the 'column density', which in this case is 0.78 $cm^{-3} \times 100$ cm = 78 cm^{-2}. We have assumed an even distribution of particles along the length of the tube, but, of course, the column density would be the same even if all the particles are crowded together at the far end. That is the problem facing the astronomer observing along a line of sight – working out where exactly along that telescope beam the particles are actually distributed. Comparative velocity distribution among molecular species becomes essential information in that analysis.

Having decided that a range of evidence points to the location of the particles exclusively towards the end of our beam, if the telescope beam is larger than the observed object itself, then the extent of 'beam dilution' will influence our estimation of particle volume density at that location. For example, observing a hot core having a dimension less than 1 or 2 arcseconds against the sky with a single dish telescope limited to perhaps 20–30 arcseconds resolution restricts column density estimates to very approximate lower limits. Given the uncertainties, we might prefer to simply express the column density of one particle species as a fraction of another that we know to exist in the same place. Since molecular hydrogen is far and away the most abundant species in dense molecular gas, the fractional abundance of all other species can be expressed in relation to it. For example, the first major detection of ethanol in the Galaxy beyond Earth's atmosphere was made towards the G34.26+0.15 hot core (Chapter 12). In this case, the column densities of ethanol (C_2H_5OH) and H_2 were deduced to be 10^{15} cm^{-2} and 10^{23} cm^{-2} respectively, hence a fractional abundance for ethanol of $10^{15}/10^{23} = 10^{-8}$. Among the following case studies, fractional abundances of other paired species, including directly related species such as chemical product to precursor, will enter the discussion.

1.21 Rotation Diagrams

One commonly used technique among observers is the rotation diagram analysis in which, assuming LTE in the first instance, at temperature T_k the column density of each populated upper level u is related to the total column density via

$$N_u = N/Z\left(g_u e^{-Eu/kTk}\right),$$

where N is the total column density of the species, Z is the partition function, g_u is the statistical weight of the level u, and E_u is the energy above ground state. Thus, all we need to deduce total column density is an observed temperature. If we have multiple transitions for the same species, then we can construct a rotation diagram relating column density per statistical weight of multiple molecular energy levels to their energy above ground state. The rotation diagram plots the natural logarithm of N_u / g_u versus E_u / k. For the LTE condition, the plot would be a straight line with a negative slope of $1 / T_k$. This derived rotation temperature is expected to equal the kinetic temperature, as we have said, for the thermalised gas case. However, most astrophysical conditions are not ideal LTE cases, and the opacities for individual transitions in given circumstances are far from known. Negotiating these and other difficulties in order to extract reliable information is a complex business, but the many rotational lines of the many molecular species in a cloud, each with their own critical density and excitation energy, in principle offer astronomers powerful diagnostic tools for interpreting physical conditions.

1.22 Radiative Transfer Modelling

In the analysis and interpretation of observations where LTE is too gross an approximation, we consider all the individual excitation and de-excitation processes, collisional and radiative, through statistical equilibrium calculations. Provided collisional data is available, line radiative transfer models permit the calculation of level populations that underpin output spectra. The codes typically compute flux or intensity of individual line emission, and output line profiles. The assumptions are that temperature, density, and abundance are constant, since these determine the collisional coefficients, the level populations, and hence the emission. In non-LTE conditions, the excitation temperatures in particular are critical in determining level populations, so alternative approximations, such as LVG (large velocity gradient) radiative transfer modelling, are often used. Just as the rate coefficients are the major uncertainty in chemical modelling, the collisional coefficients are the principal uncertainties in LVG modelling. Other methods have been developed to solve, ever more accurately, the radiative transfer problem, and the interested reader is invited to start with Williams and Viti (2013) and the references therein.

Further Reading

Molecular Astrophysics, A. G. G. M. Tielens, Cambridge University Press (CUP) 2021: The most comprehensive account of molecular astrophysics available, connecting molecular physics, astronomy, and physical chemistry. The best current standard textbook.

Observational Molecular Astronomy, D. A. Williams & S. Viti, CUP 2013: An excellent introduction to observational techniques at millimetre and submillimetre wavelengths and the extraction of useful astronomical information from raw telescope data.

The Physics & Chemistry of the Interstellar Medium, A. G. G. M. Tielens, CUP 2006: An overview of the theoretical and observational understanding of the ISM, offering greater detail than *Case Studies* on the microscopic physical and chemical processes that influence macroscopic interstellar structures.

An Introduction to Star Formation, D. Ward-Thompson & A. P. Whitworth, CUP 2011: An excellent overview of the wider aspects of star formation, comprehensive and concise.

PART II

Low-Mass Star Formation (LMSF)

2

Two LMSFR Surveys Using IRAM and ALMA

2.1 Introduction

Having presented an overview of star formation studies in Chapter 1, in this chapter we present two surveys of low-mass star formation regions (LMSFRs) before individual source locations are considered in Chapters 3, 4, and 5. Low-mass young stellar objects (YSOs) can be classified as Class 0 protostars deeply embedded in the natal envelope, Class I protostars undergoing envelope collapse and accretion into a circumstellar disk, and Class II protostars that have cleared their envelope through jets and outflow actions while an accretion disk remains to feed material onto the star surface (Lada, 1987; Andre et al., 1993; Dunham et al., 2014). While early surveys made good use of ubiquitous small molecular and molecular ion emission (e.g. CO, CS, HNC, H_2O, HCO^+), the surveys of recent years have focused more on the evidence for, and the evidence provided by, COMs (complex organic molecules having at least six atoms) as particular sources of interest and as useful evolutionary markers. These molecules help us to delineate the physical parameters of the protostellar environment, while also testing our understanding of COMs formation chemistry in a reciprocal exchange between observation and chemical modelling.

2.2 IRAM COMs

The first survey we will look at is that towards 16 deeply embedded YSOs using the IRAM 30 m telescope (Bergner et al., 2017). The IRAM (Institute for Radio Astronomy in the Millimeter Range) facility, located nearly 3,000 metres high in the Sierra Nevada in Spain, close to the Pico Veleta peak, is one of the most sensitive single-dish telescopes in the world and has been in operation for nearly four decades. While this modest survey reinforces some equivalences between the chemistries of hot corinos associated with low-mass YSOs and the hot cores we will later study associated with high-mass star formation regions (HMSFRs), it does help distinguish between the conditions in which hot and cold COMs form. From the cold, dense , collapsing cloud to warming and eventually almost 'hot' conditions, COMs form during the earlier stages of low-mass star formation well before protostellar ignition. Because of their location, not least within both middle and outer regions of protostellar disks, if they survive, we may also imagine they ultimately become incorporated into protoplanetary disks (Visser et al., 2009, 2011) and hence potentially into planetesimals, perhaps seeding planets with complex organic material (Jorgensen et al., 2012). In support of such speculations, the observational results of recent surveys and subsequent modelling of COMs in low-mass protostellar environments are not inconsistent with general observations of the composition of solar system comets (Crovisier et al., 2004; Mumma & Charnley, 2011; Cordiner & Charnley, 2021). However, such speculations are a step outside the case studies' remit, so we will not pursue them here.

Of COMs in the IRAM survey, earlier observations of the sources had not offered clear evidence of hot corino activity prior to the 2017 detections. However, in the new observations, eight molecular species were identified: CH_2CO, CH_3CHO, CH_3OCH_3, CH_3OCHO, HC_3N, HNCO, CH_3OH, and CH_3CN. From the observations, median relative abundances were compared with a chemical model of the source envisaged as a simple core plus envelope structure, with temperature and gas densities varying as functions of radial distance from a central protostar. .

Where and how exactly are COMs thought to be forming? It appears they do so as reaction products within the ice mantles that consist initially of atomic and small molecular particles frozen out and coating interstellar dust grains (see Herbst & van Dishoeck, 2009). During dense cloud collapse, initially it is the hydrogenation of atoms or small molecules by atomic hydrogen on the grain surface that produces saturated species such as CH_4, NH_3, and H_2O (Brown, Charnley, & Millar, 1988), and small COMs such as CH_3OH (so-called 'zeroth-generation' species). Photolysis or radiolysis may then break these molecules into constituent radicals that remain on the grain surface (see Oberg, 2016). As the protostellar cloud collapses and warms, diffusion and recombination reactions generate larger complex species (called 'first-generation'). It is also likely that some of the smaller COMs, including CH_3OH, are desorbed at this stage and further processed in the warming gas. As a useful rule of thumb, modelling puts the temperature at which this diffusion and recombination process becomes efficient as ~30 K (Garrod & Herbst, 2006). As temperatures continue to increase, particularly once above ~100 K, the larger molecules also desorb and are free to react in the gas phase to form ‘second-generation’ COMs.

The initial ice composition is determined by the initial dense cloud gas composition, which is a function of the local evolutionary stage of the source. We should not expect local conditions to be the same everywhere, and ice compositions have been shown to vary significantly between sources (Oberg et al., 2011a, 2011b). Additionally, the locally distinct stellar radiation field will process sources differently. Observed variations in COMs’ chemical richness will, therefore, reflect variations in both the evolutionary stage and the initial conditions of each source. The common thread is that low-mass YSOs always evolve from cold to warming prestellar cores and thence to disk and warm inner core or envelope. These developments are evident in first-generation COMs, with potentially a hot corino if the central source is warm enough, sufficient, that is, to sublimate H_2O ice and promote first- and second-generation interactions as well as direct gas-phase enrichment (Caselli & Ceccarelli, 2012).

2.3 IRAM Observations

Sixteen Class 0/I YSOs, identified by their Spitzer-derived IR spectral indices, were selected in the northern hemisphere for the IRAM survey, 10 of them in the direction of Perseus (Boogert et al., 2008; Bergner et al., 2017). A measurable ice composition, detected as H_2O ice column density, was also a key part of the selection criteria. Beam sizes ranged from 27″ at 92 GHz to 21″ at 117 GHz. Table 2.1 shows a range of data for all 16 embedded sources in the sample.

As an example of the observations, Figure 2.1 shows the spectral window containing the CH_3CN 6–5 ladder and its detection or non-detection in each source. As one might expect, there is a dispersion in line richness between these sources, and Figure 2.2 shows this graphically, with the brightest at the top.

For the line emission across the 92–117 GHz range, the brightest sources are at the top, the least bright at the bottom. Not all detected species were seen in all sources. Taking them in the following order: CH_2CO, CH_3CHO, CH_3OCH_3, CH_3OCHO, CH_3CN, HNCO, and HC_3N, each was detected in 4, 6, 2, 2, 7, 13, and 12, of the 16 sources, respectively.

Table 2.1 *Source information on the 16 sample sources (Bergner et al., 2017).*

Source	R.A.	Dec	Cloud	L_{bol}	M_{env}	αIR	$N(CH_3OH)$	$N(H_2O_{ice})$	$X_{CH3OH(ice)}$	$X_{NH3(ice)}$	rms
	(J2000)	(J2000)		$L_\odot$	$M_\odot$		10^{13} cm^{-2}	10^{18} cm^{-2}	%H_2O	%H_2O	(mK)
B1-a	03:33:16.67	31:07:55.1	Perseus	1.3	2.8	1.87	10.21(3.24)	10.39(2.26)	<1.9	3.33(0.98)	3.6
B1-c	03:33:17.89	31:09:31.0	Perseus	3.7	17.7	2.66	1.69(0.51)	29.55(5.65)	<7.1	<4.04	5.5
B5 IRS 1	03:47:41.61	32:51:43:8	Perseus	4.7	4.2	0.78	1.77(0.46)	2.26(0.28)	<3.7	<2.09	7.0
HH 300	04:26:56.30	24:43:35:3	Taurus	1.27	0.03	0.79	0.24(0.10)	2.59(0.25)	<6.7	3.46(0.09)	5.8
IRAS03235	03:26:37.45	30:15:27.9	Perseus	1.9	2.4	1.44	1.17(0.08)	14.4(2.26)	4.2(1.2)	4.71(1.00)	4.2
IRAS03245	03:27:39.03	30:12:59.3	Perseus	7.0	5.3	2.70	1.54(0.29)	39.31(5.65)	<9.8	<4.40	3.7
IRAS03254	03:28:34.51	31:00:51.2	Perseus	-	0.3	0.90	-	3.66(0.47)	<4.6	6.66(1.37)	3.9
IRAS03271	03:30:15.16	30:23:48.8	Perseus	0.8	1.2	2.06	0.42(0.04)	7.69(1.76)	<5.6	6.37(1.86)	4.8
IRAS04108	04:13:54.72	28:11:32.9	Taurus	0.62	-	0.90	1.04(0.44)	2.87(0.4)	<3.5	4.29(1.03)	4.0
IRAS23238	23:25:46.65	74:17:37.2	CB244	-	-	0.95	2.19(1.01)	12.95(2.26)	<3.6	<1.24	2.7
L1014 IRS	21:24.07.51	49:59:09.0	L1014	-	-	1.28	0.88(0.56)	7.16(0.91)	3.1(0.8)	5.20(1.43)	2.8
L1448 IRS1	03:25:09.44	30:46:21.7	Perseus	17.0	16.3	0.34	0.23(0.04)	0.47(0.16)	<14.9	<4.15	3.7
L1455 IRS3	03:28:00.41	30:08:01.2	Perseus	0.32	0.2	0.98	1.46(0.88)	0.92(0.37)	<12.5	6.21(3.51)	3.9
L1455 SMM1	03:27:43.25	30:12:28.8	Perseus	3.1	5.3	2.41	1.48(0.76)	18.21(2.82)	<13.5	<8.29	4.1
L1480 IRS	04:04:43.07	28:18:56.4	Taurus	3.7	0.1	1.10	0.60(0.14)	4.26(0.51)	4.0(1.5)	5.42(0.06)	5.4
SVS 4–5	18:29:57.59	01:13:00.6	Serpens	38	-	1.26	11.19(4.29)	5.65(1.13)	25.2(3.5)	~4.3	3.9

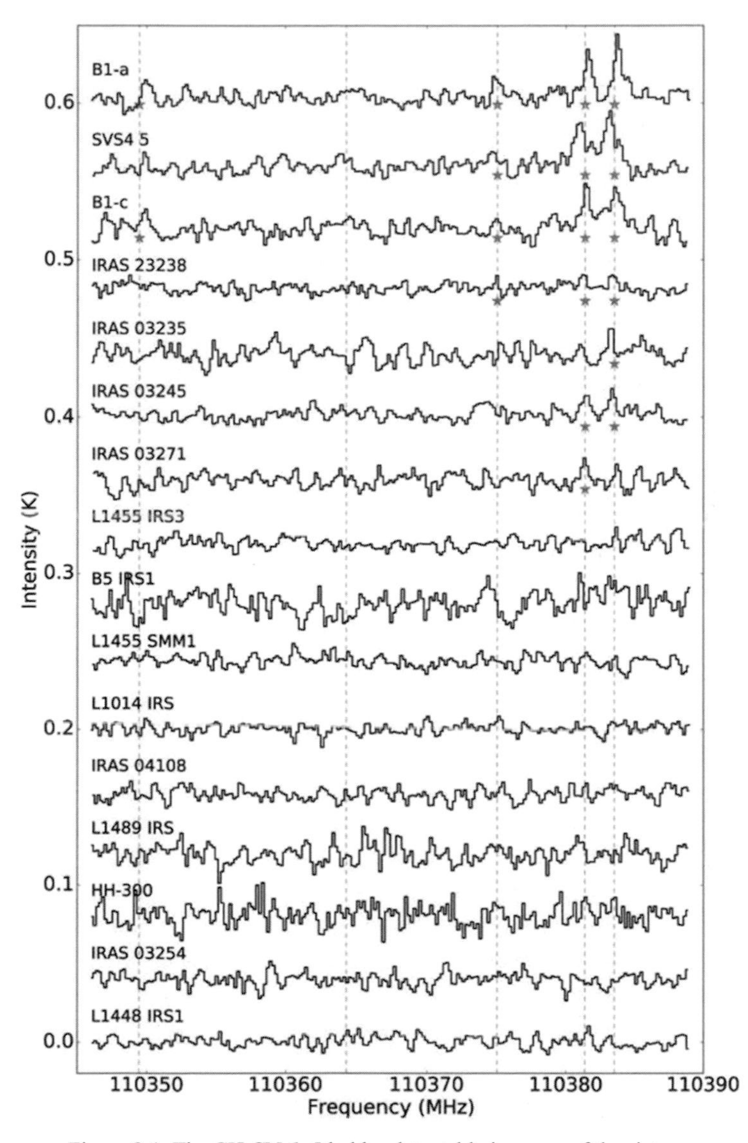

Figure 2.1 The CH_3CN 6–5 ladder detectable in seven of the sixteen sources (*top half*), with individual lines marked with stars (Bergner et al., 2017).

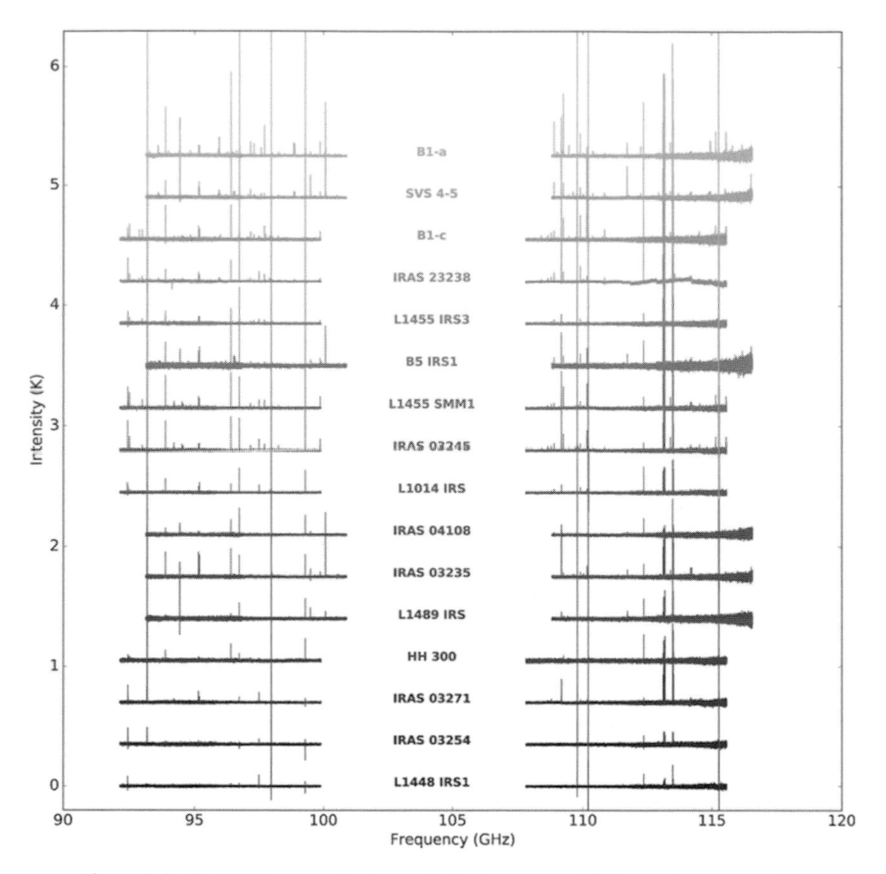

Figure 2.2 IRAM 30 m spectra towards the low-mass YSO sample, in ascending order of richness (Bergner et al., 2017).

For comparisons with chemical modelling, column densities and derived temperatures for oxygen-bearing COMs and nitrogen-bearing COMs are shown separately in Tables 2.2 and 2.3. Uncertainties in both are expressed here as the bracketed values. The temperature values, T_{rot}, were determined from rotation diagram analysis where multiple lines were detected; otherwise, sample-averaged values for that species are adopted.

2.4 COMs Correlations

To glean useful information from these observations, a common technique is to explore correlations between COMs species as indicative of possible chemical relationships. Strong positive correlations would be expected between

Table 2.2 *Column densities and rotational temperatures for oxygen-bearing COMs. Uncertainties are listed in parentheses (Bergner et al., 2017).*

Source	CH_2CO		CH_3CHO		CH_3OCH_3		CH_3OCHO	
	N_{tot} (cm^{-2})	T_{rot} (K)	N_{tot} (cm^{-2})	T_{rot} (K)	N_{tot} (cm^{-2})	T_{rot} (K)	N_{tot} (cm^{-2})	T_{rot} (K)
B1-a	$8.6(11.8) \times 10^{12}$	8(2)	$5.2(0.8) \times 10^{12}$	9(1)	$8.1(1.3) \times 10^{12}$	22(5)	$5.9(6.2) \times 10^{12}$	17(11)
B1-c	-	-	$3.6(3.4) \times 10^{12}$	4(1)	1.0×10^{13}	17(5)	$<9.5 \times 10^{12}$	16(1)
B5 IRS 1	$<1.4 \times 10^{12}$	8(2)	$<1.1 \times 10^{12}$	8(2)	$<1.0 \times 10^{13}$	17(5)	$<9.4 \times 10^{12}$	16(1)
HH 300	-	-	$<4.5 \times 10^{11}$	8(2)	$<7.1 \times 10^{12}$	17(5)	$<6.5 \times 10^{12}$	16(1)
IRAS03235	$1.0(1.4) \times 10^{12}$	8(2)	$<6.1 \times 10^{11}$	8(2)	$<6.4 \times 10^{12}$	17(5)	$<4.1 \times 10^{12}$	16(1)
IRAS03245	-	-	$6.4(4.9) \times 10^{11}$	8(2)	5.6×10^{12}	17(5)	$<5.8 \times 10^{12}$	16(1)
IRAS03254	-	-	$<6.4 \times 10^{11}$	8(2)	$<8.4 \times 10^{12}$	17(5)	$<5.3 \times 10^{12}$	16(1)
IRAS03271	-	-	$<4.8 \times 10^{11}$	8(2)	$<8.5 \times 10^{12}$	17(5)	$<4.9 \times 10^{12}$	16(1)
IRAS04108	$1.4(1.9) \times 10^{12}$	8(2)	$<7.6 \times 10^{11}$	8(2)	$<7.2 \times 10^{12}$	17(5)	$<6.0 \times 10^{12}$	16(1)
IRAS23238	-	-	$1.6(1.0) \times 10^{12}$	6(2)	$<5.1 \times 10^{12}$	17(5)	$<3.7 \times 10^{12}$	16(1)
L1014 IRS	-	-	$<3.1 \times 10^{11}$	8(2)	$<3.9 \times 10^{12}$	17(5)	$<3.2 \times 10^{12}$	16(1)
L1448 IRS1	-	-	$<4.2 \times 10^{11}$	8(2)	$<6.1 \times 10^{12}$	17(5)	$<4.1 \times 10^{12}$	16(1)
L1455 IRS3	-	-	$<4.3 \times 10^{11}$	8(2)	$<8.1 \times 10^{12}$	17(5)	$<4.8 \times 10^{12}$	16(1)
L1455 SMM1	-	-	$1.1(0.3) \times 10^{12}$	11(7)	$<6.7 \times 10^{12}$	17(5)	$<4.7 \times 10^{12}$	16(1)
L1480 IRS	$<1.4 \times 10^{12}$	8(2)	$<8.7 \times 10^{11}$	8(2)	$<8.1 \times 10^{12}$	17(5)	$<6.4 \times 10^{12}$	16(1)
SVS 4–5	$9.0(12.4) \times 10^{12}$	8(2)	$6.8(1.1) \times 10^{12}$	8(1)	$1.2(0.5) \times 10^{13}$	12(3)	$7.0(4.4) \times 10^{12}$	15(5)

Table 2.3 *Column densities and rotational temperatures for nitrogen-bearing COMs. Uncertainties are listed in parentheses (Bergner et al., 2017).*

Source	CH_3CN		HC_3N		HNCO	
	N_{tot} (cm^{-2})	T_{rot} (K)	N_{tot} (cm^{-2})	T_{rot} (K)	N_{tot} (cm^{-2})	T_{rot} (K)
B1-a	$4.9(1.1) \times 10^{11}$	33(9)	$4.2(1.2) \times 10^{12}$	12(2)	$7.7(4.3) \times 10^{12}$	14(3)
B1-c	$3.5(0.6) \times 10^{11}$	18(2)	$4.2(3.4) \times 10^{12}$	14(3)	$8.8(4.9) \times 10^{12}$	14(3)
B5 IRS 1	$<1.7 \times 10^{11}$	27(7)	$3.2(1.2) \times 10^{12}$	11(2)	$3.50(2.0) \times 10^{12}$	14(3)
HH 300	$<1.2 \times 10^{11}$	27(7)	$<1.3 \times 10^{11}$	14(3)	$<6.6 \times 10^{11}$	14(3)
IRAS03235	$1.6(0.9) \times 10^{11}$	27(7)	$3.9(1.0) \times 10^{12}$	13(2)	$1.8(1.0) \times 10^{12}$	14(3)
IRAS03245	$1.9(1.5) \times 10^{11}$	33(28)	$3.6(2.9) \times 10^{12}$	14(3)	$2.8(1.6) \times 10^{12}$	14(3)
IRAS03254	$<1.0 \times 10^{11}$	27(7)	$<9.2 \times 10^{10}$	14(3)	$<6.1 \times 10^{11}$	14(3)
IRAS03271	$1.9(1.1) \times 10^{11}$	27(7)	$1.7(1.4) \times 10^{12}$	14(3)	$1.4(0.8) \times 10^{12}$	14(3)
IRAS04108	$<1.2 \times 10^{11}$	27(7)	$<1.0 \times 10^{11}$	14(3)	$7.0(4.0) \times 10^{11}$	14(3)
IRAS23238	$1.4(0.3) \times 10^{11}$	33(6)	$3.2(2.6) \times 10^{12}$	14(3)	$4.6(2.6) \times 10^{12}$	14(3)
L1014 IRS	$<6.2 \times 10^{10}$	27(7)	$4.4(3.6) \times 10^{11}$	14(3)	$1.7(0.9) \times 10^{12}$	14(3)
L1448 IRS1	$<9.4 \times 10^{10}$	27(7)	$<8.7 \times 10^{10}$	14(3)	$<4.4 \times 10^{11}$	14(3)
L1455 IRS3	$<9.1 \times 10^{10}$	27(7)	$6.5(5.3) \times 10^{11}$	14(3)	$1.7(0.9) \times 10^{12}$	14(3)
L1455 SMM1	1.1×10^{11}	27(7)	$2.4(1.9) \times 10^{12}$	14(3)	$2.4(1.3) \times 10^{12}$	14(3)
L1480 IRS	$<1.4 \times 10^{11}$	27(7)	$3.5(2.4) \times 10^{11}$	20(13)	$7.7(4.7) \times 10^{11}$	14(3)
SVS 4–5	$5.2(0.9) \times 10^{11}$	17(2)	$1.1(0.3) \times 10^{13}$	13(2)	$7.0(3.9) \times 10^{12}$	14(3)

Table 2.4 *Column density correlations between selected COMs. Brackets indicate the number of sources with detections for both molecules. The dash indicates a pair of molecules with fewer than three sources in common (Bergner et al., 2017).*

	CH_3OH	CH_2CO	CH_3CHO	CH_3CN	HC_3N
HNCO	0.69[13]	0.98[4]	0.79[6]	0.79[7]	0.67[12]
HC_3N	0.78[12]	0.57[3]	0.82[6]	0.71[7]	
CH_3CN	0.90[7]	0.99[3]	0.96[5]		
CH_3CHO	0.90[6]				
CH_2CO	0.99[4]				

two species if one was the reaction product of another, or if they were both dependent on the same density and temperature conditions for their formation. Correlations as functions of column density are interpreted with care since they are invariably positive where molecules increase in abundance with increasing total column density (provided formation rate ≫ destruction rate). However, Table 2.4 shows some correlation figures for those COMs detected in at least three sources, and all but one of the sample 16 available (from Graninger et al., 2016). Methanol (CH3OH) was observed in all sources. The number of sample sources in which both paired species were detected is in brackets.

The ubiquitous COMs molecule CH_3OH correlates strongly with CH_2CO, CH_3CHO, and CH_3CN, and more weakly, but positively, with HNCO and HC_3N. The higher positive correlation of CH_3OH with oxygen-bearing species, compared with nitrogen-bearing molecules, supports a scenario in which CH_3OH is the parent species for these derivatives. Within the nitrogen-bearing group there are no strong correlations and HC_3N is the least correlated, perhaps unsurprising since by formation it is essentially a carbon chain molecule. While the CH_3OCH_3 and CH_3OCHO correlation is detected in only two sources, drawing conclusions would be unsound. Additionally, the correlations involving CH_2CO arise in just four of the sources.

As a supplement to the chemical correlations considered, there are observations we might make between column densities and the physical parameters of envelope mass and bolometric luminosity listed in Table 2.1. With more molecules along the line of sight where envelope mass is greatest, we expect a positive relationship, which is clear for all the species seen and most evident in the nitrogen-bearing species. More than this, with the strongest correlation being for HC_3N, and then for HNCO, we are confirming our expectation, since HC_3N is thought to form through gas-phase chemistry in the cold envelope and should therefore scale directly in abundance with envelope mass. In contrast, HNCO is thought to form most efficiently in the warm inner region (although also to a lesser extent in the cold envelope), so should be less responsive to envelope mass alone. The same function of temperature appears true for CH_3CN in this sample survey, based on a rotational temperature analysis. The fact that CH_3CHO and CH_2CO are distinctly cold cloud emitters but show little correlation with envelope mass supports a view that these molecules, having formed mainly on grain surfaces, remain as ices in the cold envelope rather than forming continuously through gas-phase reactions.

Negative results in research can be as informative at times as positive ones. In this survey, comparisons with the bolometric luminosity values shown in Table 2.1 uncover no significant correlations at all. This may indicate that thermal desorption and excess excitation have affected all molecules equally at the resolutions of this study, helping define the physical limits of these sources. However, one other interesting correlation to be explored is that of the relative abundances of COMs species against the observed ice compositions, available for 10 of the 16 sources in this survey. Given that CH_3OH is the likely initiator of oxygen-rich COMs, and NH_3 the possible precursor to their nitrogen-rich counterparts (although N_2, with zero dipole moment and therefore unobservable, remains a contender), it is these two species to which we look for COMs comparisons. Unfortunately, of the 10 sources that show both CH_3OH and NH_3 ices, not all are rich in COMs, which limits the conclusions that can be drawn. In fact, in the sample, correlations of any kind are seemingly absent, against both CH_3OH, NH_3, and even abundant H_2O ice. Only in the case of CH_3CN and HNCO are there weak but positive correlations. Perhaps the very weakness suggests that it is undetected N_2, rather than NH_3, that initiates the bulk of complex cyanide chemistry.

2.5 Low-Mass YSO Chemical Model

Despite the limitations of the IRAM 30 m single-dish telescope in resolving some of the detailed clues to COMs formation in the survey sources, it is important to compare the observational results with contemporary theoretical predictions generated by modelling. In this case, the research team used a three-phase chemical kinetics code, MAGIKAL (Garrod, 2013; Garrod et al., 2017), to simulate low-mass YSO chemistry. The model incorporates both gas-phase and gas-grain chemistries, quantifying selected ice mantle formation and destruction reactions. The restriction here is that COMs are taken as an exclusively grain-surface product, a first approximation likely to be close to reality.

The time-dependent single-point model is run multiple times at a range of radial distances from the core centre. From the grid of results the authors produce a composite, spatially extended description of the chemistry. The results show that for representative 1 $L_\odot$ and 10 $L_\odot$ protostar simulations, variations in temperature (±10% over a 10^4 AU radial distance) produce little change in peak molecular abundances. Figure 2.3 shows the radial profiles for the model density and temperature (*top row*), and those of COMs fractional abundances for 1 $L_\odot$ (*left*) and 10 $L_\odot$ (*right*) simulations, respectively. The cold emitters are shown in the two middle plots, and the warm emitters in the bottom pair of plots. In both cold and warm simulations, CH_3OH is included for comparison. The shaded regions around each line represent temperature ±10% over the fiducial run.

While the black-and-white reproduction prevents easy identification of each individual molecular species, what the model results do show are average abundances across the sample that agree within an order of magnitude with the observations of Tables 2.2 and 2.3. In other words, column densities of 10^{12} to 10^{11} cm^{-2} are typically reproduced, with the single exception of HNCO. Given the observational uncertainties and the modelling simplifications, this is a reasonable level of agreement that allows conclusions to be drawn, with modest confidence, regarding likely formation and destruction reaction networks described in the following subsection. A significant feature of the modelled results is that they demonstrate how critical is the desorption abundance of CH_3OH to the peak abundances achieved by almost every COMs species.

2.6 COMs Formation

The formation of *CH_2CO* (ketene) in ices has been shown experimentally and formation mechanisms suggested as either C_2O (dicarbon monoxide) hydrogenation (Garrod et al., 2008; Maity et al., 2014) or CH_2CHCOH (ethanol/vinyl alcohol) formation and decomposition (Hudson & Loeffler, 2013). The strong correlation between CH_2CO and CH_3OH (Table 2.4) along with its low rotational temperature (Table 2.2) supports a mechanism in which ketene forms via atom addition reactions in simple ices (i.e. during the 'zeroth-generation' phase of core collapse) – that is, $CO + C \rightarrow C_2O$, followed by hydrogenation to CH_2CO.

There is a strong correlation between *CH_3CHO* (acetaldehyde) and CH_3OH. With a low excitation energy, it is likely that CH_3CHO is another low-temperature 'zeroth-generation' COM. Its emission does not correlate strongly with envelope mass, in contrast to the gas-phase product HC_3N, reinforcing the

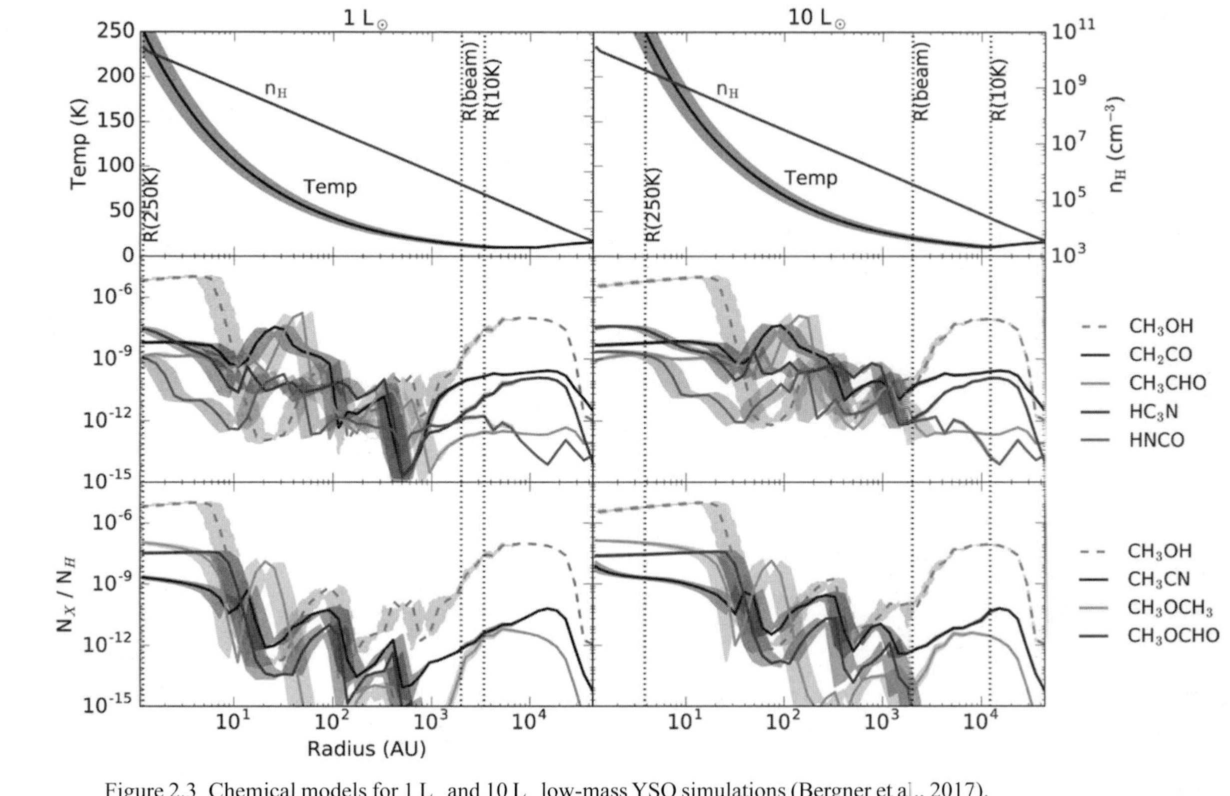

Figure 2.3 Chemical models for 1 L$_\odot$ and 10 L$_\odot$ low-mass YSO simulations (Bergner et al., 2017).

idea of ice product origin. Both experiment (Bennett et al., 2005; Oberg et al., 2009a) and modelling (Garrod et al., 2008) have shown efficient pathways to acetaldehyde from the radical decomposition products of methanol, CH_3, and HCO. The CH_3 radical is also derived from dissociation of CH_4, and the HCO radical from dissociation of H_2CO, but the tight correlation of this survey's results certainly suggests that the CH_3OH relationship may dominate.

In this survey, CH_3OCH_3 (dimethyl ether) and CH_3OCHO (methyl formate) have warm rotational temperatures, suggesting formation closer to the central protostar. Both are thought to form from methanol dissociation radicals, and they appear in greatest abundance in the survey sources with high CH_3OH column densities. However, the detection sample is small, and the emission locations too deeply embedded, requiring much higher spatial resolution for any certainty.

Isocyanic acid (*HNCO*) has a more obscure formation chemistry. Experiments have shown reactions in ices between CO and NH radicals (the latter a product of NH_3 dissociation) or simultaneous UV irradiation and hydrogenation of NO in $CO/H_2CO/CH_3OH$ ices (Fedoseev et al., 2015, 2016) that can generate HNCO. Kinetics modelling has shown efficient formation in the gas phase as a consequence of the destruction of other COMs such as $(NH_2)_2CO$ (urea) (Garrod et al., 2008), as well as the grain-surface reaction of CO and NH (Belloche et al., 2017). The former route is perhaps supported by the consistently warm rotational temperature derived for HNCO and the relatively strong correlation with envelope mass. However, given HNCO is seen across all the survey sources including those where other species are not evident, it seems confirmation of the likelihood that HNCO is a zeroth-generation COM with certainly no formation dependence on second-generation complex molecules. The correlation between HNCO and CH_3OH supports a scenario in which both form in the cold envelope, and the absence of HNCO column density increase when CH_3OH abundance increases (in B1-a and SVS 4–5) suggests that HNCO may be destroyed more readily in the denser core gas.

Acetonitrile (*CH_3CN*), also known as methyl cyanide, shows the highest rotational temperatures of all COMs observed (averaging 27 K), while also failing to show a correlation with envelope mass. Both observations suggest a centrally concentrated emission and a 'lukewarm' formation pathway. While a gas-phase formation pathway has long been assumed under cold, dense conditions ($HCN + CH_3^+ \rightarrow CH_3CNH^+$, which $+ e^- \rightarrow CH_3CN + H$), the presence of CH_3CN in the hot, dense gas associated with high-mass star formation (the hot cores we will meet in Part II) has long been observed and also modelled through the reaction of $CH_3 + CN$ on grain surfaces (Garrod et al., 2008). In the IRAM sample survey, CH_3CN correlates well with CH_3OH, encouraging us to think their formation chemistries have similarities, particularly within ices. Against this, there is no obvious correlation with the NH_3 observations, even though it is another zeroth-generation species. The fact is that all models of the past three decades have underproduced CH_3CN against observed abundances in higher-temperature conditions, typically by several orders of magnitude, and the formation mechanism has remained a significant unknown. Since we will continue to meet this widely observed molecule in many of the chapters that follow, we will leave it to a more detailed discussion in Chapters 12 and 15 to draw together all the strands of what is, and what is still not, known about it.

Cyanoacetylene (C_3N) shows the most scatter against CH_3OH, not unexpected since cyanopolyyne observations are generally well produced by exclusively cold, dense gas-phase models (Herbst & van Dishoeck, 2009). The tight correlation with envelope mass and the lack of correlation with NH_3 ices further reinforce that formation route, as is the case for most hydrocarbon species.

2.7 Increasing Resolution with ALMA

Having established a degree of agreement between observation and theory for the six COMs described, one of the key conclusions of the 2017 IRAM survey is that although hot corinos can be as rich in their variety of COMs as many hot cores (IRAS 16293 in the following chapter being a good corino-distinct example), as well as having greater numbers of species generated under the longer cold-duration timescales of low-mass protostellar evolution, the richness does not appear necessarily to be typical. Given the surveyed variations, generalising COMs as evidence for a single phase of low-mass protostellar evolution may be an overly reductive categorisation. The higher sensitivity and spatial resolution of ALMA observations, however, do allow us to identify COMs in the disks of Class II YSO sources in much greater detail (as also the IRAS 16293 example will show) (Oberg et al., 2015; Walsh & Ilee, 2020; Bergner et al., 2018; Loomis et al., 2018, Favre et al., 2018). As a follow-up to the IRAM survey, five Class 0/I protostellar disk candidates have been targeted using the much higher resolution of ALMA (Bergner et al., 2019).

As an adjunct to star formation, until recent years planet formation was assumed to develop first during the YSO Class II stage. Growing evidence, however, suggests planetesimal formation may begin earlier and the detection of COMs has become a contributing part of that research. High-resolution submillimetre continuum observations reveal the fact that dust sub-structure does appear to be already common in Class II disks (Andrews et al., 2018).

One explanation suggests this sub-structure results from interactions between material masses equivalent to Solar system planets within the disk beginning in the Class 0/I stage (Zhang et al., 2018; ALMA Partnership et al., 2015). The incorporation of COMs into the formation of planetesimal and planetary material begins then to seem a plausible possibility.

As for LMSF itself, in the ALMA survey of 2019, observations towards five low-mass candidates, all in the Serpens cluster at distances ~440 pc, detected five typical COMs species across the sample, but not in every source and with relative abundances spanning two orders of magnitude. As with the IRAM survey, this again shows a high degree of diversity at the hot corino stage. Each of CH_2CO, CH_3CHO, CH_3OCH_3, CH_3OCHO, HC_3N, and HNCO were shown to have median abundances with respect to CH_3OH (~5–10%), while CH_3CN showed an order of magnitude lower in column density and relative abundance. Table 2.5 shows the key properties characterising the five sources in Serpens, and Figure 2.4 shows the observed distribution of dust, $C^{18}O$, and CH_3OH across each.

Using COMs emission lines to unequivocally identify disk structure, even at ALMA resolution, was not possible in the 2019 survey. However, the variety of possible spectral line profiles one looks for is shown in Figure 2.5, a schematic based on H_2O observations across nearly 30 Class 0/I low-mass sources (Kristensen et al., 2012). Comparing the modelled profile with selected COMs spectra from 2019, one can see hints of structure in the spectral line shapes, as shown in Figure 2.6. For example, Ser-emb 17 shows many such double-peaked line profiles in several COMs spectra, commonly taken as a rotating disk signature (Beckwith & Sargent, 1993). Some of the lines in Ser-emb 8 also appear to show an inverse P-Cygni profile (the IPC panel in Figure 2.5) with a red-shifted absorption feature, indicative of infall (Di Francesco et al., 2001). Line profiles for NH_2CHO in Ser-emb 8 and 17 also show a slight

Table 2.5 *Properties of the five sources, with uncertainties in (Bergner et al., 2019).*

Source		R.A.	Dec.	Class	T_{bol}	L_{bol}	M_{env}	Est.M_{disk}
		(J2000)	(J2000)		(K)	($L_{\odot}$)	($M_{\odot}$)	($M_{\odot}$)
Ser-emb	1	18:29:09.1	0:31:30.9	0	39(2)	4.1(0.3)	3.1(0.05)	0.28
Ser-emb	7	18:28:54.1	0:29:30.0	0	58(13)	7.9(0.3)	4.3(0.4)	0.15
Ser-emb	8	18:29:48.1	1:16:43.7	0	58(16)	5.4(6.2)	9.4(0.3)	0.25
Ser-emb	15	18:29:54.3	0:36:00.8	I	101(43)	0.4(0.6)	1.3(0.1)	0.15
Ser-emb	17	18:29:06.2	0:30:43.1	I	117(21)	3.8(3.3)	3.6(0.4)	0.15

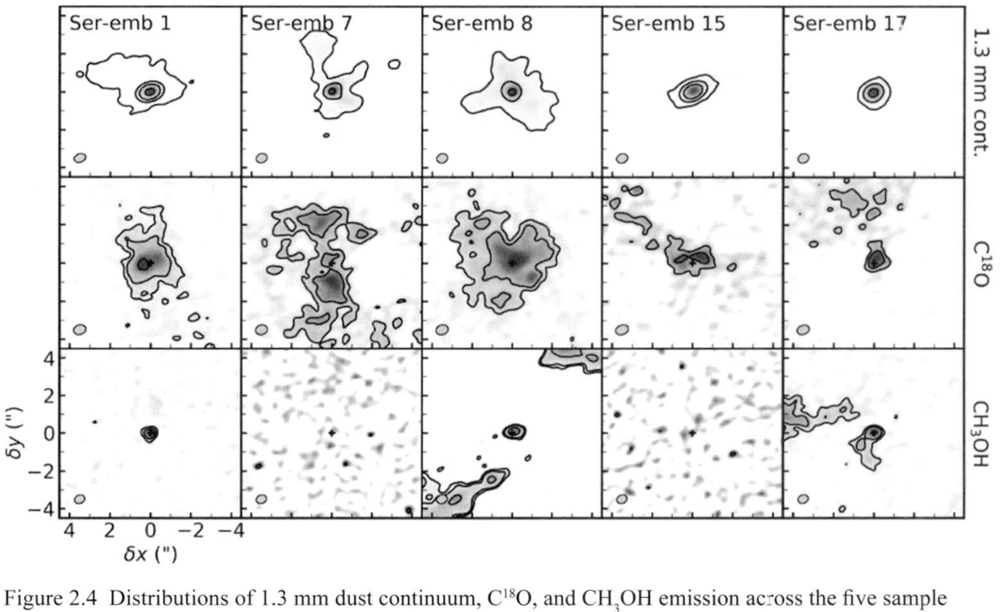

Figure 2.4 Distributions of 1.3 mm dust continuum, $C^{18}O$, and CH_3OH emission across the five sample YSOs. The synthesised beam size is shown in the bottom left of each panel (Bergner et al., 2019).

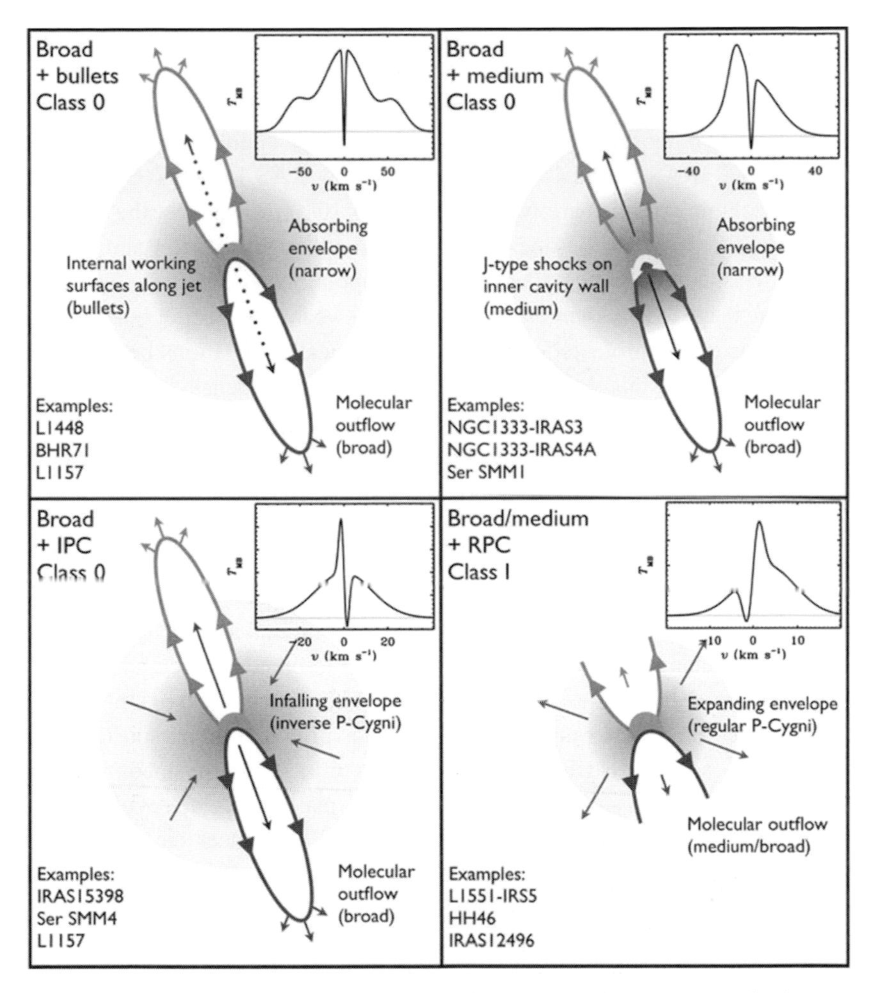

Figure 2.5 Schematics of various protostellar configurations and the indicative line profiles identified in H_2O emission (Kristensen et al., 2012).

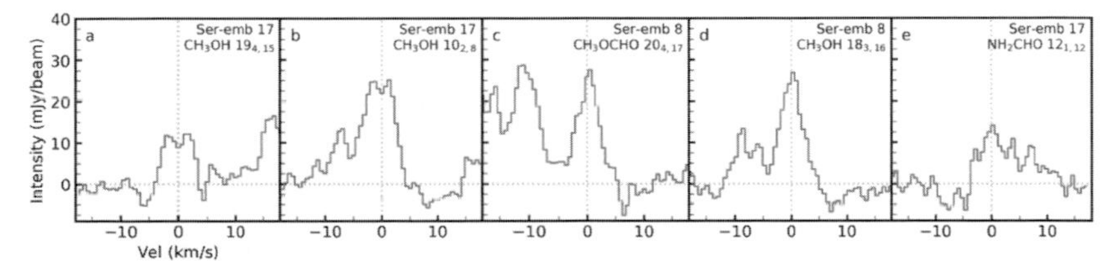

Figure 2.6 Examples of COMs spectra with line profiles hinting at rotation (a, b), infall (b, c, d), and outflow (e) (Bergner et al., 2019).

broadening at red-shifted velocities, suggesting the possible presence of a jet or outflow (Codella et al., 2017).

In using CH_3OH as a critical marker in corino observations, it is worth noting that where other sources have been studied at high angular resolution, such as NGC 1333 IRAS2A and 4A (see Chapter 4) and IRAS 16293 B (Chapter 3), CH_3OH column densities are typically of order 10^{19} cm^{-2}, against the Serpens sources ~10^{17}–10^{18} cm^{-2}. If one is confident about beam dilution and optical depth effects, in the absence of additional isotopologue observations, we might conjecture that Serpens sources are intrinsically weaker or smaller hot corinos, perhaps due to low protostellar source luminosities. Correlations between the brightest COMs in this sample against CH_3OH column densities nonetheless reinforce aspects of the IRAM 2017 results and those of other studies. The NH_2CHO/CH_3OH and CH_2CO/CH_3OH column density ratios are consistent across the Serpens sources and others, whereas the CH_3OCH_3 and CH_3OCHO ratios with CH_3OH are increased by at least a factor ~10 compared to other hot corinos. We can note that the range of those ratios in the Serpens and other published sources involving CH_3OCH_3, CH_3OCHO, and NH_2CHO span an order of magnitude. While different angular resolutions and analytical techniques in deducing column densities are factors to be considered particularly closely, there remains the apparent fact that abundances of oxygen-bearing COMs cover the widest range. With CH_3OH as principal precursor, the contributions to spatial distribution arise from a mix of physical components (infall, jets, accretion shocks, rotating disks). Equally, there are evolutionary timescales that may vary. Infall rate, for example, influences the duration of heating and therefore chemical processing within ices prior to sublimation of product COMs (Garrod et al., 2008). The size and age of hot corino regions themselves will influence observed gas phase abundances. The fact that the Serpens sources appear chemically similar to one another, albeit in contrast to many other observed hot corinos, confirms that local and initial conditions are always likely to play an important role.

2.8 Summary

Disentangling hot corino, infall, disk, and jet/outflow components, whether distinct or overlapping, will be the subject of ever higher-resolution observations of key COMs species in the years ahead. At submm wavelengths we can probe the gas-phase COMs reservoirs that link to the ice-phase reservoirs through the variety of desorption processes and reaction networks. Gas phase formation of saturated molecules is inefficient at these temperatures and densities. Their production must be almost exclusively within the ice mantles. Gas phase emission from more evolved, warmer and denser regions reveal the many ice sublimates. Molecular astronomers interested in planetary formation will also be looking to the presence of COMs in Class II disks for comparison against cometary detections, and the associated gas-phase observations which are believed to reflect pristine ices. While reprocessing between the many stages is inevitable, including shock heating during infall and accretion on to the disk (Oya et al., 2016), at least some of the more complex material in the outer Solar system could be inherited from a protostellar formation phase, according to this tentative evidence (Mumma & Charnley, 2011; Altwegg et al., 2017).

3
IRAS 16293 in Ophiuchus

3.1 Introduction

Having considered aspects of two low-mass star formation surveys in the previous chapter, particularly as they relate to COMs as indicators of both physical and chemical conditions, as a first example of an individual low-mass star formation region let us look at what is probably the most studied example of a warm core surrounding a binary protostellar source. About 120 pc away in the L1689 N cloud, the IRAS 16293–2422 source is chemically rich with emissions from abundant simple as well as complex molecules, and many unusual deuterated isotopologues. As with each of the individual sources in the case studies, we begin by locating the site in its wider Galactic context.

3.2 Dark Clouds and Streamers

Both the sky map of Figure 3.1 and the optical image of Figure 3.2 show the central disk of the Galaxy lying diagonally northeast to southwest (top left to bottom right). With a Galactic longitude of 354° and latitude 15°, the star formation complex ρ Ophiuchus lies in the direction of the Galactic centre but sits significantly above the equatorial plane. On the sky map, the ρ Ophiuchi cloud complex is in the bottom right corner, tucked into the angle of the constellation boundary with Scorpius, just above Antares and the M4 globular cluster. Figure 3.3 shows a closer infrared image from NASA's Wide-Field Survey (WISE) in which the major reflection nebula (associated with HD147889) is in the upper centre, and stars σ Scorpii (on the right) and τ Scorpii (far bottom left) show most clearly, with α Scorpii (Antares) just left of centre. We are actually seeing the M supergiant α Scorpii (the more luminous of the Antares binary) about 5 pc to the south of the cloud complex's centre, with σ Scorpii a little closer at 4 pc to the southwest.

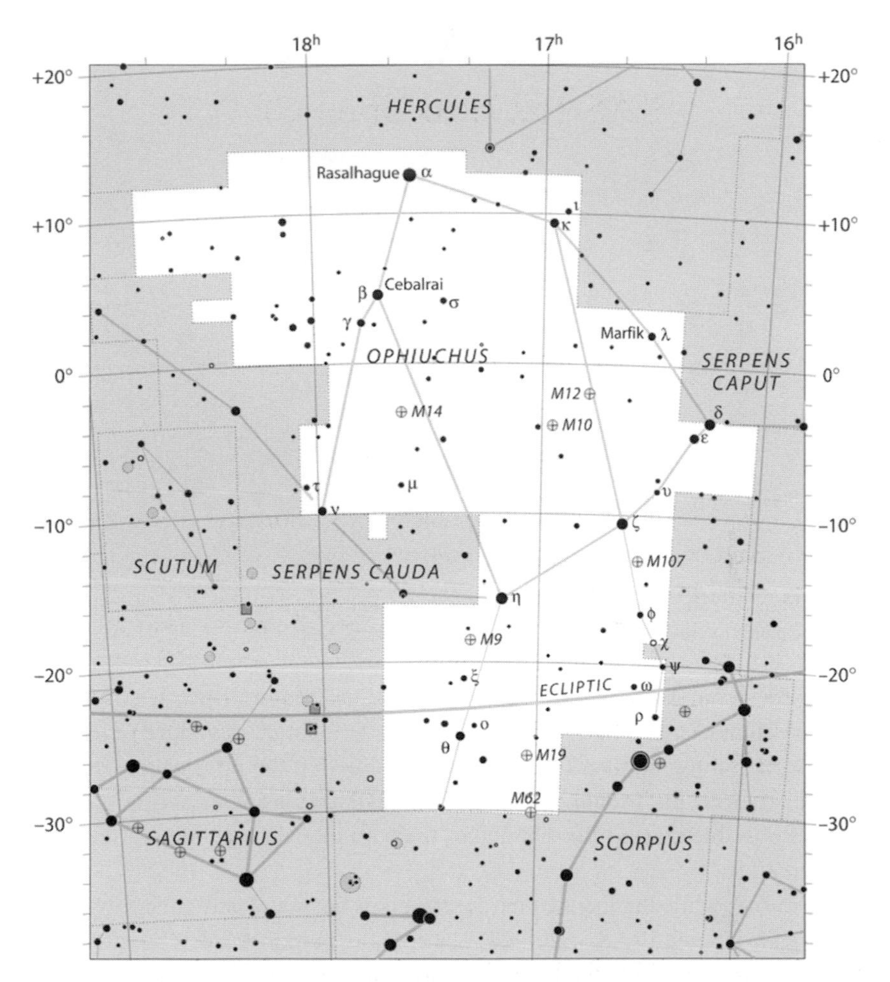

Figure 3.1 On the sky map the ρ Ophiuchus cloud complex is in the bottom right corner, tucked into the angle of the constellation boundary, just above Antares (IAU/Sky & Telescope).

The warm dust of the infrared image in Figure 3.3 shows the central cold dark cloud complex streaming east in which many young low-mass stars are embedded and where we will find IRAS 16293-2422. The quadruple star cluster of the ρ Ophiuchi group, giving its name to the complex, is a degree north of the HD147889 reflection nebula which sits almost centrally in this image. The three dark clouds L1688, L1709, and L1689 are then shown in close-up in Figure 3.4. The central dark cloud, Lynds 1688, is about 140 pc from Earth and is a location in which much star formation activity has been

Figure 3.2 The Galactic disk lies northeast to southwest across this image with the ρ Ophiuchus complex on the right (ESO/Stéphane Guisard).

observed. From L1688, the L1709 streamer extends northeast, while the other streamer stretching due east is L1689. This latter is a more quiescent star formation region than L1688. If we assume that the physical conditions of both regions are broadly similar, their differences in formation activity and the possible reasons for that makes them an interesting comparative subject for study.

Therefore, while the focus of our interest will be on the particular low-mass protostellar source IRAS 16296-2422 which lies within the L1689 streamer, we can briefly consider the morphology of all three streamers on the larger scale which pose their own questions and are of evolutionary significance for both the L1688 and L1689 cloud complexes. One can first make a plausible case for the joint influence of both α Scorpii as an O9 star (even, perhaps, supernova activity in the nearest elements of the Upper Sco OB2 association) irradiating the L1688 region from the southwest, coupled with that of the ρ Ophiuchi cluster itself impinging from the north (Ladjelate et al., 2020). High-mass star activity, let alone supernova demise, offers the potential for widespread shock compression via expanding shells of gas over parsec scales powered by stellar winds or explosive dispersal. One such analysis identifies possible pressure fronts and compression arcs in the L1688 cloud complex as shown in Figure 3.5.

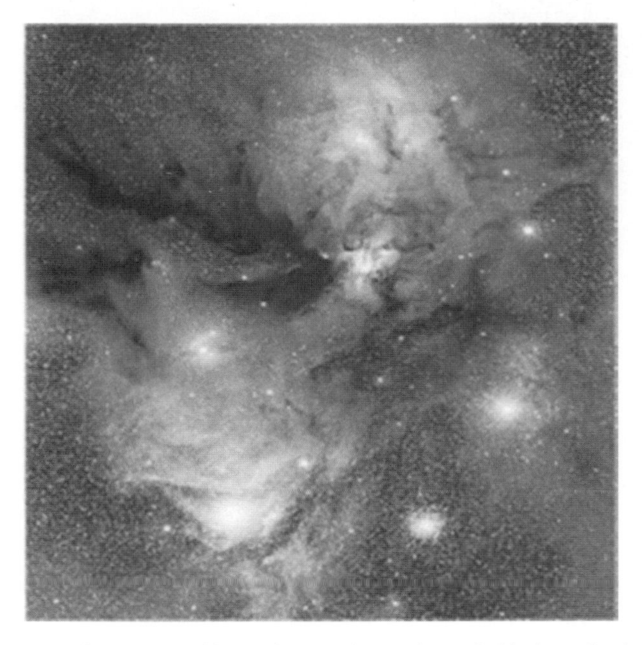

Figure 3.3 This IR image shows the ρ Ophiuchus cloud complex with the HD147889 reflection nebulae upper centre and three stars beneath: σ Scorpii and τ Sco, right and left respectively, each surrounded by reflection nebulae, and α Sco (Antares) the bright star lower left of centre, separate from major nebulae (Judy Schmidt/NASA WISE).

3.3 Filaments and Cores

Closing in to look for stellar formation activity within these dark clouds, the clear delineations at parsec scales shown in Figure 3.6 and Figure 3.7 begin to reveal an L1689 region consisting of three highly compact dust locations, each having their identifiable prestellar gas and dust cores. The sensitivity of the Herschel survey across many parts of the Galaxy leaves astronomers in no doubt that within large-scale streamers generally, such as those of ρ Ophiuchus, there is an inextricable link between dense filaments and compact prestellar cores as progenitors of low-mass protostars (André et al., 2010, 2014). Prestellar cores typically string out along elongated filamentary structures. Sensitive temperature measurements are also able to distinguish between the prestellar cores and their unbound starless core counterparts. In L1689, for example, detections have shown 35 warm examples of the former and 79 cold examples of the latter, some of which are circled in Figure 3.7.

It has become clear from both observational and theoretical research in recent years that most, and quite possibly all, stars are born in multiple

Figure 3.4 A magnified optical image showing dark cloud L1688 dead centre, L1709 streaming northeast, and L1689 the lesser complex due east (Adam Block/ Seward Observatory/University of Arizona).

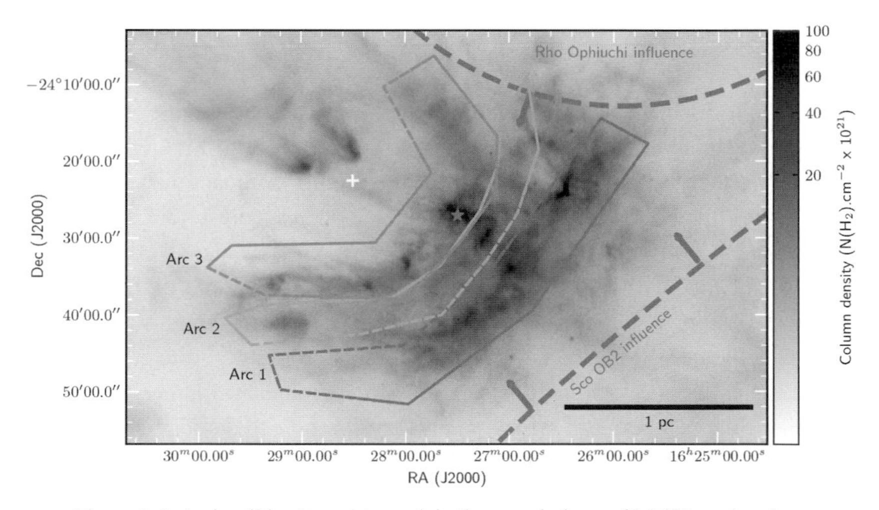

Figure 3.5 A plausible attempt to explain the morphology of L1688, and perhaps L1689, as compression filaments under the influence of the O9V star α Scorpii about 4pc to the southwest and skewed also by the ρ Ophiuchi cluster 'bubble' to the north (Ladjelate et al., 2020).

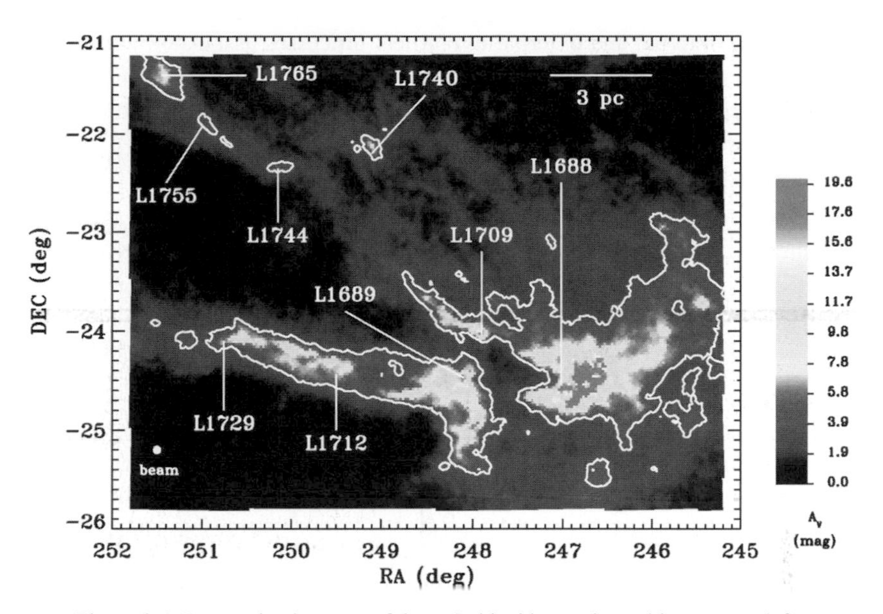

Figure 3.6 Dust extinction map of the ρ Ophiuchi complex, with contours defining Av $\geq$ 3mag (Ridge et al., 2006).

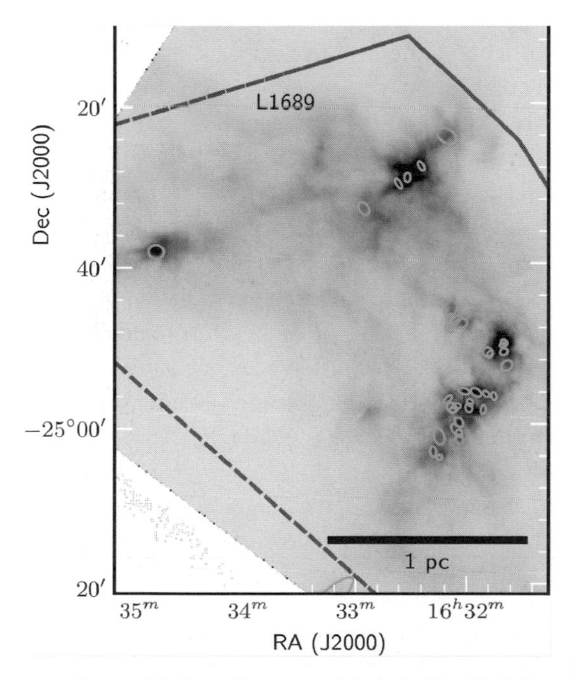

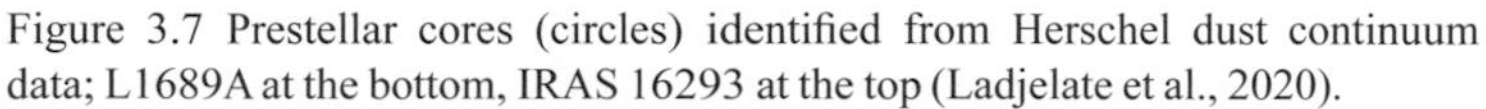
Figure 3.7 Prestellar cores (circles) identified from Herschel dust continuum data; L1689A at the bottom, IRAS 16293 at the top (Ladjelate et al., 2020).

systems. The internal and external dynamical processes associated with these systems then determines the emergent field population of single, binary, or multiple stars. As we stated in Chapter 1's overview of the star formation process, the dynamics associated with low- and intermediate-mass stars, unlike those driven by the highly energetic jets, outflows, and radiation fluxes of high-mass stars, make for a distinctively different environment in which physical and chemical evolutionary tracers offer their own complex puzzles for solution. IRAS 16293-2422 offers a well-observed research target through which many LMSF answers have been gleaned.

3.4 IRAS 16293–2422

As an archetypal example, embedded within the northern cluster of the L1689 cloud, the low-mass star forming core IRAS 16293–2422 was first targeted in the late 1980s when the focus was on trying to distinguish between circumstellar material falling inwards towards a conjectured low-mass protostar, as distinct from material rotating in a surrounding disk-like structure. In the 1990s IRAS 16293 became the classic example of a Class 0 protostar, one thought of as having accreted less than half its final mass (André et al., 1993). Mapping the envelope material with single-dish telescopes showed infall on the scale of a few thousand AU and, at the even more focused scales of interferometric observation, IRAS 16293 became the first ever protostar identified as a binary system, its A and B components resolved at ~5″, about 600 AU apart (Wootten, 1989; Mundy et al., 1992). Figure 3.8 shows a more recent dust continuum separation of the two components, A in the south showing greater complexity of dust distribution but both sources being within a larger circumbinary envelope (Jorgensen et al., 2016). Dust continuum emissions at scales down to ~100 AU indicate compact disk-like structures surrounding both protostar sites (Looney et al., 2000). The differences in character of the continuum fluxes have subsequently enabled identification of emission from A as a combination of dust and shocked ionised gas, while that from B is predominantly optically thick dust emission. IRAS 16293A is probably no more than a few 10^4 years old, and it is partly through the longer wavelength emission that this component has been resolved into its own tight binary system, with sources separated by just 1″, which at this distance is a separation ~120 AU (Chandler et al., 2005).

Having identified three separate stellar cores (A1, A2, and B), the interest for the molecular astronomer is in the physical and chemical distinctions between them. Figure 3.9 shows a schematic distribution of some of the observed

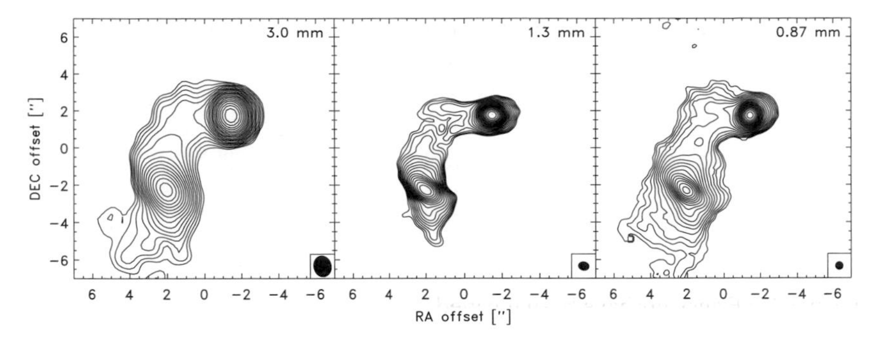

Figure 3.8 Dust continuum location of IRAS16293 A and B with the Atacama Millimetre Array (ALMA) (Jorgensen et al., 2016).

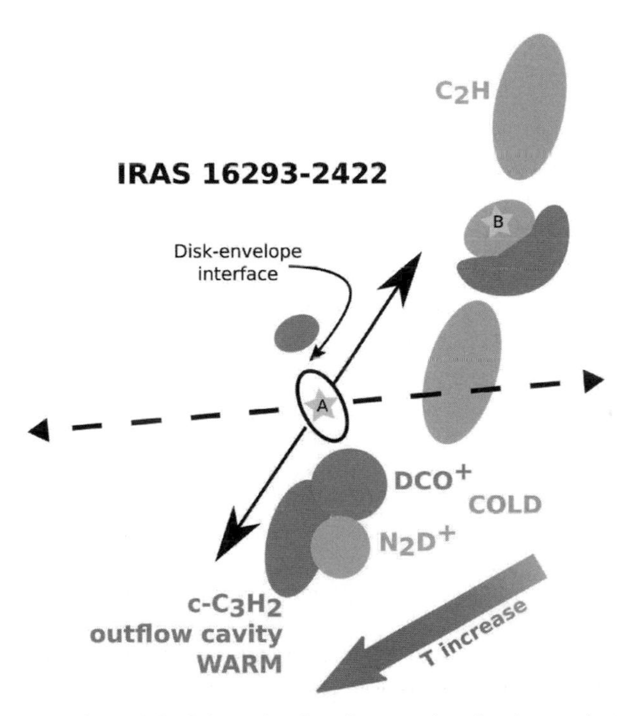

Figure 3.9 Schematic of outflows and molecular species around IRAS16293A and B (Murillo et al., 2018).

molecules associated with sources A and B, which give us interesting information about aspects of the circumstellar conditions. Firstly, the schematic shows the principal molecular outflow directions in relation to what is observed to be an obliquely angled accretion disc around source A. The dynamics around A and B, let alone A1 and A2 as separate entities, have yet to be entirely well resolved, and researchers currently face some rather confusing velocity gradients and uncertain contributors to overall luminosity. Nonetheless, the basic outflow structure has been studied in different molecular tracers and at different angular resolutions. There appears to be a quadrupolar outflow structure originating from source A, with one collimated pair of lobes in the north-east to south-west direction and one less collimated (and less well aligned) in the east–west direction. These are shown in Figure 3.10, although unfortunately in mirror image orientation in this figure. The upper two CO molecular outflow contour sets (positive v_{LSR}) in this diagram are red-shifted and the lower two (negative v_{LSR}) are blue-shifted, with the arrows delineating the bipolar directions of each pair. The central white triangle identifies IRAS 16293A and the square marks a DCO^+ (3-2) emission peak that is offset for reasons we will shortly consider.

The initial capability of low-mass protostars to warm much of their surrounding mass of gas and dust is limited. Average densities and temperature in the IRAS 16293 circumbinary envelope gas as a whole are ~10^4 cm^{-3} and ~10 K, values typical of cold, dense, prestellar molecular gas. This is the cold envelope in which we expect zeroth-generation COMs to be forming within

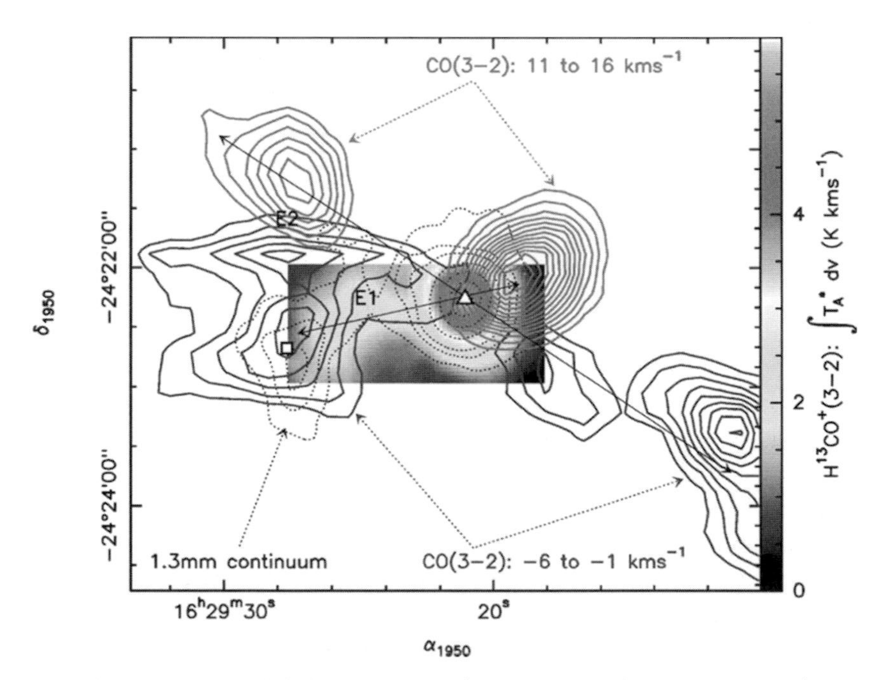

Figure 3.10 Red-shifted (positive kms^{-1}) and blue- shifted (negative kms^{-1}) molecular outflows: IRAS16293A (Lis et al., 2002).

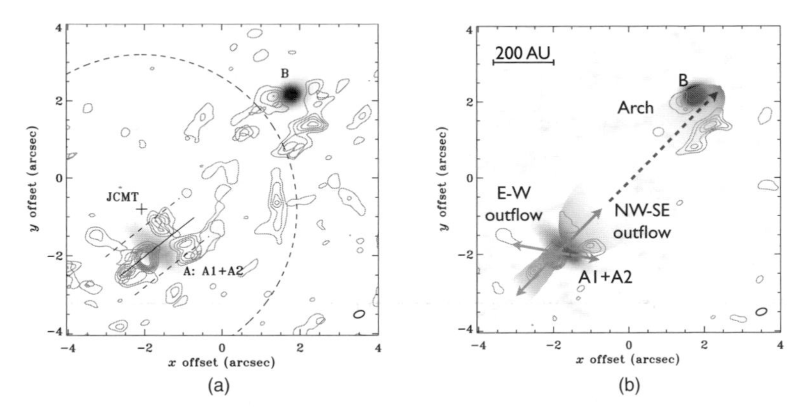

Figure 3.11 Carbon monoxide (CO) emission contours (panel (a)) showing A1 and A2 binary sources. The single-dish JCMT beam (dotted circle with centre marked) contrasts with the ALMA beam (small oval, extreme bottom right). The schematic in panel (b) offers an interpretation of what these observations suggest may be occurring (Kristensen et al., 2013).

grain ices. Interferometer detections, however, are capable of identifying localised hotspots associated with the dense cores, and the most evident are those associated with A1 through a variety of molecular line emissions. It also appears that a blue-shifted outflow bow-shock from IRAS 16293A might be impacting B, as shown in Figure 3.11, entraining material over a ~100 AU distance before deceleration through impact with the envelope (Loinard et al., 2013; Kristensen et al., 2013).

3.5 A Gas Bridge

Looking in more detail at the gas around and between IRAS 16293A and B, we can reflect further on the fact that the majority of stars form as binary or multiple systems, each starting at a different time and evolving under its own local circumstances. With a separation between protostars A and B of ~600 AU, both embedded in a single ~10^4 AU scale envelope, molecular line observations have been made of the associated gas in a variety of species, with an ALMA spatial resolution at its best to date down to ~60 AU (van der Wiel et al., 2019). Figure 3.12 shows these species detected (including CO, H_2CO, HCN, CS, SiO, and C_2H) in the ALMA-PILs 329 to 362 GHz spectral survey. Figures 3.14 and 3.15 show a representative selection of velocity maps for some of these species and others.

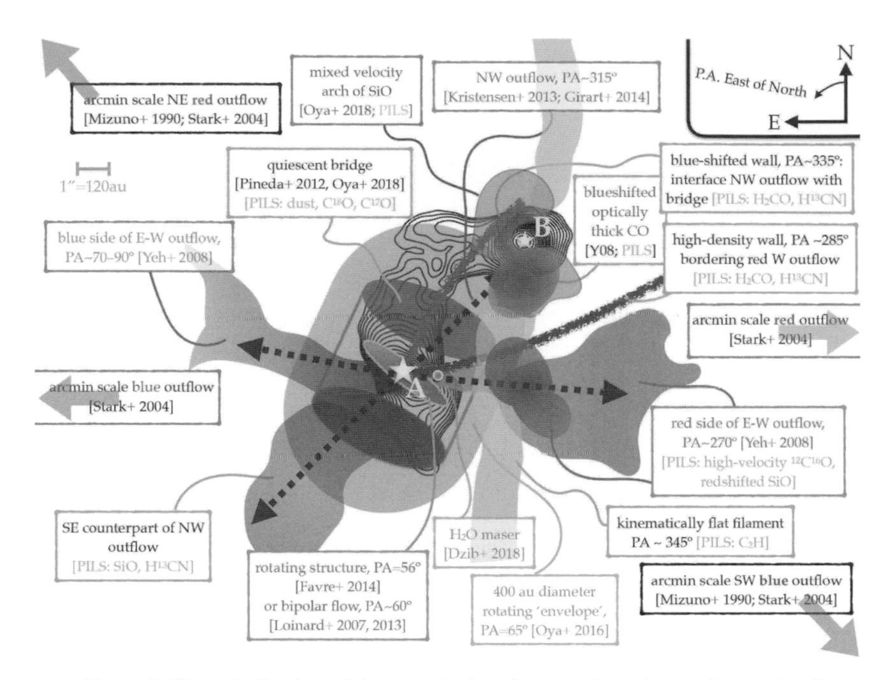

Figure 3.12 An indication of the complexity of gas motions detected around and between the two sources A and B (van der Wiel et al., 2019).

The velocity maps for both optically thick and thin molecular lines observed in recent decades indicate a quiescent bridge of dust and gas spans the region between the two sources, with an inferred density ranging from 10^7 cm^{-3} closest to the sources to 10^4 cm^{-3} around a 'mid-point' (actually closer to source B). As we have noted, a variety of outflows are apparent from source A, evidence of high-density, warm gas; none of the outflow material coincides spatially or kinetically with the bridge.

Simplifying some of the visual complexities of Figure 3.12, Figures 3.13, and 3.14 show the distributions of various observed species. Were Figure 3.13 in the colour coding of the original research paper, we would see the v_{LSR} values on the right-hand scale translated into dark red regions to the south associated with source A, indicating gas at red-shifted velocities between +7 and +11 km s^{-1} moving away from us along the line of sight. The darker colouration to the north associated with location B indicates blue-shifted gas at velocities between –4 and –1 km s^{-1} moving towards us. The SiO velocity map traces both sides of the NW–SE outflow pair most clearly, extending out to ~8″ from source A, blue-shifted to the north-west, red-shifted to the south-east. The gas and dust between protostars A and B is traced in submillimetre dust continuum emission and cold C^{17}O gas, showing itself as kinematically quiescent with a central velocity at

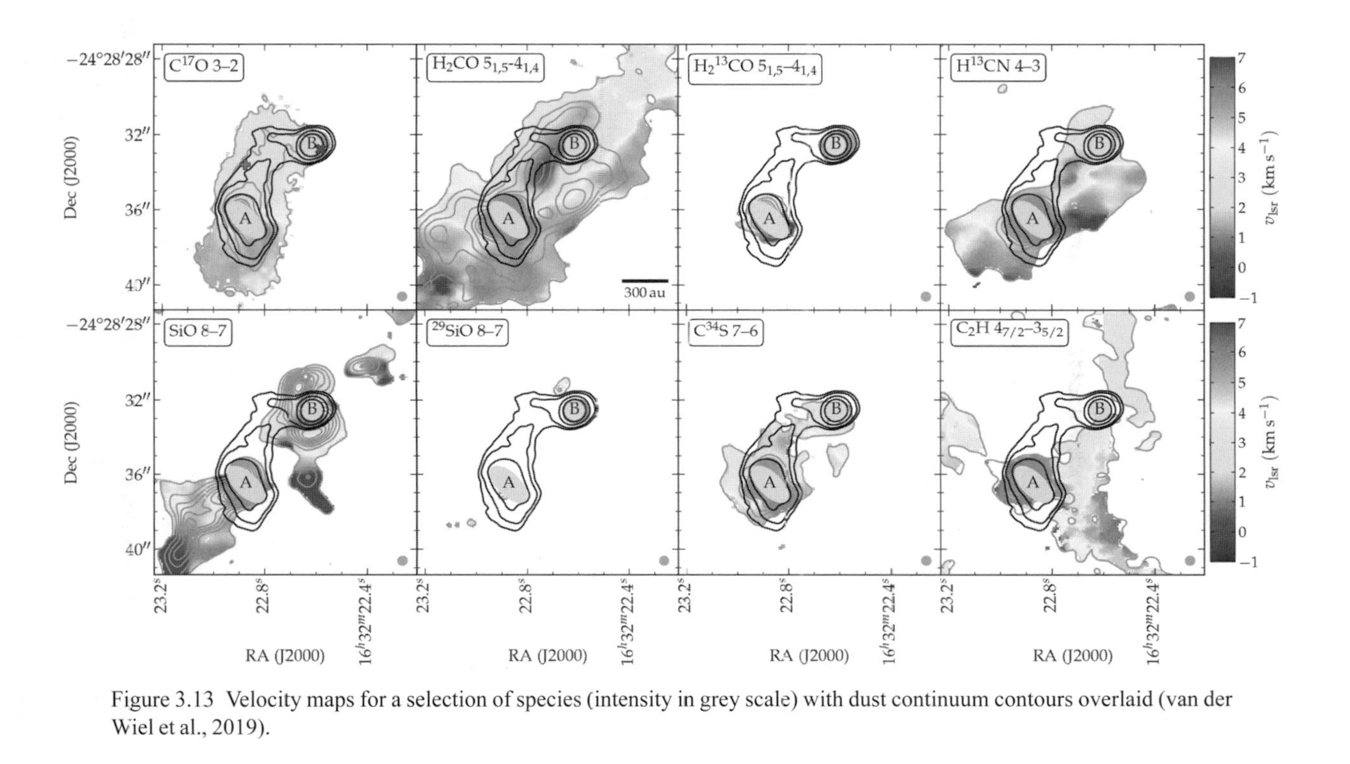

Figure 3.13 Velocity maps for a selection of species (intensity in grey scale) with dust continuum contours overlaid (van der Wiel et al., 2019).

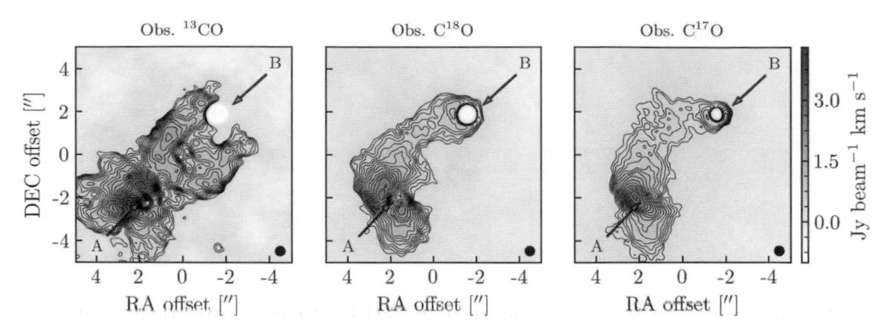

Figure 3.14 Maps of ^{13}CO, $C^{18}O$, and $C^{17}O$ emission from the PILS observations (Jacobsen et al., 2018).

v_{LSR} = +3 km s^{-1}, consistent with the velocities of the individual protostars (+2.7 and +3.1 km s^{-1}), therefore unlikely to be associated with the outflows.

From all the observations it does seem likely that a quiescent bridge of gas could be a remnant of filamentary substructure in the protostellar envelope within which the two sources A and B separately coalesced. The absence of detected outflow signature associated with source B suggests it is in an earlier evolutionary stage than source A. With A and B presenting different orientations of likely accretion disks with respect to our line of sight, A inclined to it and B almost face on, there is not yet direct evidence even of accretion at source B. Even if it is accreting, it appears to be among the youngest low-mass protostars known. Within the complex dynamics inside the larger envelope, one slow-moving outflow possibly associated with B would suggest from its extent and speed a dynamical age of only a few hundred years. We will, however, return to the comparison of sources A and B when considering COMs observations in subsection 3.8.

3.6 Deuterated Evolutionary Tracers

In comparing the evolutionary developments of A and B, the molecular line spectrum is a particularly rich source of potential information. The lack of sensitivity in early interferometer detections meant that single-dish surveys, using telescopes such as the Caltech Submillimetre Observatory (CSO) and the James Clerk Maxwell Telescope (JCMT) equipped with sensitive submillimetre receivers, were employed as alternatives. These two telescopes were both on the high, dry site of Mauna Kea in Hawaii with ready access to ρ Ophiuchus as a southern hemisphere source (Van Dishoeck et al., 1993). Such studies detected hundreds of lines and identified two dozen molecular species through rotational diagrams, non-LTE excitation calculations, and line profile analyses. The envelope structure was modelled from these single-dish observations and found to reproduce the far-infrared/submillimetre line and continuum observations, providing a useful reference for chemical modelling of infalling gas on scales of several thousand AU (Zhou et al., 1995; Narayanan et al., 1998; Ceccarelli et al., 2000; Schoier et al., 2002). The highly sensitive IRAM 30 m telescope and the JCMT then revealed a much larger variety of COMs, including HCO_2CH_3, CH_3OCH_3, and C_2H_5CN (Cazaux et al., 2003, 2011) and including the first detection of NH_2CHO in a low-mass protostar (Kahane et al., 2013). Surveys have also been made with the Herschel Space Observatory (Ceccarelli et al., 2010; Hily-Blant et al., 2010; Bacmann et al., 2010; Bottinelli et al., 2014) and on SOFIA (Parise et al., 2012), revealing deuterated hydrides, water, and high-excitation lines of heavier molecules. With more-recent high-resolution observations, such as the ALMA-PILS survey (Protostellar Interferometric Line Survey), this rich chemistry has been shown to include the first detections towards a low-mass source of CH_2OHCHO (glycoaldehyde), c-C_2H_4O (ethylene oxide), and C_2H_5CHO (propanal) (Jorgensen et al., 2016), and we will look more closely at these 'prebiotic' COMs and their exact locations within IRAS 16293 in subsection 3.10.

Observing the envelope around the protostars of IRAS 16293 as a function of depth along the line of sight, three distinct physical and chemical environments were first distinguished twenty years ago. The first is compact, turbulent, warm (over 80 K), and dense (~$10^7 cm^{-3}$), rich in silicon- and sulphur-bearing molecules as well as the ubiquitous CH_3OH and CH_3CN, eventually identified specifically with the IRAS 16293A location. Apart from the lower temperature and lesser total mass of gas and dust, environments such as this were originally designated hot corinos in recognition of their similarity (in molecular richness) to the earlier delineated hot cores associated with high-mass star formation. The second distinct environment originally identified was a quiescent (narrow line width) circumbinary envelope traced in commonly observed dense cloud molecules such as CS, HCO^+, and H_2CO. Thirdly, the colder outer envelope observed and modelled from the 1990s was identified by typical cold cloud radicals such as CN, C_2H, and C_3H_2.

The key distinction between the molecular activity that emerges in regions close to low-mass, as opposed to high-mass protostars arises from the much longer timescales that gas and dust are exposed to lower-temperature and high-density conditions in the LMSF case. This allows for extensive fractionation in gas–grain chemistry and explains the abundance of doubly and even triply deuterated molecules that we now know are common towards IRAS 16293A – species such as D_2CO, ND_2H, ND_3, CHD_2OH, and CD_3OH. In general, gas-phase deuterated species are more stable than their hydrogenated counterparts, having higher activation energy requirements in both exchange and recombination reactions that destroy them. The reactions of interest take the general form of $XH^+ + HD \Leftrightarrow XD^+ + H_2 + \Delta E$, with ΔE values typically a few hundred degrees K, so deuterium fractionation in the left to right direction is favoured where the gas is cold. Deuteration is also enhanced through cold grain-surface reactions with accreted D atoms. With high deuterium fractionation levels in the gas, deuterium exchange reactions such as H_2D^+ to DCO^+ are followed by dissociative electron recombinations that result in high atomic D-to-H abundance ratios. The same range of hydrogenation reactions with surface species are open to deuterium reactivity, although typically are critically dependent on the other species present that preferentially remove the lower-mass elemental hydrogen (in forming species such as CH_3OH, for example). The developing chemical complexity in the gas phase therefore correlates well with the length of time available for the evolution of dust surface chemistries in low-mass protostellar envelopes, to be followed by the release into the gas phase of a rich chemical mix as the protostar warms its immediate surroundings. This would account for the range of species observed in warm core/hot corino gas, including many complex hydrocarbons such as $HCOOCH_3$ that have an undoubted grain-surface origin. Additional species seen towards IRAS 16293A are CH_3OCH_3, C_2H_5CN, and the first detection of NH_2CHO (formamide) in a low-mass protostellar environment, not to mention abundant H_2O, its minor isotopologues ($H_2{}^{17}O$ and $H_2{}^{18}O$) and deuterated forms (HDO and D_2O) (Jorgensen et al., 2016). These studies of IRAS 16293 and other comparative sources make it very clear that there are localised gradients in temperature and density on the scale of just a few hundred AU that do undoubtedly affect the chemical signatures of these low-mass protostellar regions.

3.7 Disk–Envelope Interface

One interesting tracer specific to the interface between disk and envelope in IRAS 16293A is DCO^+. It occupies a position slightly offset from the protostar itself, bordering what is taken to be a disk-like structure, a positioning similarly observed towards other comparable sources. Detections in DCO^+ 5-4 and 3-2 transitions show a half-crescent shape location centred around source A, with the peak of both transitions red-shifted and close to one another about 2″ to the south-west, as shown in Figure 3.15. This asymmetric distribution, its location, and red shift are consistent with a rotating accretion disk as drawn in Figure 3.16, with DCO^+ additionally tracking the outflow extending south-east.

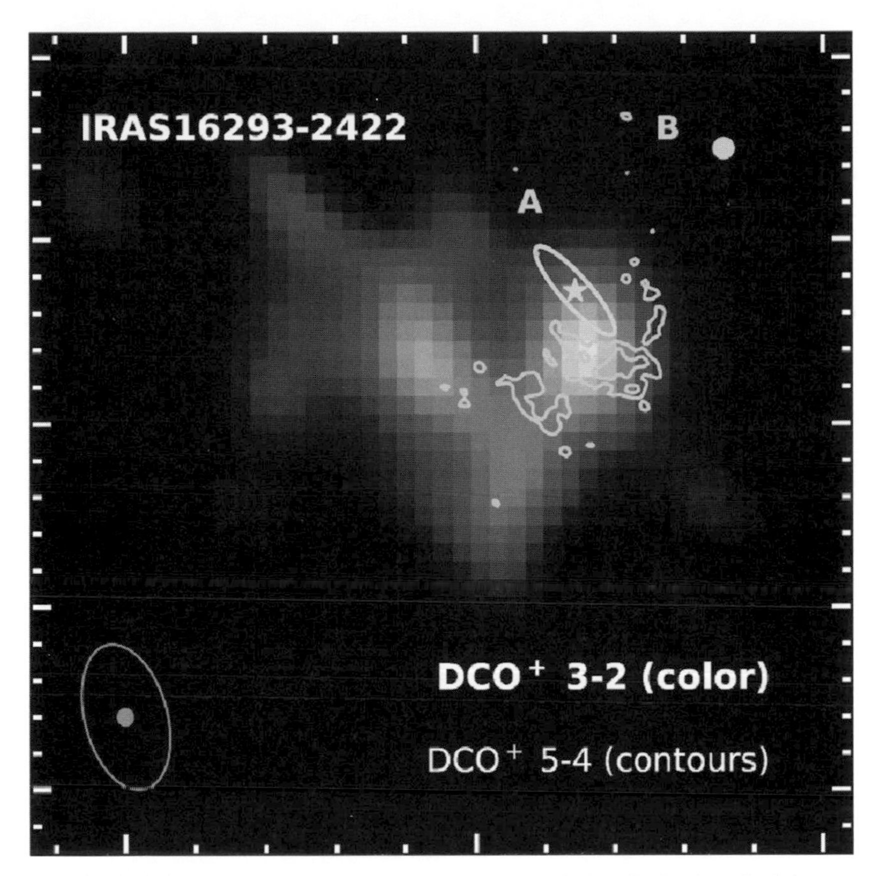

Figure 3.15 Atacama Array images of two DCO+ emission distributions, in greyscale and contours (Murillo et al., 2018).

DCO^+ formation chemistry is particularly temperature sensitive, and the DCO^+ peak emission position in both A and B sources corresponds to a drop in the temperature profile if we envisage the presence of a disk that shadows (thus cooling) the envelope. DCO^+ formation and emission would be closer to the source protostar in the disk plane than would otherwise be anticipated. This scenario ties in with observations in other sources using ALMA in which DCO^+ emission is evident in both the cold envelope and the disk–envelope interface (Murillo et al., 2018). Taking CO and H_2D^+ as precursors to DCO^+ (Figure 3.17), given the ready abundance of CO frozen out onto grain surfaces, the rate-determining step in the reaction sequence is the gas-phase formation of H_2D^+. In the case of $H_3^+ + HD \leftrightarrow H_2D^+ + H_2$, the reverse reaction has an activation energy of over 200 K. Since H_3^+ is a product of gas-phase reaction between neutral and cosmic ray–ionised molecular hydrogen, the cosmic ray

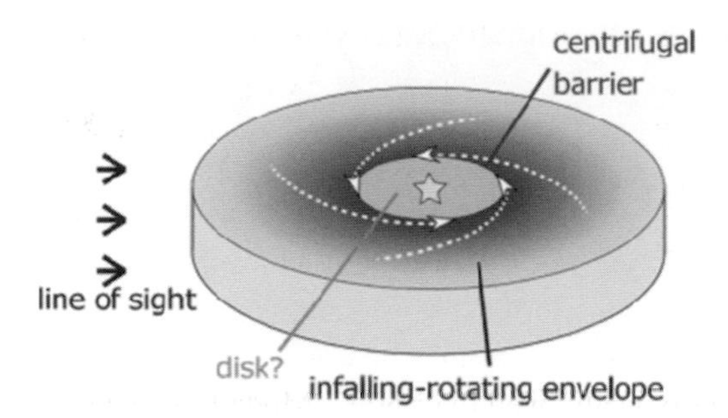

Figure 3.16 Schematic of an infalling-rotating envelope and accretion disk (Sakai et al., 2014).

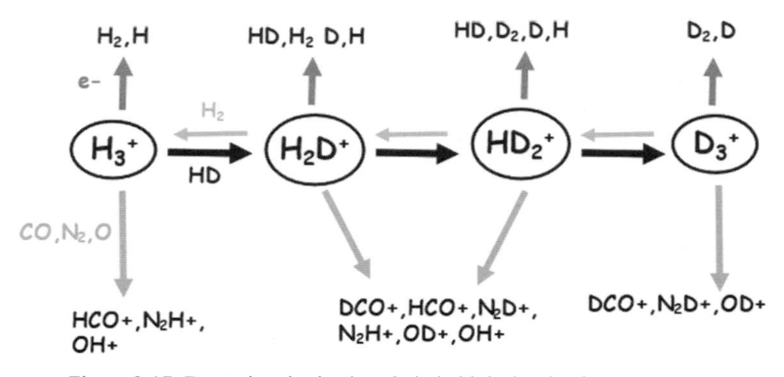

Figure 3.17 Deuterium ionisation chain in high-density, low-temperature gas (Roberts & Millar, 2006).

ionisation rate becomes a critical factor and one of several modelling variables we can explore, alongside such contributors as CO sublimation temperature and desorption density (Murillo et al., 2018). As H_3^+ is difficult to observe directly, the degree of ionisation in a cloud relies on the abundances of species further along the chain of ionisation, illustrated in Figure 3.17. For example, the DCO^+/HCO^+ ratio (R_D) is often used, and standard texts derive pragmatic expressions which can be linked to other species (e.g. HCO^+ to CO in Tielens, 2021, pp. 409–10). The R_D parameter becomes a function of the degree of ionisation (literally the abundance of anions), the CR ionisation rate divided by the density (ζ_{CR}/n_{H2}), and (in the case of HCO^+/CO) the depletion factor of CO (f_d).

With temperature a particular controlling factor in the evolution of the gas-phase chemistry here, the influence of disk shadowing on envelope temperatures is significant, as is the impact of early protostellar UV heating on outflow cavity material. A best fit from both modelling and observational analysis puts the DCO^+ emission towards IRAS 16293A in a gas environment in which particle densities may be as high as 8×10^7 cm^{-3}, while temperatures are little more than 20 K (alongside dust a few degrees less). We can compare this with a second tracer – the small hydrocarbon c-C_3H_2 – which seems to mark the boundary between the cavity engendered by the southern outflow and its ambient cloud. The formation of small hydrocarbons such as this probably follows CO sublimation from dust-grain ices and UV photodissociation freeing up atomic carbon for warm gas-phase reactions. The line emission from c-C_3H_2 is itself probably energised by UV radiation from source A within the warmer outflow where temperatures are typically well over 50 K. Modelling suggests that c-C_3H_2 follows the interface, moving out into the disk plane as the outflow cavity expands with time (Drozdovskaya et al., 2015).

3.8 Disk Reservoirs

Where disks form in the early stages of low-mass protostellar evolution, they provide a long-lived reservoir of molecules, with rotation preventing major infall, and their presence inevitably influencing the thermal structure of their surroundings. The disk will shield surrounding envelope material in the same plane from heating by the evolving protostar, as we have said, and DCO^+ at its peak emission position seems to be associated with just such a drop in temperature at the disk–envelope interface. Figure 3.16 shows that DCO^+ emission is close to source A but red-shifted and located 2″ southwest as both envelope and disk–envelope interface gas, consistent with the schematic model of Figure 3.18.

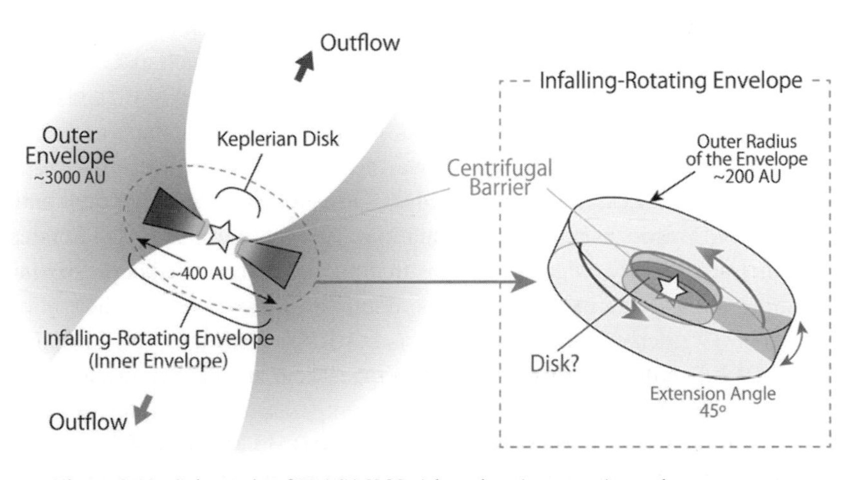

Figure 3.18 Schematic of IRAS16293-A based on Atacama Array data from selected hydrocarbons at sub-arcsecond resolution (0.″6 × 0.″5) (Oya et al., 2016).

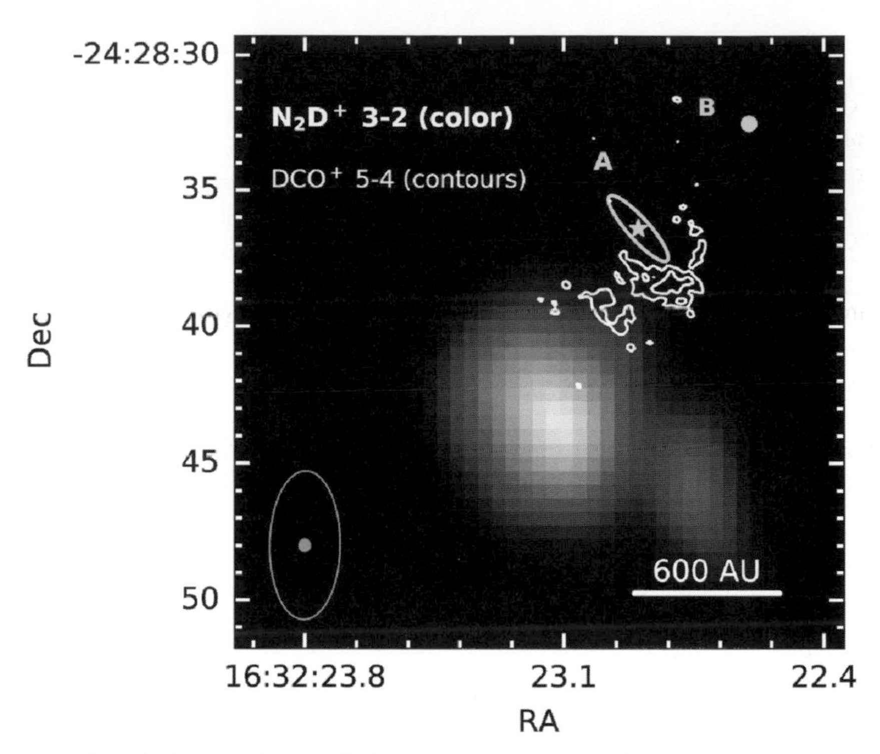

Figure 3.19 ALMA images of N2D+ against DCO+ distribution (Murillo et al., 2018).

The peak emission distribution of a second deuterated molecule N_2D^+ in spatial relation to DCO^+ is shown in Figure 3.19 and also likely to be indicative of temperature differentiation. Below 20 K, molecular nitrogen (N_2) freezes out onto grains. While other simple nitrogen-bearing species such as CN, HCN, HNC, and NO can form both in the gas and on grain surfaces, N_2H^+ and N_2D^+ form only in the gas if N_2 is present. As marked in the schematic of Figure 3.10, it looks as though N_2D^+ is tracing warmer gas probably close to the cavity boundary. As in the case of the observed HCO^+/DCO^+ abundance ratio, so with the N_2H^+/N_2D^+ ratio; as a first approximation, we may infer that the deuterated variants trace slightly colder conditions than their hydrogenated counterparts, assuming comparable particle densities, depletion factors, and CR ionisation rates. The CO derivatives are characteristic of standard dense cloud conditions ~10 K, while the nitrogenous species characterise slightly warmer gas, >20 K. It may therefore be the case that, as temperature probes, the sequence runs $HDO^+ < HCO^+ < N_2H+ < N_2D^+$ over, say, a 10–30 K range. However, the evolutionary development of IRAS16293A offers a much more complex mix of physical structure and kinematic activity than these rather simple chemical clocks can yet differentiate.

3.9 Centrifugal Barriers

The schematic of Figure 3.16 shows an idealised infalling-rotating envelope of gas and dust that we might still envisage is representative of what is occurring at the location of IRAS 16293A. In our particular case, the disk angle is more oblique to our line of sight than shown, but we may imagine an evolving Keplerian disk of radius ~100 AU within the surrounding envelope of radius ~1000 AU. Recent observations of a number of COMs have lent support to a model of the velocity structure of such an infalling-rotating envelope, in which the centrifugal barrier results in a marked distinction between chemistries in distinct zones. The centrifugal barrier is located at half the centrifugal radius, a transition zone in which the kinetic energy of infalling gas is converted to rotational energy. We expect a weak accretion shock on the protostellar side of the barrier, perhaps sufficient to disrupt grain mantles and alter the gas-phase mix within material also experiencing protostellar heating. An observed enhancement of SO, sulphur monoxide, in a different low-mass protostellar source has previously been explained in this manner (Sakai et al., 2014). Since the rotationally supported disk forms within the barrier, disk evolution is potentially traceable by chemical composition.

In Figure 3.20, ALMA observations of IRAS 16293A have placed CH_3OH and $HCOOCH_3$ in compact distributions, taken to be within the possible centrifugal barrier, with OCS exclusively in the surrounding envelope and H_2CS in both disk and envelope locations (Oya et al., 2016).

The OCS map of Figure 3.20 shows a velocity gradient around protostellar source A in a north-east to south-west direction. The darkest colouration in the greyscale to the north-east is blue-shifted gas, the lighter grey to the south-west is red-shifted. This seems to be a clear signature of infalling gas in a rotating envelope.

The comparable maps for CH_3OH and $HCOOCH_3$ in Figure 3.21 show the same direction of velocity gradient around source A, over a smaller portion of the envelope. All three species observed show no comparable velocity gradients associated with source B, which lies face-on in the plane of the sky (Oya et al., 2016).

Along both the disk-envelope and the outflow (perpendicular to it) directions (position angles 65° and 155°, respectively), shown in Figure 3.22 for OCS, position-velocity (P-V) diagrams are shown in Figure 3.23.

The P-V diagram along the disk-envelope direction shows a spin-up feature towards the protostar, and the P-V diagram along the outflow direction shows a significant velocity gradient. However, since OCS emission is concentrated around the protostar, the contribution of outflows to the velocity gradient

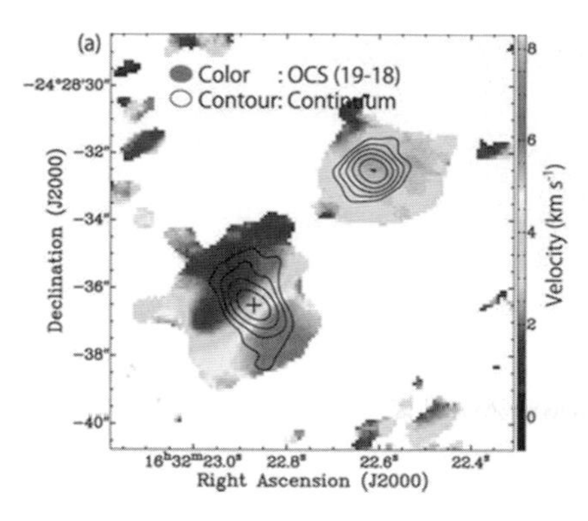

Figure 3.20 OCS velocity and distribution mapped in greyscale against continuum contours in IRAS16293A and B (Oya et al., 2016)

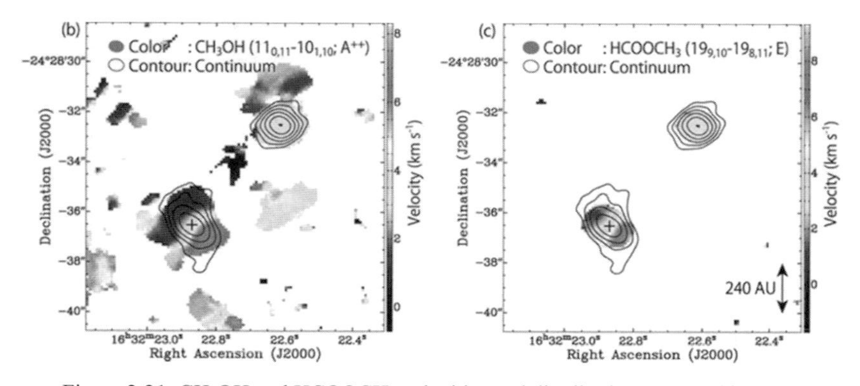

Figure 3.21 CH_3OH and $HCOOCH_3$ velocities and distributions mapped in greyscale against continuum contours in IRAS16293A and B (Oya et al., 2016)

seems much less likely than that the gradient indicates infalling gas in a rotating envelope close to the protostar. Modelling the infalling-rotating envelope scenario matches a kinematic structure reproduced for a protostellar mass of 0.75±0.25$M_{\odot}$, a disk radius of 50±10AU, and an angle of inclination of 60° (edge-on being 90°). The velocity components of other emissions (e.g. H_2CS) are equally well modelled, assuming a 0.75$M_{\odot}$ protostellar value and modelling further points to gas temperatures in the three regions (i.e. infalling-rotating envelope, centrifugal barrier, and Keplerian disk) as 95±25 K, 130±30 K, and 105±35 K, respectively. The ratios of $HCOOCH_3/CH_3OH$, both of which species are confidently believed to form efficiently on grain surfaces, equally increase from two to five to nine progressively towards the protostellar centre.

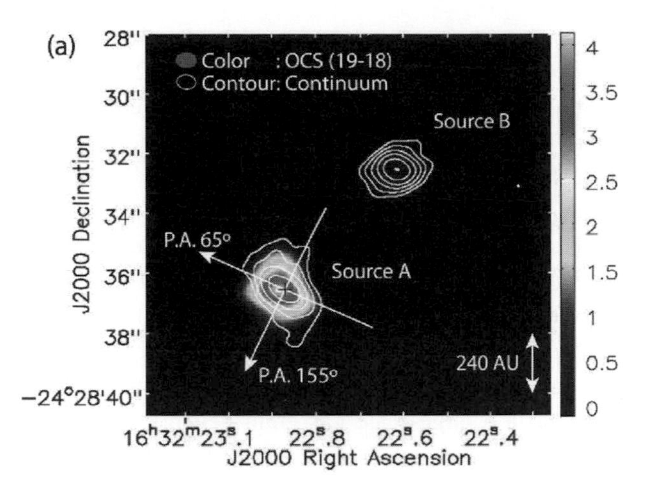

Figure 3.22 The disk-envelope and outflow directions of protostellar source IRAS 16293A (Oya et al., 2016).

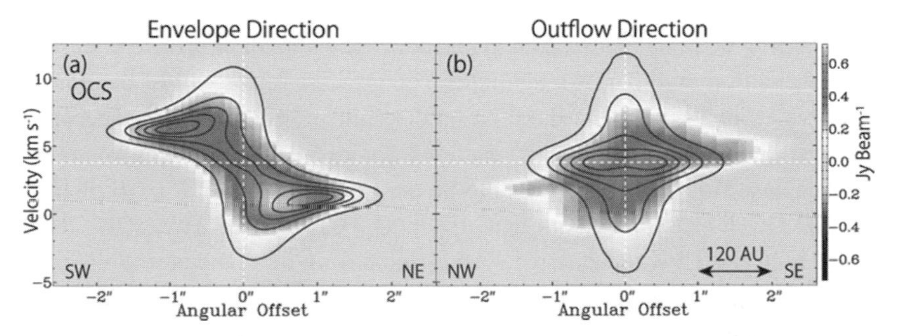

Figure 3.23 The position-velocity diagrams for OCS along each direction shown in the previous figure (Oya et al., 2016).

In short, the physical conditions associated with the centrifugal barrier and a rotationally supported disk are intimately related to the observable chemical complexities on either side of, and within, it.

3.10 Prebiotics

With the resolution of the ALMA observations, the PILS survey has led to detections of ever more-complex species around solar-type protostars, some of which are considered precursors to biologically significant molecules, and therefore of particular interest to those studying protoplanetary disks, planet formation, and transient Solar system bodies. Towards IRAS 16293A, these have included NHDCHO and NH_2CDO (deuterated formamide) and HNCO (isocyanic acid) (Coutens et al., 2016); CH_3Cl (methyl chloride) (Fayolle et al., 2017); CH_3NCO (methyl isocyanate) (Ligterink et al., 2017); NH_2CN (cyanamide) (Coutens et al. 2018); CH_3NC (methyl isocyanide) (Calcutt et al., 2018); and HONO (nitrous acid) (Coutens et al., 2019). Towards the hot corino in IRAS 16293B, detections have also been made of C_2H_5CHO (propanal) (Lykke et al., 2017) and CH_3CCH (methyl acetylene) (Calcutt et al., 2019). In addition, the unsaturated radicals C_2H_3CHO and C_3H_6, both thought to be significant precursors to saturated prebiotic COMs, have been detected towards IRAS 16293B and modelled in a large-scale gas-grain network (Manigand et al., 2021).

C_2H_3CHO and C_3H_6 were detected along with other species, listed in Table 3.1, and including some non-detection column density upper limits for those species thought to be related in the chemical network and for which abundance constraints contribute.

Modelling was then undertaken to reproduce the abundances observed, successfully within an order of magnitude. The greatest sensitivity of the model is, as we would expect, the duration of the cold prestellar phase, especially where successive hydrogenation reactions dominate the formation of saturated species. This certainly appears to be the case for CH_3CCH, HC(O)CHO, and CH_3OCHO, as precursors to C_3H_6, C_3H_8, $CH_2(OH)$, CHO, $(CH_2OH)_2$, and CH_3OCH_2OH, with saturated species increasing in abundance while precursors decrease the longer the cloud remains cold.

In modelling the warmer conditions, successive hydrogenation reactions on the ices prior to desorption highlight that of C_3O, forming HCCCHO, C_2H_3CHO, and C_2H_5CHO, and the radical-radical additions of HCO and C_2H_3 (or C_2H_5) equally contributing to C_3-species abundances. However, the modellers suspect that there are missing destruction pathways for C_3 and C_3O since the overall production of C_3-species generally seems so efficient. One possibility would be the production and growth of polyaromatic hydrocarbons (PAHs) on the grain surfaces, which is not included in the model (Manigand et al., 2021).

Finally, Figure 3.24 shows some abundance comparisons observed between prebiotic COMs in the two protostellar sources A and B. The differences are partially spatial and functions of sublimation temperatures, with the more compact resolved hotter corino region detected towards source A.

Table 3.1 *Column densities (including non-detection upper limits) and excitation temperatures for COMs towards IRAS 16293B. (†) The authors chose a conservative estimation of the relative uncertainty on the column density of 50 per cent. (a) The excitation temperature is fixed to the mean excitation temperature of C_2H_3CHO and C_3H_6 and consistent with the excitation temperature of CH_3CCH (Manigand et al., 2021).*

Species	T_{ex} (K)	N_{tot} (cm^{-2})
C_2H_3CHO	125±25	$3.4 \pm 0.7 \times 10^{14}$
C_3H_6	75±15	$4.2 \pm 0.8 \times 10^{16}$
HCCCHO	100[a]	$<5.0 \times 10^{14}$
n-C_3H_7OH	100[a]	$<3.0 \times 10^{15}$
i-C_3H_7OH	100[a]	$<3.0 \times 10^{15}$
C_3O	100[a]	$<2.0 \times 10^{13}$
cis-HC(O)CHO	100[a]	$<5.0 \times 10^{13}$
C_3H_8	100[a]	$<8.0 \times 10^{16}$
CH_3CCH	100±20	$1.1 \pm 0.2 \times 10^{16}$†
C_2H_5CHO	125±25	$2.2 \pm 1.1 \times 10^{15}$†
CH_3CHO	125±25	$7.0 \pm 3.5 \times 10^{16}$†
CH_3COCH_3	125±25	$1.7 \pm 0.8 \times 10^{16}$†
$C_2H_5OCH_3$	100±20	$1.8 \pm 0.2 \times 10^{16}$†

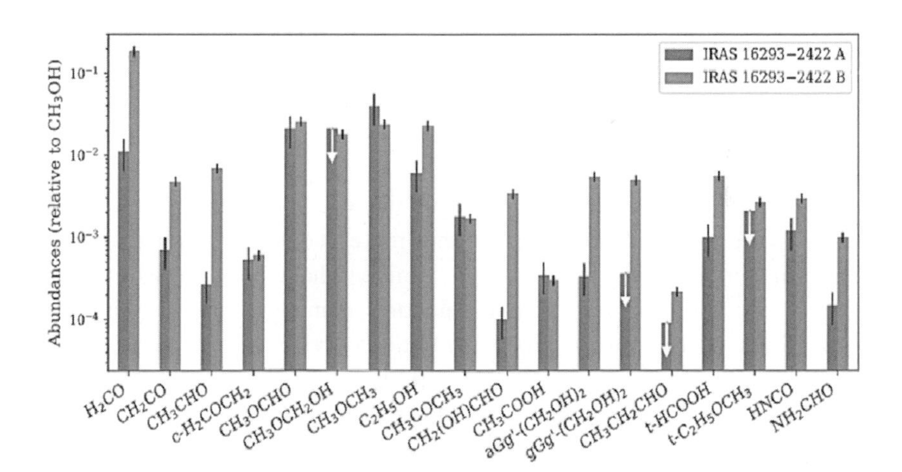

Figure 3.24 Abundance of the main isotopologues relative to CH_3OH in IRAS 16293A and B chemical modelling. The left-hand side in each pair is protostellar source A, and the right-hand side is source B (Manigand et al., 2020).

3.11 Nitriles

The ALMA-PILS survey also detected a variety of molecular species in the nitrile (R-CN) family, with derived abundances broadly similar for both sources A and B, with one notable exception. That exception, CH_3CN, was observed along with five of its isotopologues, including CHD_2CN (a first detection in the ISM), and the three COMs: C_2H_5CN (ethyl cyanide), C_2H_3CN (vinyl cyanide), and HC_3 N (cyanoacetylene). LTE spectral emission modelling produced excitation temperatures for all nine species spanning 100–160 K, and abundances relative to CH_3OH and CH_3CN (Calcutt et al., 2018).

Most species peak in abundance at source A, although on the smaller scale the abundance differences between those of A and B are less noticeable, particularly when relative to CH_3CN. Figure 3.25 shows those relative abundances (with Sgr B2 (N2) as a canonical reference; see Chapter 7). The exception to that close equivalence is C_2H_3CN, vinyl cyanide, which is found exclusively in source B. It has been proposed as an evolutionary tracer in high-mass hot core conditions (Caselli et al., 1993; Fontani et al., 2007), with higher abundances relative to its saturated daughter species ethyl cyanide (C_2H_5CN) indicating a more evolved object. Vinyl cyanide could form early in low-mass star formation on dust grains from the hydrogenation of HC_3N, before further hydrogenation results in ethyl cyanide. On the other hand, modelling has also found C_2H_3CN abundance only being significant after the warm-up phase as a destruction product of C_2H_5CN through protonation and dissociative recombination (Garrod et al., 2017). Both scenarios suggest that observed differences between sources A and B at least indicate differences in evolutionary stage, whether the result of timescale or temperature. In either schema, source B is either more evolved or has had a longer warm-up period for much more vinyl cyanide to have been formed (the difference between observed and non-detection abundance ratio is a factor of ten). How does this square with the additional evidence of source B being much younger, not least in its lack of outflow activity compared with source A and its lower apparent luminosity? One possibility is that, while both protostars could be of roughly similar age, source A could have a much higher accretion rate. This would account for both luminosity and outflows, and result in warm-up timescales sufficiently short that vinyl cyanide is not efficiently formed.

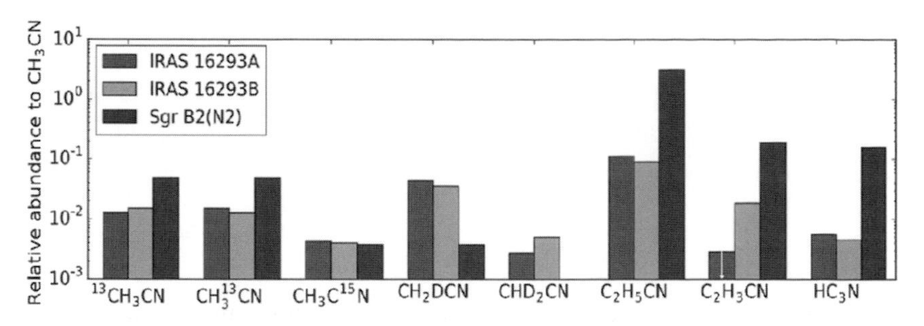

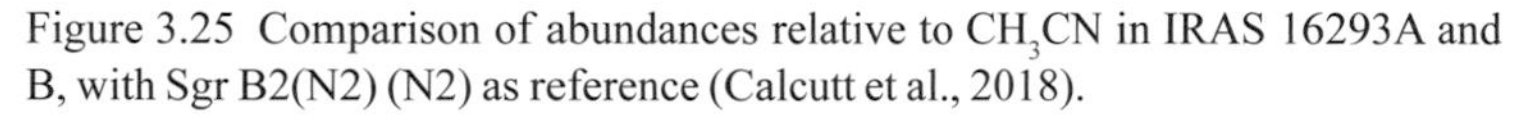

Figure 3.25 Comparison of abundances relative to CH_3CN in IRAS 16293A and B, with Sgr B2(N2) (N2) as reference (Calcutt et al., 2018).

3.12 Summary

IRAS 16293-2422 offers a detailed view of the principal components associated with low-mass star formation, from dense cloud filament to starless core to prestellar core to protostellar envelope, accretion disk, and warm/hot corino chemistry. It illustrates the conditions in which the chemical signatures involving COMs help us to define the structure of disks and envelopes on scales of ~100–1000 AU. Both COMs and deuterated species, particularly the ratios of deuterated species to their hydrogenated counterparts, trace gas and dust temperatures and densities, and compositionally dependent gas–grain interactions, through comparisons with chemical modelling. The following chapter takes us to a second low-mass star formation region and extends the questions about chemical composition as indicator of protostellar evolution, reinforcing some conclusions derived from IRAS 16293 studies while identifying distinctions between what appears to be hot corino chemistry and a warm carbon-chain chemistry (WCCC) alternative.

4
NGC 1333 in Perseus

4.1 Introduction

Having touched on COMs formation in previous chapters and what these species can tell us about the low-mass protostellar environment, we can now look to alternative formation mechanisms and attempt to quantify distinctions between corino and warm core chemistries. The protostellar molecular content within the Perseus Molecular Cloud (PMC) has been probed with observations at mid-IR (Spitzer: Enoch et al., 2009), far-IR (Herschel: Sadavoy et al., 2014), sub-millimetre (JCMT: Sadavoy et al., 2010), and radio wavelengths (VLA: Tobin et al., 2016; Tychoniec et al., 2018). A total of 94 Class 0/I protostars and Class II objects are known to populate the entire cloud (Tobin et al., 2016). Focusing on the most active low-mass star formation cluster gives plenty of scope for comparative sampling.

4.2 Envelope and Accretion

The PMC is almost invisible in the optical but for the two major cluster nebulae, IC348 and NGC 1333, both shown in Figure 4.1. The latter, located in the western part of the PMC, is illuminated by a B8 star (BD+30°549), and the NGC 1333 label these days denotes both the reflection nebula and an associated young stellar cluster evident in numerous Hα emission lines. The molecular mass of the region is estimated to be about 450 $M_\odot$, and there is evidence of continuing infall observed through red-shifted HCO^+ at an estimated rate of 10^{-4} $M_\odot$ yr^{-1} over a 0.4 $parsec^2$ region (Walsh et al., 2006).

The sky map of Figure 4.2 shows the principal visible stars of the Perseus constellation, in relation to which the PMC lies north-east to south-west behind Atik (ζ Persei) in the constellation's far south. NGC 1333 alone

Figure 4.1 The principal features of the Perseus Molecular Cloud in the optical (Adam Block/Bally et al., 2008).

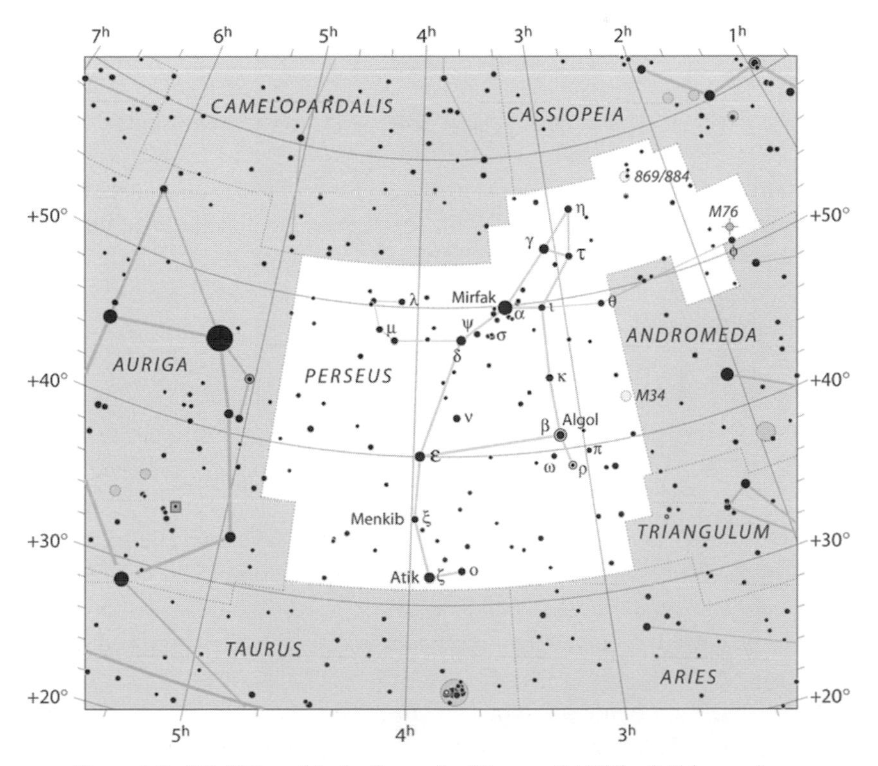

Figure 4.2 Atik (ζ Persei) in the far south of Perseus (IAU/Sky & Telescope).

marks the location for hundreds of newly formed stars, all less than a million years old, forming at an average rate around one Solar mass star every ten thousand years.

Lying at the western end of the PMC shown in Figure 4.3, an exact distance to NGC 1333 remains rather uncertain but is thought to be between 300 and 350 pc. This puts it closer even than the Orion features to be considered in later chapters that are typically thought of as near neighbours to the Solar system. Importantly, the proximity of NGC133 brings the structures of envelopes and accretion disks around the low-mass stars in the feature within range of detailed interferometric observation. The Spitzer Space Telescope observations of the first decade of the century identified about 40 low-mass protostars and 100 pre-main-sequence stars with disks (Jorgensen et al., 2007). Given their roughly uniform distribution within a 0.3 pc radius, a typical separation

Figure 4.3 The western PMC region (Jeff Lunglhofer/Bally et al., 2008).

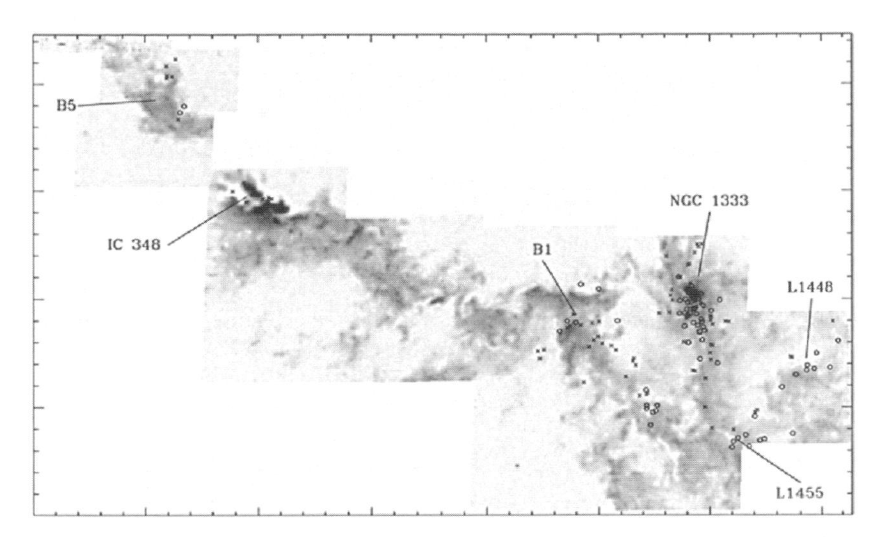

Figure 4.4 The Perseus Molecular Cloud in ^{13}CO emission, identifying the location of NGC1333, with small circles marking Herbig–Haro objects (jet impact reflection nebulae), clear indicators of star formation activity (Bally et al., 2008).

of about 0.045 pc seemed likely. In contrast to observations at the north-eastern end of the PMC, the stellar population of NGC 1333 and its south-western neighbours, L1448 and L1455, are more recently formed. Of the embedded protostars in NGC 1333, its southern half contains the brightest Class0/I objects, including the two sources IRAS2 and IRAS4 and their individual components, which will be our focus. Figure 4.4 shows a ^{13}CO map of the PMC and in particular shows the distribution of a large sample of shock-excited Herbig–Haro (HH) objects arising from many active outflows, all less than a million years old and all within a parsec range of one another. Figure 4.5 is from an Hα survey of the southern sector of NGC 1333, showing the specific IRAS2 and IRAS4 locations (identified also as G311.37-31.0).

4.3 Warm Carbon-Chain Chemistry (WCCC)

Before looking closely at IRAS2 and IRAS4, let us reinforce our understanding of the issue of molecular complexity in low-mass star formation environments that we met in the previous chapter. It is obvious that the first detections to be made among astronomical sources are those with the brightest emission. Subsequently, the less apparent cases in both observation and theory come to light and we begin to understand that a spectrum of possibilities may actually

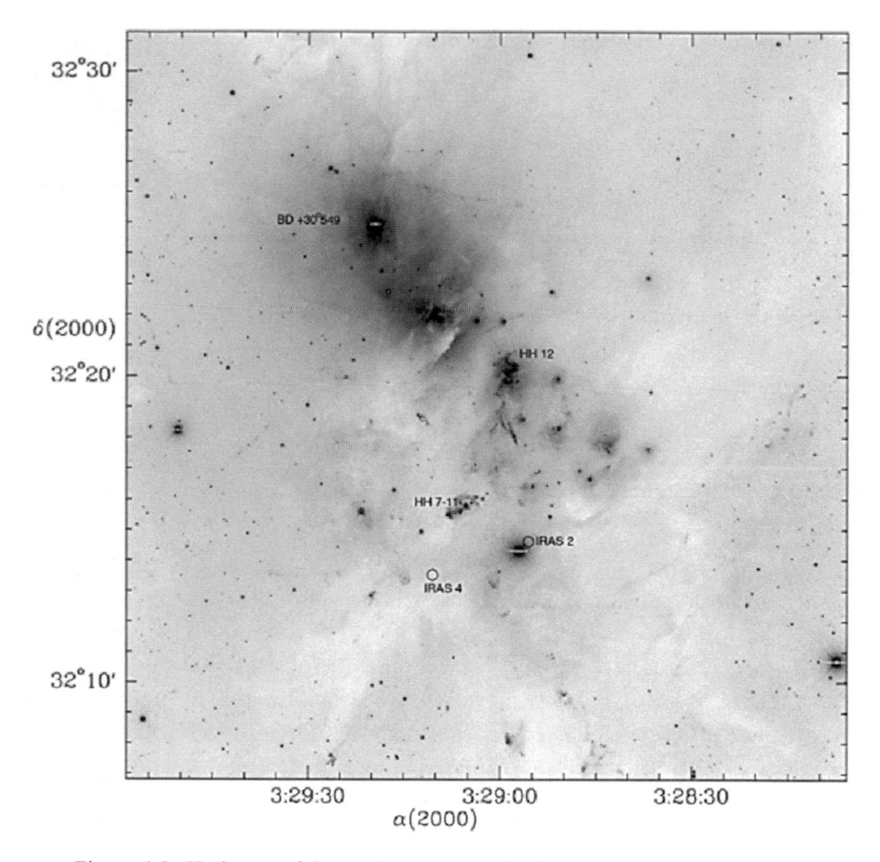

Figure 4.5 Hα image of the southern region of NGC 1333 showing IRAS 2 and IRAS 4 (Walawender et al., 2008).

be the norm. For example, in observing the immediate surroundings of low-mass Solar-type protostars, the compact (<100 AU), dense (>10^7 cm^{-3}), and hot (>100 K) sources designated hot corinos were first observed nearly twenty years ago with single-dish telescopes. These emissions revealed a relatively undifferentiated but rich gas-phase mix of COMs that we met in Chapters 2 and 3, such as CH_3OH (methanol), CH_3CHO (acetaldehyde), and $HCOOCH_3$ (methyl formate) (Ceccarelli et al., 2000, 2017; Cazaux et al., 2003; Bottinelli et al., 2004a). The corino was taken as the low-mass star formation equivalent of the hot cores associated with high-mass stars that had been observed by that time for over a decade previously, even though astronomers had little detailed understanding of the nature of either. Interferometric detections subsequently offered resolutions of hot corino sources down to just a few arc-seconds, pinpointing the inner regions of what we now take to be protostellar envelopes (Bottinelli et al., 2004b; Kuan et al., 2004). In contrast, while innermost regions were devoid of COMs, they were found to be surrounded by larger regions, out to distances ~2,000 AU, that were enriched by what became known as warm carbon chain chemistries (WCCCs). These were typified by emission from species such as CCH (ethynyl radical), c-C_3H_2 (cyclopropenylidene), and C_4H (butynyl radical) (Sakai et al., 2008; Sakai & Yamamoto, 2013). Observations of intermediate cases having both hot corino and WCCC characteristics have also since been discovered (Oya et al., 2017) and the research thinking moves on to ever more nuanced interpretation.

From the theoretical point of view, two basic scenarios for the formation of all COMs in corinos and WCCC sources are current, although we will consider some recent refinements to these later in the chapter. The first model has already been referred to in the preceding chapters, and this original schema posits a gas-phase chemistry triggered by the sublimation of dominant ice components, in particular CH_3OH, a scenario that was an obvious extension of the high-mass star formation hot core case which had been previously modelled with significant success (Millar et al., 1991; Charnley et al., 1992). In the low-mass cases, however, with ongoing laboratory experiments and developing theoretical calculations, one or two assumptions made in the early gas-phase models have been called into question. For example, the dissociative recombination of large ions is no longer taken to necessarily result in COMs formation but at least as often to their fragmentation (Geppert et al., 2006; Hamberg et al., 2010). Equally, some of the accepted ion–molecule reactions in gas-phase modelling persistently fail to generate observed abundances of common species such as $HCOOCH_3$ under corino conditions, just as we will see in high-mass sources that they still sometimes fail in the case of hot core CH_3CN (Millar, Macdonald & Gibb, 1997; Horn et al., 2004).

A second proposed explanation is therefore based on the recombination of radicals (unsaturated molecular fragments) at the surface of interstellar grains during the warm-up phase (T_k ~30–100 K) that occurs in the envelopes around Class 0 protostars (Garrod & Herbst, 2006; Garrod et al., 2008). The radicals form either as UV photodissociation products in the ices or as remnants that avoid full hydrogenation in the conversion of CO to CH_3OH during ice formation (Taquet et al., 2012). In pursuit of accurate protostellar abundance and location information with which to constrain any proposed modelling, the ratio of dominant WCCC species against ubiquitous corino CH_3OH has been one interesting area of study as we have already noted in early chapters. Species compared include $HCOOCH_3$, CH_3OCH_3, $HCOCH_2OH$, C_2H_5OH, CH_3CN, and C_2H_5CN (Taquet et al., 2015, 2019).

Single-dish telescopes with their broadband receivers can detect many transitions, so that over a relatively large beam towards these sites (say, 10–30″) the column densities of COMs can be derived. However, beam dilution is significant for hot corinos with radii of only ~5″, preventing observational separation of corino from cold envelope or even outflow sources. Fortunately, interferometers offer us angular resolutions ~1″, and multi-line observations of methanol's optically thin isotopologue ($^{13}CH_3OH$) can provide accurate column density estimates. In the PMC case, that is exactly what has been measured with the Plateau de Bure Interferometer (PdBI) towards IRAS2 and IRAS4, observations which also include, for example, $HCOOCH_3$, CH_3CN, CH_3OCH_3, C_2H_5OH, $HCOCH_2OH$, C_2H_5CN, HC_3N, $H_2{}^{13}CO$, NH_2CHO, and CH_2CO (Tarquet et al., 2015). Let us now turn our attention to these two sites of interest.

4.4 IRAS2 and IRAS4

Both IRAS2 and IRAS4 (Figure 4.6) are multiple sources. IRAS2 has two components, A and B, separated by 32″ in a north-west to south-east orientation in the plane of the sky, and it is the low-mass protostar IRAS2A whose inner envelope and disk have been probed most closely through a variety of line emissions. The observations show an east–west outflow in SiO and CO (Jorgensen et al., 2004), which has also been targeted for sulphur species, in pursuit of SO_2 or CS ratios with respect to SO, ratios that could offer potential chemical clocks for the protostellar evolutionary process (Charnley, 1997; Wakelam et al., 2005).

IRAS4 is also a multiple source, with IRAS4A a close binary whose components (IRAS4A-SE and IRAS4A-NW) are separated by less than 2″. While unresolved in the PdBI observations of COMs line emission, it is the IRAS4A-SE component that offers the detectable continuum peak and is the most interesting protostellar site of the pair. At 0.8″ resolution there are H_2O maser sites all within 100 AU of the IRAS4A-SE focus, feasibly marking the extent of the circumstellar disk.

The focus of observations, where available towards both these sites, has been on the relative abundance ratios of COMs, in the identification of offspring to precursor relationships. The CH_3OH species, which we know is widely evident in absorption in ices, shows a gas-phase jump in abundance within the hot corino radius ~0.5″ (50 AU) in IRAS4A2, where the temperature has risen to the ready ice sublimation value of 100 K. Of course, laboratory analogues for sublimation can reproduce neither the timescales of

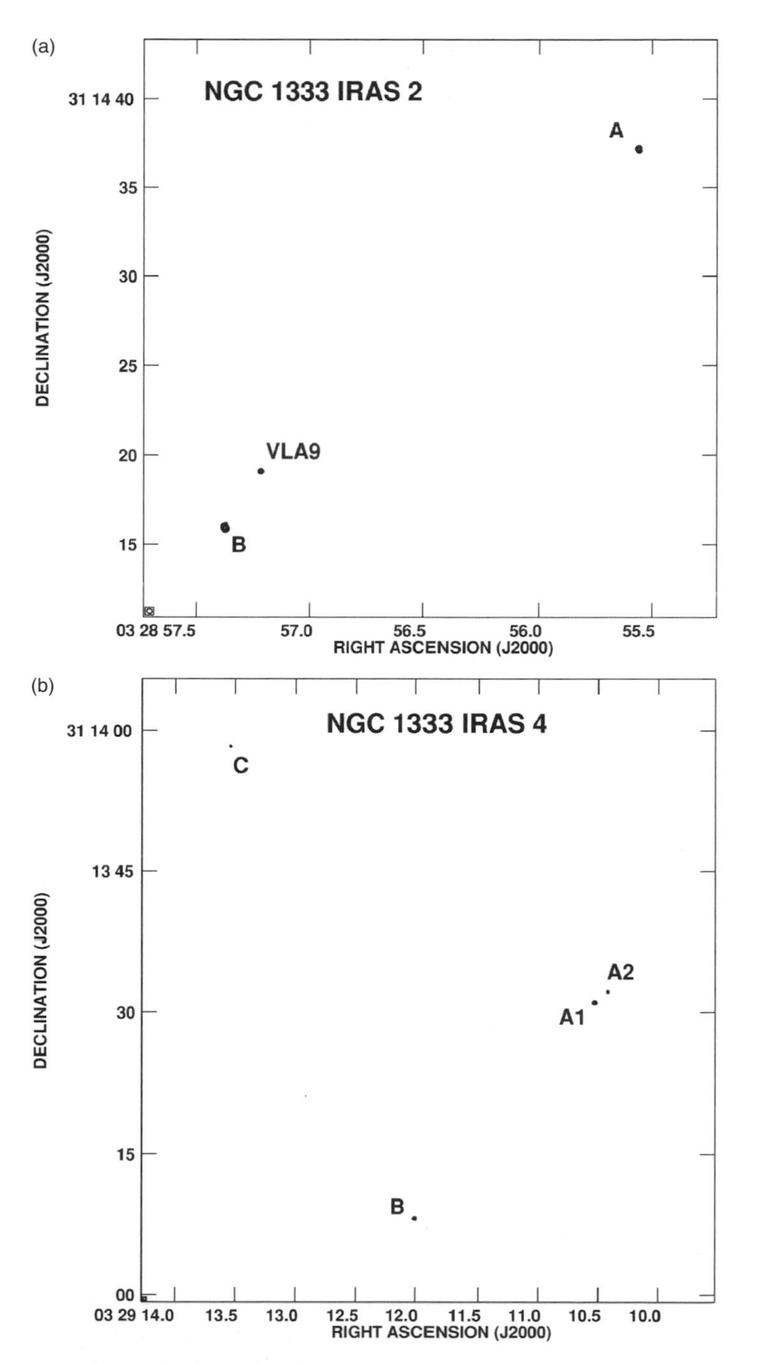

Figure 4.6 The two locations, IRAS2A and IRAS4A in NGC 1333 (VLA/Reipurth et al., 2002).

astronomical processes nor the likely complexity of mixed ice composition. However, in the case of a volatile species trapped in a less volatile matrix, such as CO in H_2O (or to a lesser degree, but extant, CH_3OH in H_2O) we expect the matrix to loosen up sufficiently for selected species to escape at about half the pure sublimation temperature.

In using the observations to characterise the physical conditions, what kind of assumptions are made in accommodating the limitations of, in this case, the PdBI interferometer? The beam size to these IRAS sources is ~2″, so the observers have an undifferentiated total flux for the emission of most transitions, delineating a hot corino within that arcsecond scale. Given multiple transitions and derived column densities, a rotation diagram analysis is used to obtain a rotation temperature for given gas densities. Plotting column densities against excitation energy gives a straight line with slope equal to rotation temperature, provided the emission is optically thin. The expectation is that rotation temperatures equal kinetic temperatures if all the levels are thermalised (LTE), a reasonable assumption in many cases. Since we know there is beam dilution in this case, the rotational diagram analysis can also reveal whether the emission is optically thin or thick. On the question of beam dilution, the modelling expectation is that COMs abundance will typically be low in a dense protostellar envelope where dust temperatures are lower than plausible sublimation temperatures. The origin of COMs observed seems inevitably to be from the hot corino region closest to the protostar. From observations of species such as $HCOOCH_3$ having a larger source size (in this case 4″) we can try to relate this to differing binding energies and resulting sublimation factors. In making such judgements there are always comparisons from elsewhere to colour one's expectations. For example, in the case of CH_3OH in IRAS2 and IRAS4, from observations in both high- and low-mass cores elsewhere it seems likely that low-energy transitions are optically thick towards the centre of a protostellar envelope, so that the rotational diagram analysis underestimates level populations. Assuming this, low-energy transitions can be excluded from the analysis.

4.5 COMs Ratios

With little scattering, the data for most COMs gives a decent best-fit line in the rotational diagram, from which derivation of both sources shows a column density for CH_3OH between 6×10^{17} cm^{-2} and 12×10^{17} cm^{-2} against the variety of COMs between 6×10^{14} cm^{-2} and 6×10^{16} cm^{-2} (Taquet et al., 2015). We wish to compare abundance ratios with respect to methanol for those COMs for which it is a likely precursor. While absolute abundance values have a high uncertainty, the relative column densities determined from the same observational limitations and analytical assumptions do offer potentially valid comparisons. From this data, the two protostellar sites do show similar chemical compositions, supporting a consistent model of hot corino activity. As for the ratios with respect to CH_3OH, methyl formate ($HCOOCH_3$) is ~3 per cent, ethanol (C_2H_5OH) ~1.5 per cent, dimethyl ether (CH_3OCH_3) ~1 per cent, and the rest less than 1 per cent. Single-dish observations of comparable low-mass protostellar sources suggest ratios several times higher, but all of these suffer from the single-dish limitations of fewer detected transitions, low numbers of upper-energy-level transitions, and considerable beam dilution – all in all, they are likely to be much less reliable than the PdBI detections. The conclusion seems to be that, in spite of their lower luminosities inducing lower temperatures and smaller sizes, nonetheless the immediate environments of low-mass protostars seem to be as chemically complex as their high-mass counterparts.

Comparing the resulting ratios in IRAS2 and IRAS4 to computational modelling results highlights several specific issues, including the fact that methyl formate, $HCOOCH_3$, as we have said, is underproduced compared with observations. In this case, it is underproduced by several orders of magnitude in both pure gas-phase and gas-grain models that include ice-surface reactions. Similarly, while gas-phase models successfully reproduce the observed abundance ratios for CH_3OCH_3 and CH_3CN, it is the model of formation within grain ices that alone reproduces C_2H_5OH, C_2H_5CN (ethyl cyanide), and $H_2COHCHO$ (glycoaldehyde) observations. All of this points to the likelihood of combinations of gas and grain reaction pathways tracing a variety of physical conditions over time (De Simone et al., 2020).

4.6 Comparisons

It is worth emphasising that none of the results described in our case studies exist in isolation when it comes to interpretation of observations. We have considered IRAS 16293-2422 in Chapter 3 as the prototypical hot corino source, but other well-studied examples include L1527, a dense core in the Taurus molecular cloud (TMC) about 140 pc from Earth. At the heart of L1527 is an embedded infrared source that interferometry has pinpointed as two continuum sources ~25 AU apart, so this is another compact binary source similar to the NGC 1333 examples, in this case having an apparent combined mass of just 2 $M_\odot$. One of these sources (IRAS 04368+2557) shows evidence of a flattened disk , and the site has become a well-studied WCCC object. After the early CO, H_2CO, and CH_3OH observations (Jorgensen et al., 2007), single-dish observations began to identify multiple carbon-chain species with up to nine carbon atoms (Sakai et al., 2008). What was not observed in the L1527 source were many saturated COMs. Pinning down the exact location of species, the PdBI interferometer was used to show unsaturated WCCC abundances peaking in the infalling envelope extending 1,000 AU out from the protostar.

Another example is IRAS 15398, a Class 0 protostar embedded in the B228 core in the Lupus-1 molecular cloud, about 150 pc from Earth. It is the only outflow source observed in Lupus-1 and has an estimated dynamical timescale of less than a few 10^4 years, with the outflow aligning close to our line of sight. The disk, perpendicular to that, is therefore close to face-on and the distribution of WCCC molecules in a defined region between 3,000 and 5,000 AU of the central star. While we will return to IRAS 15398 in Chapter 5, one of its brightest tracers is C_2H, which is also bright in NGC 1333 and therefore worth closer consideration as a key representative species of physical and chemical conditions.

4.7 C_2H

We noted earlier in this chapter that C_2H has proved a useful indicator of WCCC conditions, marking the early stages of longer carbon chain formation in warm gas. To get a handle on some competing processes, in the spirit of back-of-the-envelope approximations, let us estimate some key timescales that might influence the likely lifetime of protostellar environments and the opportunities for WCCCs and/or hot corino conditions to evolve. Within the dense envelope of gas and dust that is a protostellar core within a larger molecular cloud, molecules are well shielded from any impinging UV radiation arriving from surrounding stars. Under the steady-state conditions we might conjecture, the destruction and formation rates of competing reactions exist in dynamic balance. In one dominant process, helium is ionised by cosmic rays (CR), subsequently neutralising itself through charge transfer to CO (Sakai & Yamamoto, 2013). With the rate of He^+ formation expressed as $\zeta[He]$, where ζ is the CR ionisation rate coefficient and the rate of CO destruction is expressed as $k[He^+][CO]$, where k is the charge transfer rate constant, we have

$$\zeta[\mathrm{He}] = k[\mathrm{He}^+][\mathrm{CO}].$$

From this, the timescale for destruction (t_d) of the dominant molecular species will be

$$t_d = 1/k[\mathrm{He}^+] = [\mathrm{CO}]/\zeta[\mathrm{He}], \text{ which is about } 1 \times 10^6 \text{ years.}$$

Against this, molecular formation in dense clouds is typically driven by reaction with H_3^+ as the rate-determining step. Therefore, we need first to estimate the rate of formation of H_3^+ as it evolves through a two-step sequence of CR ionisation of H_2 followed by hydrogen exchange ($H_2^+ + H_2 \rightarrow H_3^+ + H$), while its destruction is predominantly through reaction with CO ($H_3^+ + CO \rightarrow HCO^+ + H_2$). Hence at steady state:

$$\zeta[H_2] = k[CO][H_3^+],$$

from which the timescale for formation (t_f) of molecular species is approximated as

$$t_f = 1/k[H_3^+] = [CO]/\zeta[H2] = f_{CO} / \zeta, \text{ which reduces to about } 3 \times 10^5 \text{ yr.}$$

The formation rate of CO (f_{CO}) is representative of the formation of all molecules. The approximation obviously neglects multiple reaction networks and focuses only on the dominant pathways, as it also neglects potential variations in H_2 density, so 3×10^5 yr is a minimum timescale with significant uncertainties. Nonetheless, as a first approximation, we can compare it with the standard free-fall timescale (t_{ff}) formula for dense clouds (having n $>10^4$ cm^{-3}), which is

$$t_{ff} = 4 \times 10^5 / \sqrt{n}\,(10^4 \text{ cm}^{-3})^{-1}.$$

We can also compare it with the depletion rate (t_{dep}) for molecules adsorbed onto typical dust grains, which is

$$t_{dep} = 10^5 / n\,(10^4 \text{ cm}^{-3})^{-1} \text{ yr},$$

where the sticking probability is assumed to be unity (again an approximation but a good one for the dominant species).

Figure 4.7 shows the correlation between these three timescales. What we see is that for any protostellar envelope with densities above the 10^4 cm^{-3} norm for cold, quiescent, dense clouds, and knowing that infalling envelopes will invariably have higher densities, depletion is efficient. While atomic hydrogen remains a significant addition to grain surface composition, its rapid surface diffusion results in widespread hydrogenation, not least in CO converting to the most abundant of small, saturated COMs (H_2CO and CH_3OH). With protostellar formation, the temperature of the surrounding core is gradually raised and COMs form progressively more efficiently. However, in time the warmth also puts any remaining atomic hydrogen back into the gas phase, preventing further hydrogenation and enabling smaller unsaturated species to react and form larger saturated COMs, such as $HCO + CH_3O \rightarrow HCOOCH_3$, on the vacated grain surface. Increasing heat puts even these into the gas as the hot corino

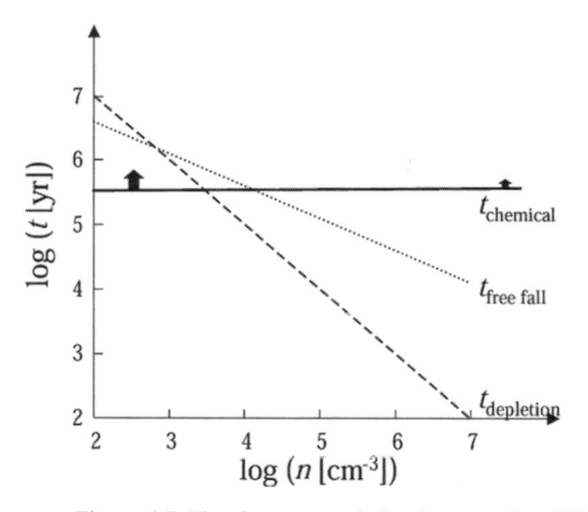

Figure 4.7 The three-way relation between free-fall time, depletion time, and chemical time. The latter can be longer depending on cloud density (Sakai & Yamamoto, 2013).

phase, and there is no reason to suppose that further chemical complexity does not continue to evolve there (Balucani et al., 2015; Taquet et al., 2015).

In the case of a gravitational collapse being as fast as free fall, the starless core phase is shorter than the chemical timescale. In this case, much of the original atomic carbon may be depleted onto grains before conversion to CO, where it is saturated efficiently with atomic hydrogen to form CH_4. As the protostar warms the grains and species return to the gas phase, carbon-chain molecules evolve first through ion-neutral reaction with C^+ to form protonated acetylene ($CH_4 + C^+ \rightarrow C_2H_3^+ + H$), and then through electron recombination to C_2H_2 and C_2H, with longer chains forming again through interaction with C^+. It is this abundance comparison we make then between the simplest carbon chain molecules that show bright lines in WCCC conditions against CH_3OH as a precursor in corino gas. Figure 4.8 summarises the two basic hot corino versus WCCC scenarios.

4.8 Wider Sampling

In seeking sources across a comparable range of sites, the wider PMC has also proved useful, given all are essentially the same distance away and they share the same beam dilution constraints for a given telescope. Figure 4.9 shows the six sites shown previously in Figure 4.4 as dust continuum maps in greyscale. The C_2H/CH_3OH abundance ratios scale with the radii of greyscale circles marking each specific site.

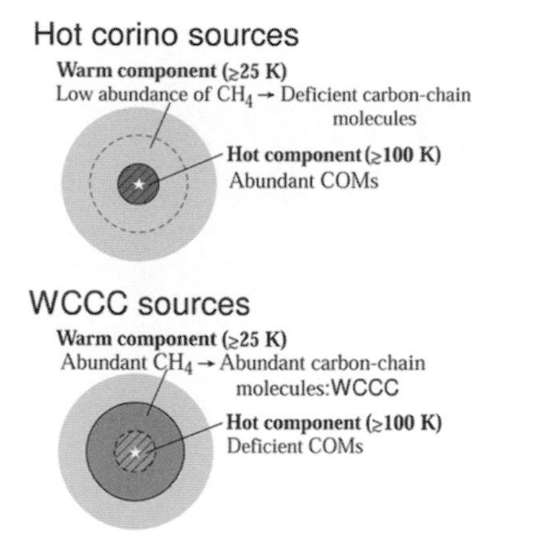

Figure 4.8 The two protostellar scenarios, hot corino and warm carbon-chain chemistry (Sakai & Yamamoto, 2013).

Since C_2H and C_3H_2 will form by comparable pathways in WCCC conditions, in a sample of three dozen protostellar sources we might expect a correlation between their beam-averaged column densities, which is indeed evident in Figure 4.10b. In contrast, if the bulk of CH_3OH occupies hot corino conditions and shares no commonality in its formation pathways with carbon-chain molecules, we would not expect a correlation between its abundance and that of either C_2H or C_3H_2, and that is also evident in Figure 4.10a.

While these observations in the PMC sources using the IRAM 30 m and Nobeyama 45 m telescopes cannot probe scales below 1,000 AU and so cannot directly distinguish conditions on hot corino or WCCC scales, the chemical signatures do suggest precisely that such a distinction does exist. The fact that higher excitation lines are used, tracing the high-density conditions closest to the protostar, effectively excludes any line-of-sight emission from the larger surrounding cloud. Comparisons with other sources also contributes to the confidence of attribution as, for example, the high CH_3OH abundance ratio in the NGC 1333 IRAS4A hot corino and its converse in the L1527 WCCC location.

4.9 Summary

In conclusion, the ratios of molecular species in IRAS2 and IRAS4 locations in NGC 1333 provide observational evidence for distinctions between conditions favouring COMs or WCCC production in the immediate neighbourhoods of low-mass protostars. Chemical modelling that explores the range

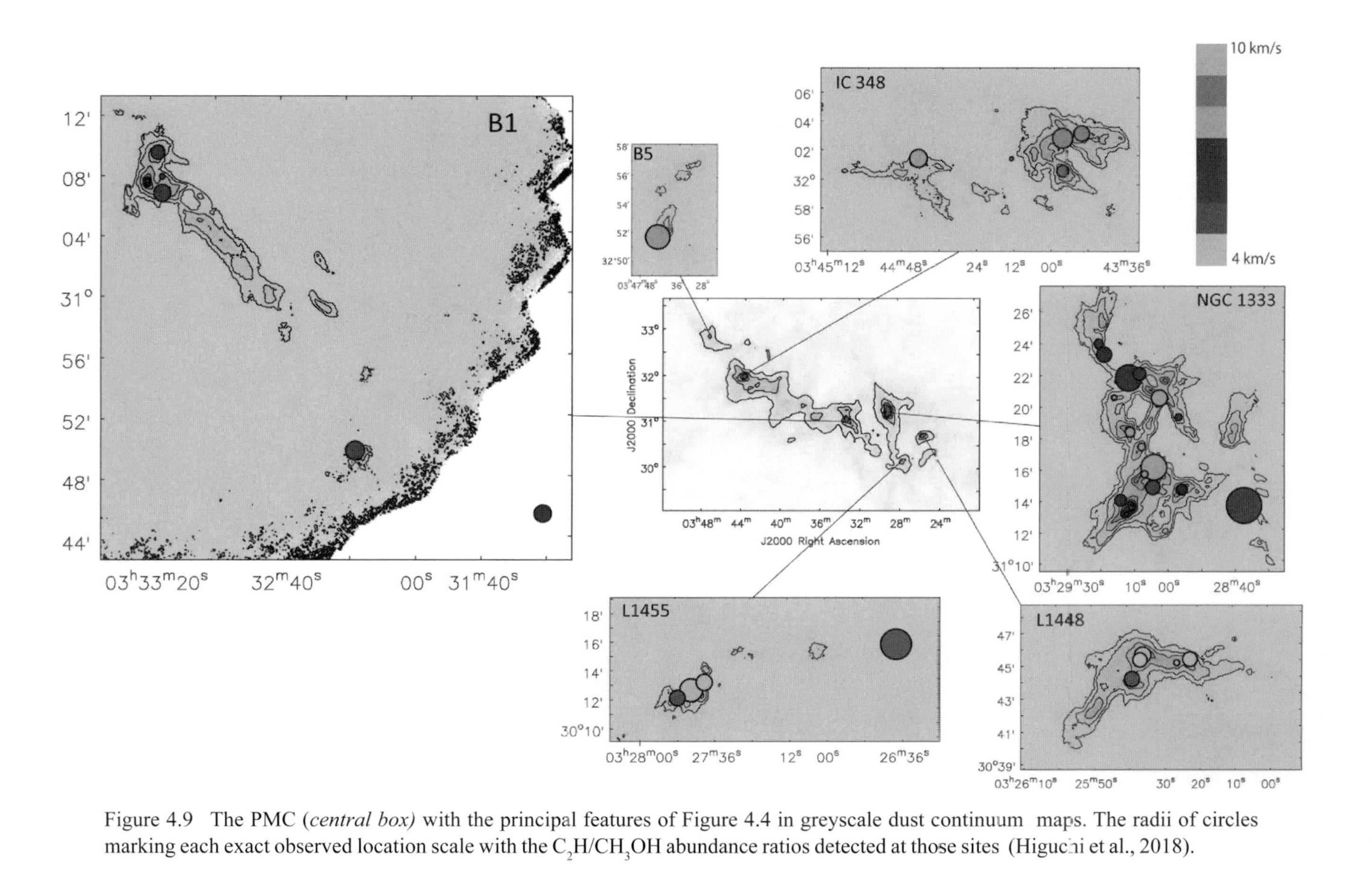

Figure 4.9 The PMC (*central box*) with the principal features of Figure 4.4 in greyscale dust continuum maps. The radii of circles marking each exact observed location scale with the C_2H/CH_3OH abundance ratios detected at those sites (Higuchi et al., 2018).

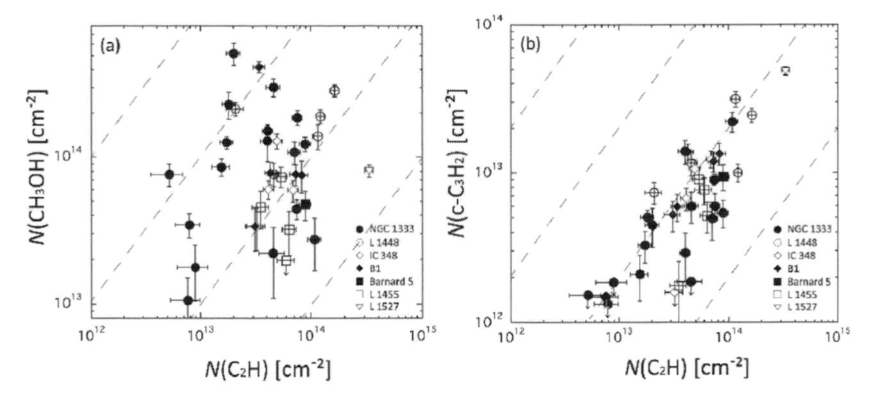

Figure 4.10 Correlation plots between the respective beam-averaged column densities, C_2H:CH_3OH in (a), C_2H:C_3H_2 in (b) (Higuchi et al., 2018).

of possibilities in conjunction with progressively higher-resolution observation using the latest generation of millimetre and submillimetre instrumentation will undoubtedly enhance our understanding in the next few years. Let us briefly look at one further low-mass star formation site, that of the previously mentioned IRAS 15398 in Lupus, specifically to see how protostellar outflows and the complexities within them may be traced through molecular emission.

5
IRAS 15398 in Lupus

5.1 Introduction

Another among the closest and largest low-mass star formation complexes seen in Earth's southern sky is that of Lupus, taken to be about 155 pc distant but with a line-of-sight depth of perhaps an additional 50 pc (Lombardi et al., 2008). The Lupus complex is immersed in one of many very large-scale structures associated with massive star clusters towards the Galactic centre. At the interface between the Galactic plane and halo these 'loop' structures are observed in atomic hydrogen (HI) emission, gamma rays, and synchrotron radio emission, having driving forces which may originate in Galactic centre outflows or perhaps ancient supernovae supplemented by recent multiple OB cluster activity (Vidal et al., 2015). The Lupus complex itself is firmly embedded in the expanding HI shell of the Upper Scorpius (USco) OB cluster, and the resulting large-scale compression shocks impacting the Lupus cloud likely trigger its multiple star formation. Figure 5.1 shows the location of Lupus on the larger scale, through a total column density distribution (derived from a Planck all-sky dust extinction map) with both the Lupus and neighbouring Ophiuchus cloud complexes marked.

Both the Ophiuchus and Lupus molecular complexes are associated with this particular loop (also called the North Polar Spur and marked as a dashed ellipse in the image), and both cloud complexes also feature as associated with the Gould Belt of massive star clusters (Pattle et al., 2015; Mowat et al., 2017), part of which is shown in Figure 5.2.

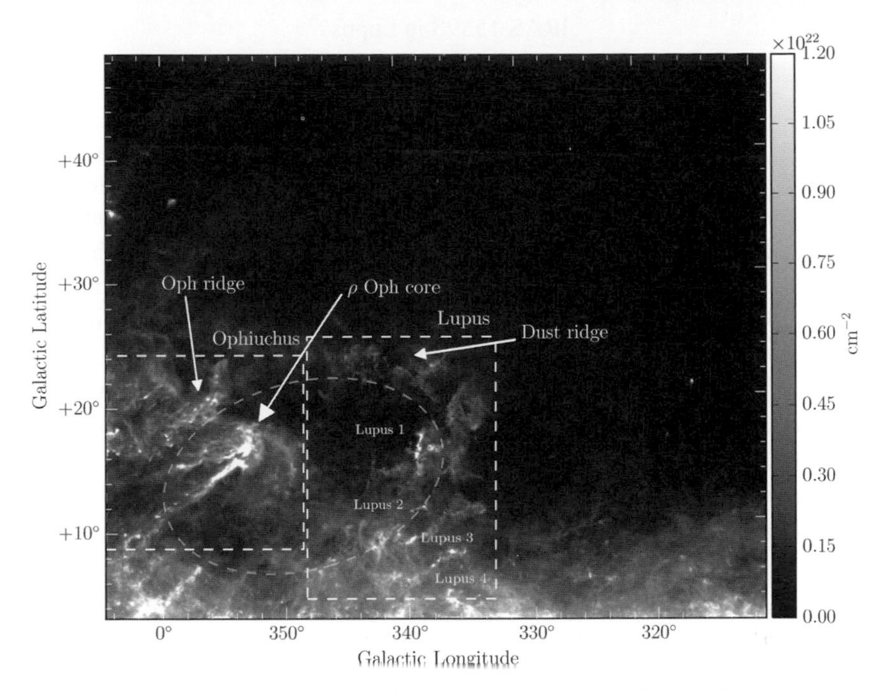

Figure 5.1 The upper Lupus complex components (1–4) in relation to Ophiuchus, traced in total gas column density derived from Planck dust optical depth mapping (Robitaille et al., 2018).

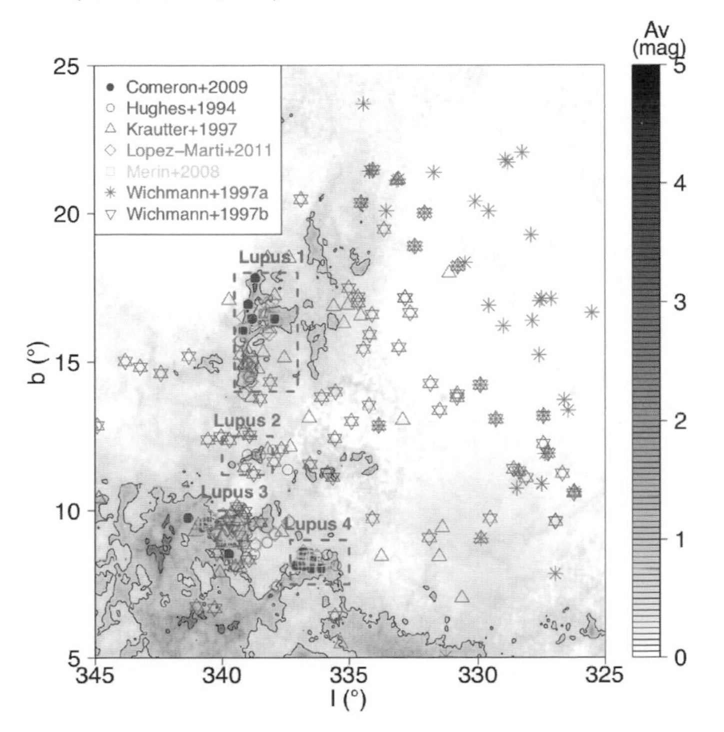

Figure 5.2 Location of the Lupus clouds 1–4 within the Gould Belt, showing star distributions from successive optical surveys overlaid on dust extinction in grey-scale (Galli et al., 2020; Dobashi et al., 2005).

5.2 Nine Clouds

Figure 5.3 focuses in on the Lupus 1 component of the four clouds, white contours delineating the cloud traced in Figure 5.1 but against an HI emission map which clearly shows a lesser ridge (albeit massive) within the wider Galactic bubble that emerges from the Galactic plane into the halo. The schematic diagram of Figure 5.4 then places the Lupus and Ophiuchus clouds in relation to that giant bubble and the lesser ridge. The central cavity of the bubble is filled with hot, ionised, X-ray-emitting gas, and it is thought that hot gas outflows have breached the cavity to the west and north-west, probably through the fragmented Lupus complex (Robitaille et al., 2018).

Dust extinction and CO maps subdivide the whole Lupus complex traditionally into nine distinct molecular clouds (Dobashi et al., 2005), and the stellar activity of each is quite diverse. Lupus 3, for example, has dense concentrations of T Tauri stars (intermediate mass, pre-main sequence), while Lupus 7, 8, and 9 appear completely quiescent. Clouds 1–4 appear the most

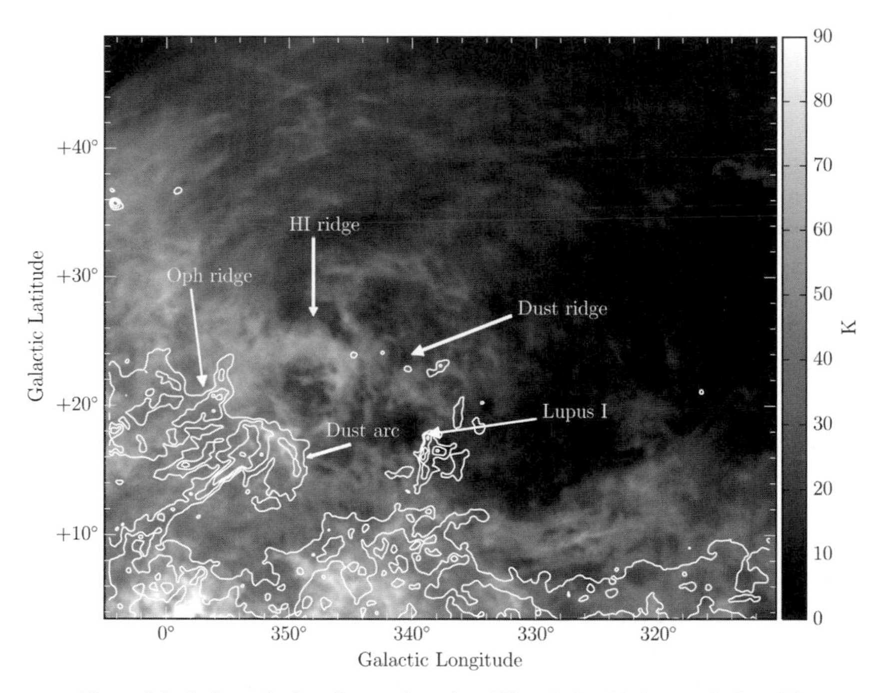

Figure 5.3 A closer look at Lupus 1 against HI emission background, the white contours delineating the cloud complexes identified through dust extinction in Figure 5.1 (Robitaille et al., 2018).

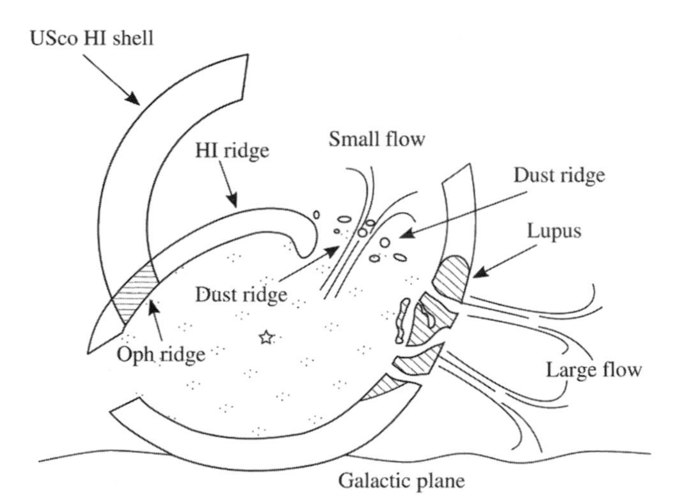

Figure 5.4 Schematic of the bubble at the Galactic plane and halo interface, placing the Lupus molecular cloud complex and the Ophiuchus ridge in their wider context (Robitaille et al., 2018).

active, although YSOs have recently also been detected in 5 and 6 (Melton, 2020). At the higher Galactic latitude, the dense gas and dust of Lupus 1 renders it the most clearly separated of these clouds from its surroundings. It is dominated by the B228 ridge ('B' after Barnard, 1927) extending northwest to south-east over 2° (~5 pc), which lies along the trailing edge of the HI shell and is the site of the most concentrated low-mass star formation activity. Dust extinction and CO mapping suggest a material mass for Lupus 1 of ~10^4 $M_\odot$ (Mowat et al., 2017), with a dust distribution along the ridge peaking in the centre, dropping away in the south-east and remaining lower but constant across the northwest 'plateau' (Tothill et al., 2009). Figure 5.5 shows the B228 ridge and the three distribution locations in ^{13}CO emission. A far higher abundance of Class 0 and Class I stars is observed in Lupus 1 than expected from comparable formation regions (Mowat et al., 2017), making this a particularly rich source for study. Where CO (4-3) emission intensity peaks in the centre of the ridge is the site of IRAS15398-3359 (also known as Lupus I-1, the Roman numeral now replacing the Arabic for the initial cloud designation, the latter now for the sub-site), sitting about 0.5 pc behind the ridge's steep edge. As we shall see, this source is a Class 0 YSO with a particularly compact young outflow. Figure 5.6 shows Herschel 250 μm dust images of the IRAS 15398 source and its immediately surrounding starless cores (named Lupus I-2/3/4/5/6/7/8/9/11).

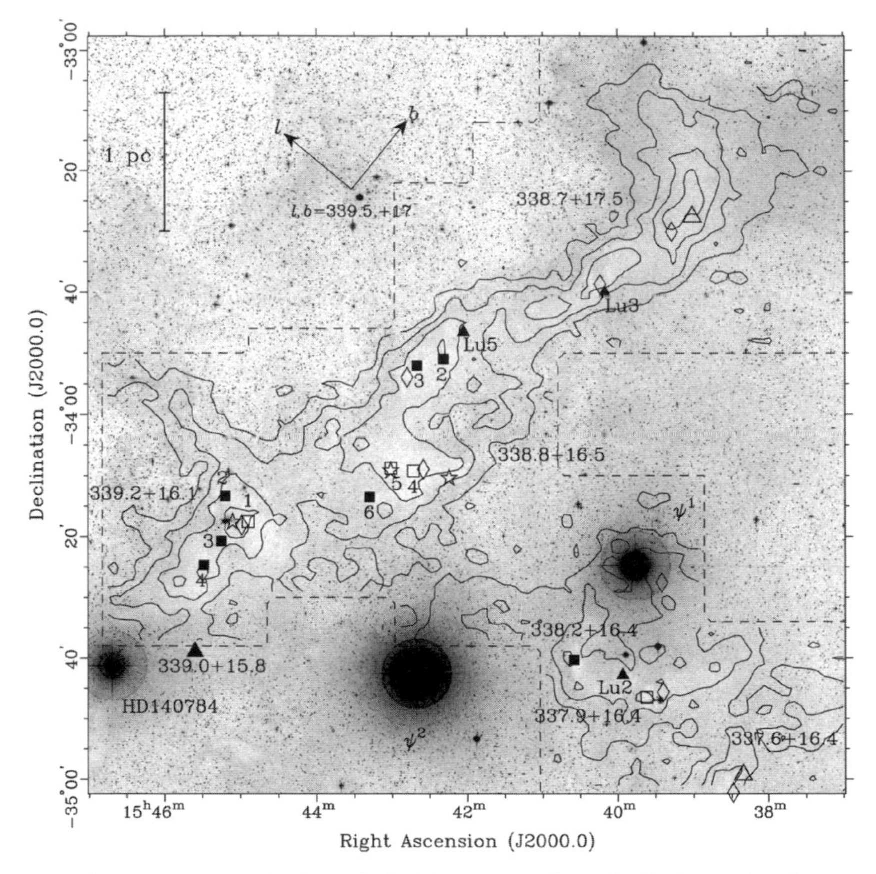

Figure 5.5 Lupus 1 in the optical with contours of CO distribution tracing the cloud behind, showing the central ridge of B228, the south-east locus, and north-west plateau (Tothill et al., 2009).

5.3 IRAS 15398–3359 (Lupus I-1)

Recent years have shown that the three physical structures always associated with early star formation (rotating protoplanetary disk, collapsing protostellar envelope, and bipolar molecular outflow) show complex chemistries that are quite distinct. Obviously, conditions even within these separable spatial locations also change over time, with both thermal and shock heating competing with freeze-out processes determining the dominant chemical reaction networks. The example of IRAS 15398 is interesting as both a warm carbon-chain chemistry (WCCC) source and as a young outflow source offering close comparative evidence of outflow shock chemistry with its neighbours. The young

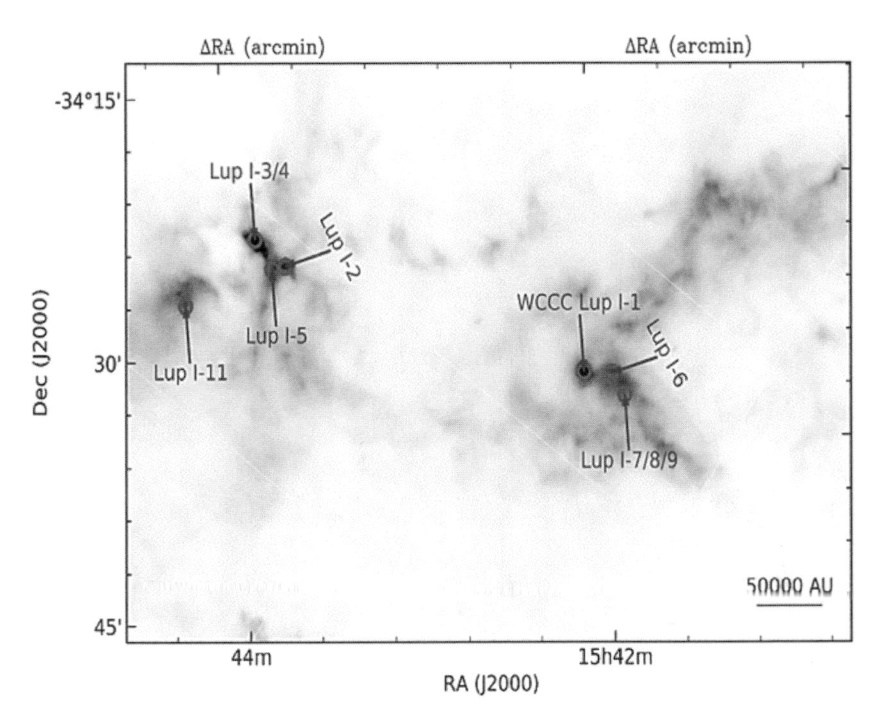

Figure 5.6 The starless cores of Lupus 1 observed in Herschel dust extinction with IRAS15398 marked as WCCC Lupus I-1 (Wu et al., 2019).

Class 0 object at the locus of this source (a_{2000} +15^{h}43^{m}02^s.16, d_{2000} -34°09′.0″) shows a fairly low bolometric temperature of ~48 K and luminosity ~1.8 $L_{\odot}$ (Kristensen et al., 2012; Jorgensen et al., 2013) but, in addition to the expected protostellar radiation, it is characterised as having undergone a burst of accretion activity during the last 100–1,000 years. This would have increased short-term stellar luminosity by a factor of 100, raising the temperature of the dust and gas envelope and generating conditions favourable, perhaps, to WCCC formation.

Figure 5.7 shows the distribution of observed species from an ALMA survey in which a principal component analysis (PCA) was applied across all detected molecular emission lines (Okoda et al., 2020). This is a statistical method for cleaning the data in order to identify the most significant variables, and results in plots in which positively correlated species according to the variable under consideration appear grouped. The upper plot in Figure 5.7 distinguishes molecules that appear to trace compact distributions around the protostar (Group 1) and those tracing extended regions, particularly the outflow (Group 2). The lower plot then further distinguishes the Group 1 species

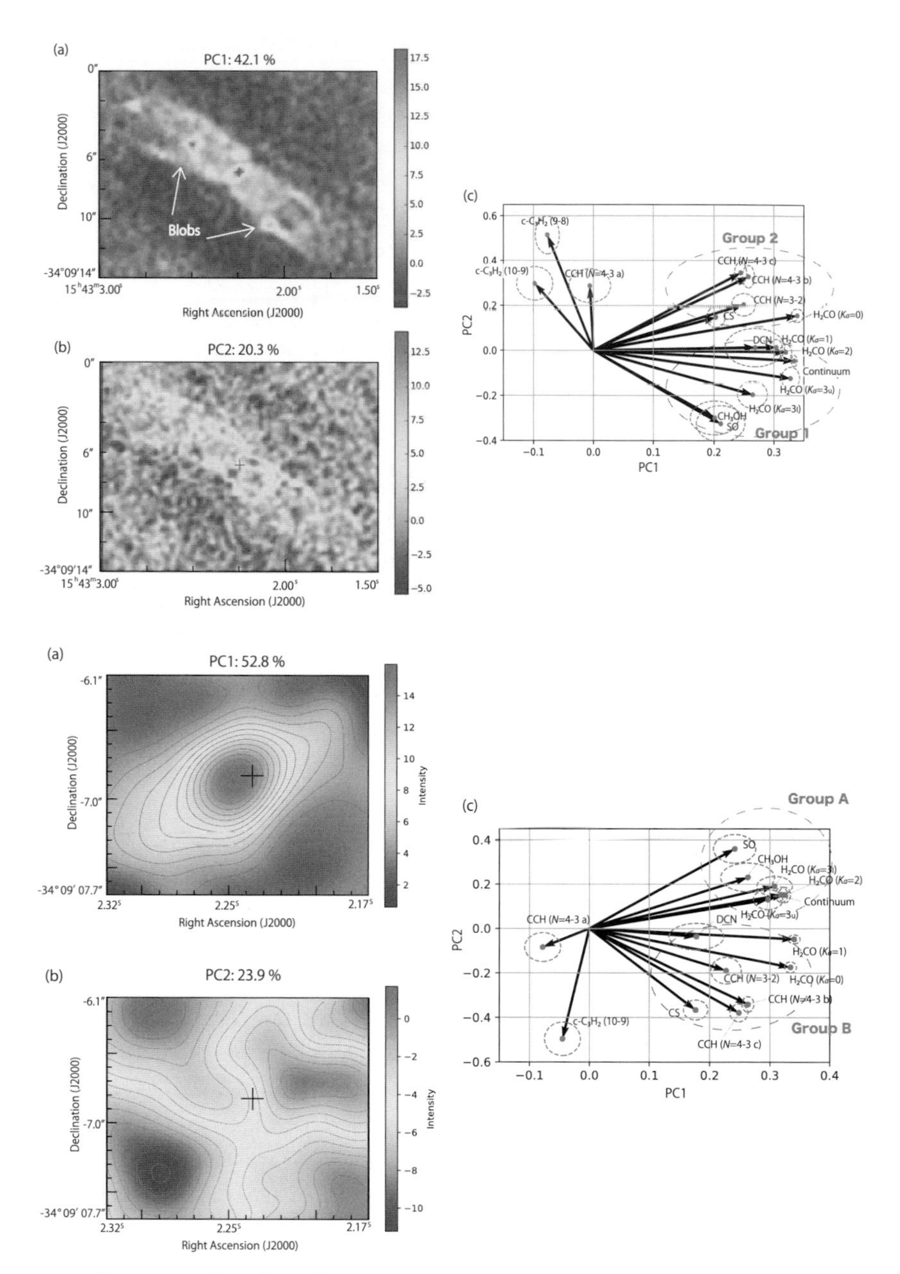

Figure 5.7 PCA plots of the observed emission lines for species in the three protostellar environments around IRAS 15398 (Okoda et al., 2020).

into those showing closest to the protostar in a likely disk distribution (Group A) and those most likely a step away in the infalling envelope (Group B).

Single-dish telescope observations first identified IRAS 15398 as the second clear WCCC source (after L1527) just over a decade ago (Sakai et al., 2008, 2009; Sakai & Yamamoto, 2013), rich in carbon-chain molecules such as CCH, C_4H, and CH_3CCH, all on the scale of a few thousand AU, surrounding the stellar source. Among the subsequent sub-arcsecond resolution observations is revealed, for example, a 'ring' distribution of $H^{13}CO$ on the scale of 150–200 AU (Jorgensen et al., 2013), and a Keplerian disk structure based on SO line emission at an angular resolution ~0.2″ (Okoda et al., 2018).

5.4 The Bipolar Outflow

Mass loss through molecular outflows from young stars is likely a magneto-centrifugal process, related to accretion and the conservation of angular momentum. Observing accretion infall directly is not yet possible, but indirectly we do see red-shifted components in self-absorbed, optically thick emission lines enhanced against their blue-shifted counterparts in the innermost protostellar regions. As for the outflow in IRAS 15398, ^{13}CO emission has shown two lobes (Bjerkeli et al., 2016a) with wide enough wind angles for cavities observable, for example, in C_2H (Jorgensen et al., 2013). The absence of HCO^+ close to the protostellar source has been linked to accretion shocks (Oya et al., 2014), or at shocked locations along the inward-facing outflow edge (Podio et al., 2014). Where HCO^+ is seen in both lobes, it is both red- and blue-shifted, and locations along the cavity walls are where velocities perpendicular to the outflow axis are expected to be greatest. Figure 5.8 shows the integrated emission from a variety of molecular species observed in the outflow lobes. With an inclination angle of 20° (almost parallel to the plane of the sky), the bipolar outflow is marked by the top left to bottom right diagonal dotted line. The associated red- and blue-shifted emission is shown as line contours in the upper-left and lower-right regions, respectively.

The two principal CO isotopologue distributions shown in Figure 5.8 are present close to the outflow origin as well as throughout the length of each lobe. HCO^+ distribution, as we have said, is both evident and absent. As for N_2H^+, it is not a commonly found shock tracer and is known to be destroyed in reaction with CO in dense conditions. With $C^{18}O$ emission peaking close to the protostar, it could be that here N_2H^+ is largely destroyed by that reaction, hence the absence of its emission, while further downstream where CO density should be lower, the N_2H^+ abundance is detectably present. The N_2D^+ emission

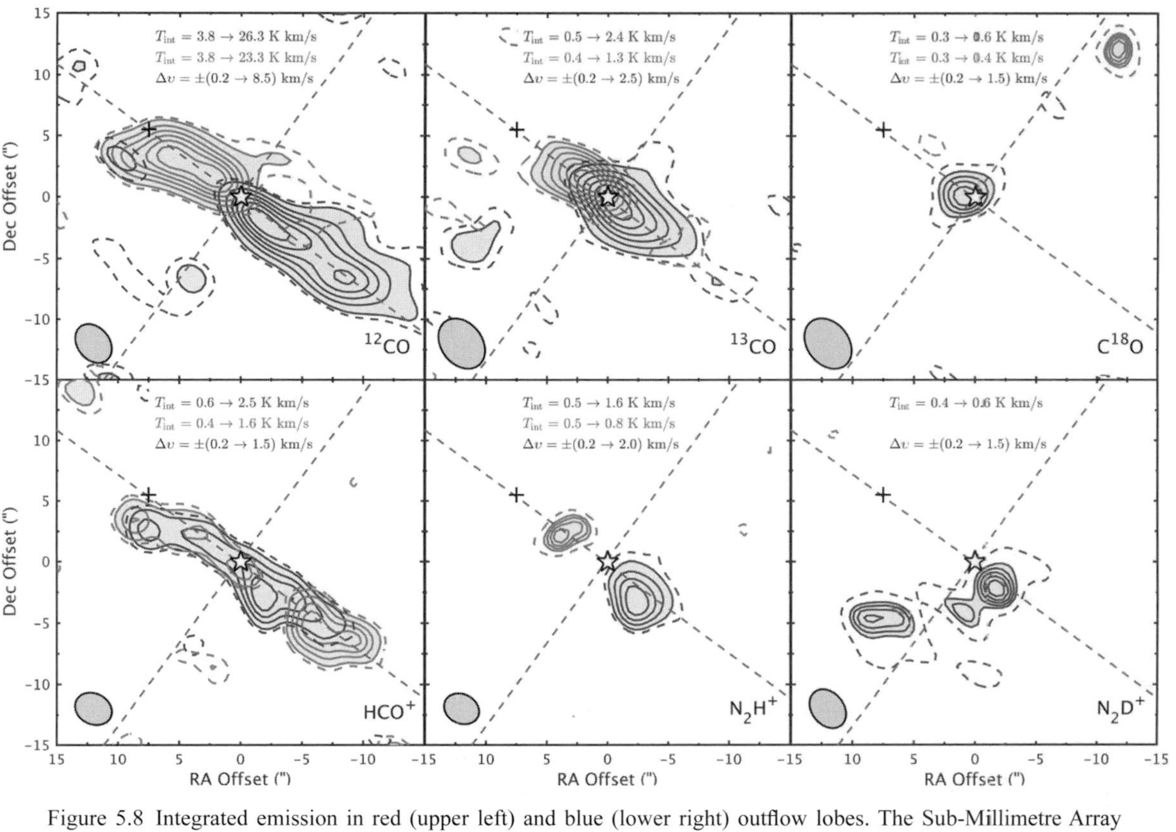

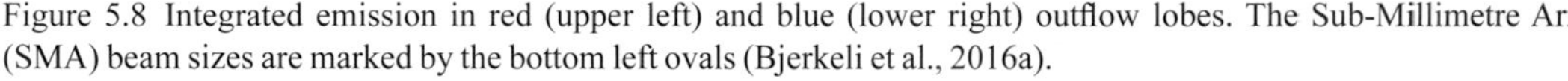

Figure 5.8 Integrated emission in red (upper left) and blue (lower right) outflow lobes. The Sub-Millimetre Array (SMA) beam sizes are marked by the bottom left ovals (Bjerkeli et al., 2016a).

is exclusively south and south-east and only at blue-shifted velocities, so it may well be as much associated with the envelope as the outflow. While the outflow morphology may appear clearly defined here, it is worth noting that with this very young, low-velocity outflow, we can find ourselves mixing low-velocity emission lines from both outflow and envelope in our interpretations, as seems likely from the evidence of radiative transfer modelling (Bjerkeli et al., 2016b).

Figure 5.9 shows the two outflow cavities seen in C_2H emission with ALMA, and with the radiative transfer modelling dimensions superimposed. The modelled lobe as envisaged is shown in the schematic of Figure 5.10.

Before summarising the model parameters, we can note that additional corroborating observations also show the outflow structure traced in H_2CO, CS, and C_2H and Figure 5.11 shows local peaks ('blobs') in that emission, reflecting shocked regions that experience the localised impact of the outflow on

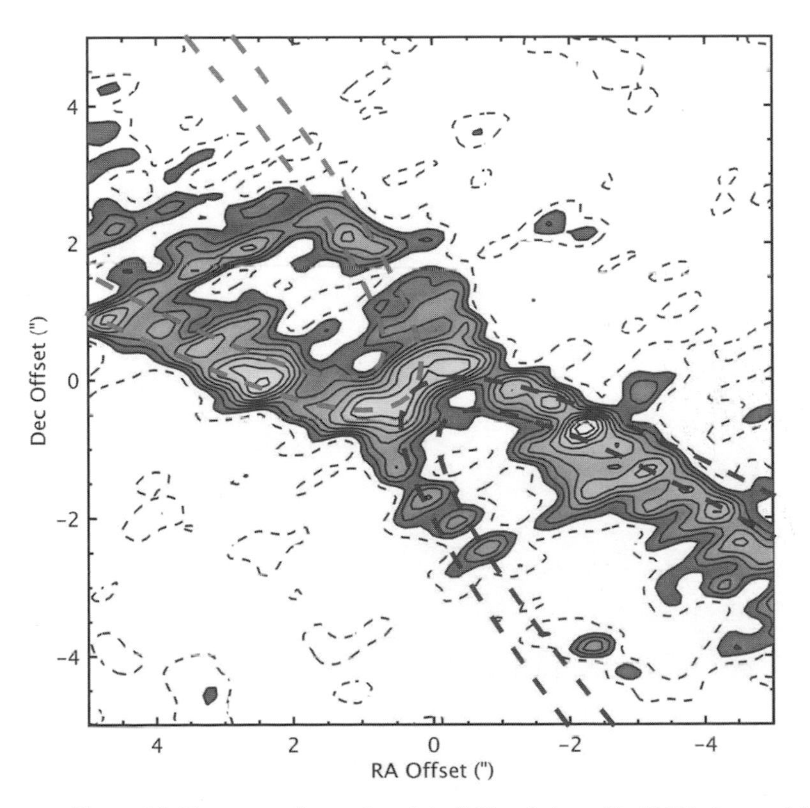

Figure 5.9 The two outflows plotted in C_2H emission with ALMA (greyscale) with modelled cavities superimposed (dotted lines) (Jorgensen et al., 2013; Bjerkeli et al., 2016a).

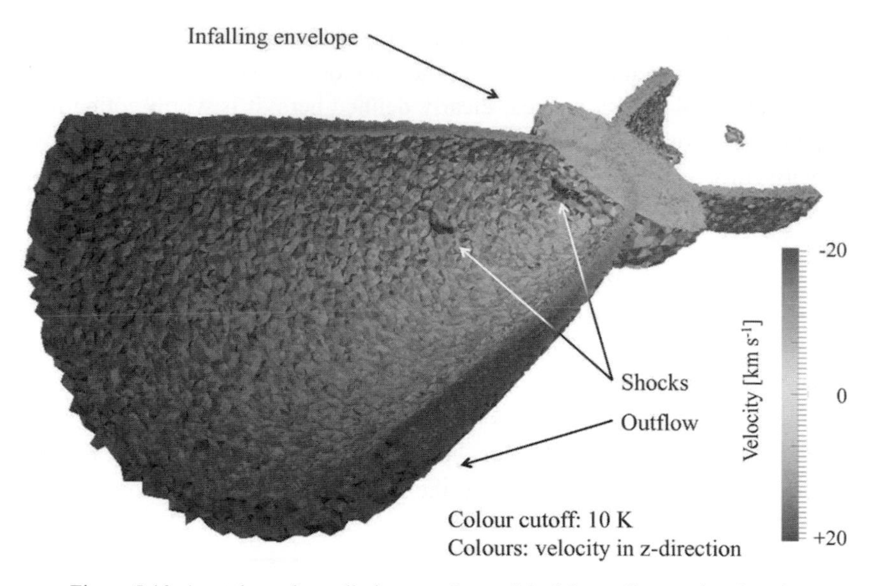

Figure 5.10 A cut through a radiative transfer model of the outflow cavity, showing shocked regions along the jet axis (Bjerkeli et al., 2016a).

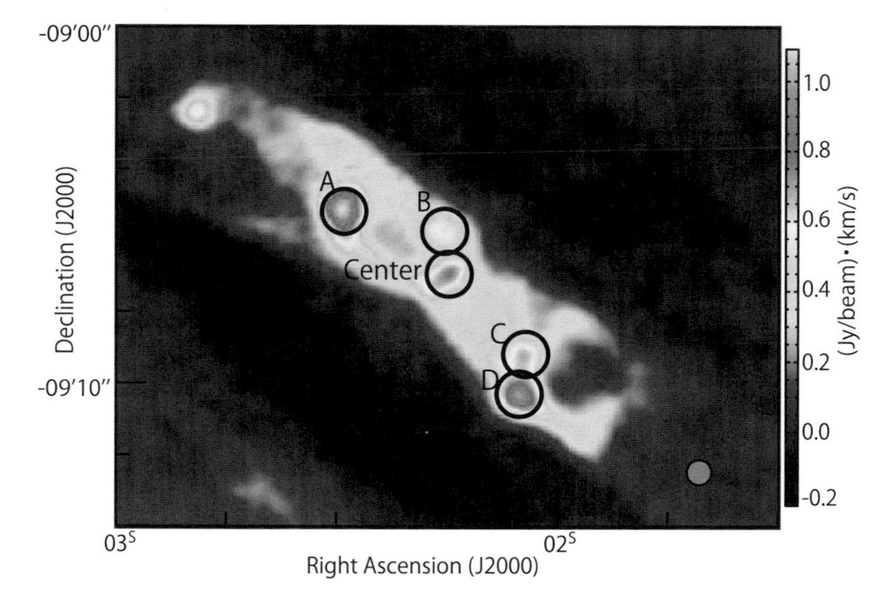

Figure 5.11 Localised shocked 'blobs' in the outflow emission of H_2CO in IRAS 15398, with 'centre' indicating the protostellar source (Okoda et al., 2020).

Table 5.1 *Derived parameters for each location (Okoda et al., 2020).*

Blob	Column density (10^{14} cm^{-2})	T_{gas} (K)
Centre	0.87 ± 0.05	55 ± 2
A	0.83 ± 0.0	63 ± 2
B	0.66 ± 0.06	43 ± 3
C	0.48 ± 0.04	54 ± 4
D	0.78 ± 0.06	45 ± 2

Table 5.2 *Physical parameters of the outflow derived from ^{12}CO emission maps (Bjerkeli et al., 2016a).*

	Red lobe	*Blue lobe*
M_{lobe} ($M_{\odot}$)	2×10^{-4}	2×10^{-4}
L_{lobe} (pc)	8×10^{-3}	1×10^{-2}
R_{lobe} (pc)	3×10^{-3}	4×10^{-3}
v_{char} (km s^{-1})	5.1	7.4
v_{max} (km s^{-1})	6.0	8.0
t_{dyn} (yr)	4×10^{2}	5×10^{2}
Momentum ($M_{\odot}$ km s^{-1})	2×10^{-4}	4×10^{-4}
Kinetic energy (erg)	2×10^{40}	5×10^{40}
Momentum rate ($M_{\odot}$ km s^{-1} yr^{-1})	4×10^{-7}	1×10^{-6}
Mechanical luminosity ($L_{\odot}$)	3×10^{-4}	1×10^{-3}
Wind mass-loss rate ($M_{\odot}$ yr^{-1})	3×10^{-8}	4×10^{-8}

ambient gas. The H_2CO map is shown in Figure 5.11, with the column densities and kinetic temperatures derived from it shown in Table 5.1.

As for key physical properties of the outflow, Table 5.2 shows parameters for the two outflow lobes derived directly from the ^{12}CO emission map of Figure 5.8. As for the model, Table 5.3 lists the initial parameters adopted, based on both observations and the best available theoretical assumptions. In the absence of better data, a typical excitation temperature of 100 K in both flows is adopted (see e.g. van Kempen et al., 2009) and a CO/H_2 ratio of 10^{-4} (van Dishoeck & Black, 1987), which suggests an H_2 column density ~3×10^{20} cm^{-2} (Bjerkeli et al., 2016a). The lobes extend ~3,700 AU in length, with a median width ~700 AU, all of which combines to give a total outflow mass ~3×10^{-4} $M_{\odot}$. Given the protostellar mass estimate of 4×10^{-2} $M_{\odot}$ (Oya et al., 2014), this indicates a relatively small amount of gas ejection to date, reinforcing the very young nature of IRAS 15398 compared with other Class 0 sources (Dunham et

Table 5.3 *Modelling parameters adopted from the available observational data and recent theoretical assumptions (Bjerkeli et al., 2016a).*

Outflow parameters	
Inclination angle (i)	20°
Temperature (T)	100 K
Structure parameter of the shell (C)	1.0 arcsec^{-1}
Velocity parameter of the shell (v_0)	1.1 km s^{-1} arcsec^{-1}
Length of lobe (L_{blue})	2,350 AU
Length of lobe (L_{red})	1,400 AU
Envelope parameters	
Radius of envelope (R_{env})	4,900 AU
Velocity at r_{vo} (v_0)	0.1 km s^{-1}
Radius where $v = v_0$ (r_{v0})	1,000 AU
Velocity exponent (p_v)	0.5
H_2 density at r_{no} (n_0)	2 × 10^9 cm^{-3}
Radius where $n = n_0$(r_{n0})	6.1 AU
Density exponent (ρ_n)	1.4
Temperature at r_{T0}(T_0)	30 K
Radius where $T = T_0$ (R_{T0})	175 AU
Temperature exponent (p_T)	1.5
Shock parameters	
Maximum velocity (v_{max})	20 km s^{-1}
Temperature (T)	1,000 K
H_2 density (n_{H2})	1 × 10^8 cm^{-3}
Shock distance from source (d_{shock})	600 and 1400 AU
Surrounding cloud parameters	
H_2 column density (N_{H2})	1 × 10^{20} cm^{-2}
Temperature (T)	10 K
Shell thickness (I_{shell})	5,000 AU

al., 2014; Bjerkeli et al., 2013). The maximum observed flow velocities in red- and blue-shifted lobes respectively are 6 km s^{-1} and 8 km s^{-1}, which we can deproject to obtain an outflow dynamical time $t_{dyn} = L_{lobe}$ cos $(i)/v_{max}$, where L_{lobe} is 3,700 AU and the inclination angle i is 20°. The deprojected red- and blue-shifted velocities are 18 km s^{-1} and 23 km s^{-1} respectively, and the outflow timescale derived between 400 to 500 years (Bjerkeli et al., 2016a). Other parameters such as mass-loss rate, momentum rate, and kinetic energy for IRAS 15398 have been deduced and found to be equally lower than the norm for Class 0 stars. This is surprising since younger outflow sources are typically more energetic than their evolved counterparts (e.g. Bontemps et al., 1996). Obviously, a reduced entrainment of gas and dust (i.e. low outflow mass) limits the energetic parameters, but if outflow is a function of accretion, the values perhaps correlate with not only variable accretion rates but potentially long periods in a low accretion-rate mode.

What the modelling does show is that the observed emission profiles cannot be modelled without a combination of outflow and infalling gas, so the observations are tracing the complex dynamics at the interface between competing processes involving a wide-angle wind outflow, a circumstellar envelope, and surrounding cloud components.

5.5 H_2O with ALMA

As observations and radiative transfer modelling make clear towards IRAS 15398, the accretion process is significantly obscured by the dense surrounding environment and researchers have yet to clarify what are the key drivers of accretion in young stars generally. Observed luminosities of protostars are typically an order of magnitude lower than inferred from standard accretion models, hence the conjectured lengthy quiescent phases followed by short-lived accretion bursts (Kenyon et al., 1990; Evans et al., 2009). If the luminosity varies, then so do the 'snow line' boundaries in the surrounding gas and dust, which define the freeze-out or sublimation temperatures for molecules adsorbing onto, or desorbing from dust grains (Rogers & Charnley, 2003; Johnstone et al., 2013). Observations at 0.5″ with the Atacama Large Array (ALMA) identify a hole in $H^{13}CO^+$ emission (Jorgensen et al., 2013), alongside peak emission for $C^{17}O$ and CH_3OH, very close to the protostar, so tracing either the accretion disk or the inner envelope. If we match dynamical events in the outflow, such as CO peaking on outflow timescales of a few hundred years (Bjerkeli et al., 2016a), and relate outflow activity directly to accretion events (Raga et al., 1990), then the associated luminosity increase would release H_2O ices (as well as the mixed contents they harbour). The destruction of HCO^+ plausibly follows from the efficient reaction with water, in which $HCO^+ + H_2O \rightarrow H_3O^+ + CO$.

The HCO^+ hole is observed to extend 150–200 AU out from the protostellar locus, perhaps much larger than the likely extent of water ice sublimation (100 K for pure ice), but also perhaps not, given partial sublimation of impure water ice is efficient enough at temperatures as low as 50 K. Choosing isotopologues $H_2{}^{18}O$ and HDO as alternatives to the essentially invisible low-energy transitions of H_2O from ground-based telescopes, fresh observations with ALMA have looked to distinguish differences in thermal structure close to the protostar (Bjerkeli et al., 2016b). The emission transitions chosen for these two isotopologues probe gas at very different excitations, E_{up} = 390 K and 22 K respectively. Significantly, one of them, $H_2^{18}O$, was not detected at all. HDO, on the other hand, while closely correlated with the continuum peak (i.e. line of sight to the protostar itself) and gaseous CH_3OH, is present precisely where HCO^+ is absent, as the contour maps of Figure 5.12 make clear. CH_3OH is the clear indicator of ice sublimation, and the presence of sublimated water where HCO^+ is absent is consistent with the recent accretion burst scenario in which the former species destroys the latter.

The morphology of HDO distribution does seem to show an additional presence in the outflow, although not at distances beyond 500 AU from the protostar. Estimates of gas thermal pressure based on the non-detection excitation limits for $H_2^{18}O$, coupled with narrow linewidths (low-velocity gas) and the constrained distribution, do, however, argue against a high-temperature shock origin through gas-phase reactions, thermal desorption, or even sputtering from ices. The only additional idea tested with modelling to date is that of an infalling envelope density profile varied, for example, where the inner parts of the outflow cavities have a lower density than the outer (Bjerkeli et al., 2016a; Bjerkeli et al., 2016b). In such cases the snow line can extend out to the distances where HDO is observed. Unfortunately, we do not yet have particle densities for the cavity gas derived from observation to compare with the modelling. What the modelling alternatively suggests is that a similar gas density in both envelope and outflow shows a 100 K snow line within 150 AU. This ties in neatly with the thermal desorption of H_2O, the location of the HCO^+ hole, and the almost static velocity relative to the systemic observed for HDO closest to the protostar, which rises very little (to just ~1 km s^{-1}) when entrained in the low-velocity outflow of the blue-shifted lobe (Bjerkeli et al., 2016b).

5.6 Summary

What is clear from the molecular emission line observations made towards IRAS 15398 is that we can distinguish between molecules that are particularly prevalent in identifiably distinct regions, either compact or extended. These are molecular emissions from close to the protostar as well as gas spreading in outflow material. Within the latter are found distinguishable localised components that show likely shock-enhanced chemistry. As is the case for IRAS 16293 and NGC 1333, we can separate disk from envelope through both characteristic

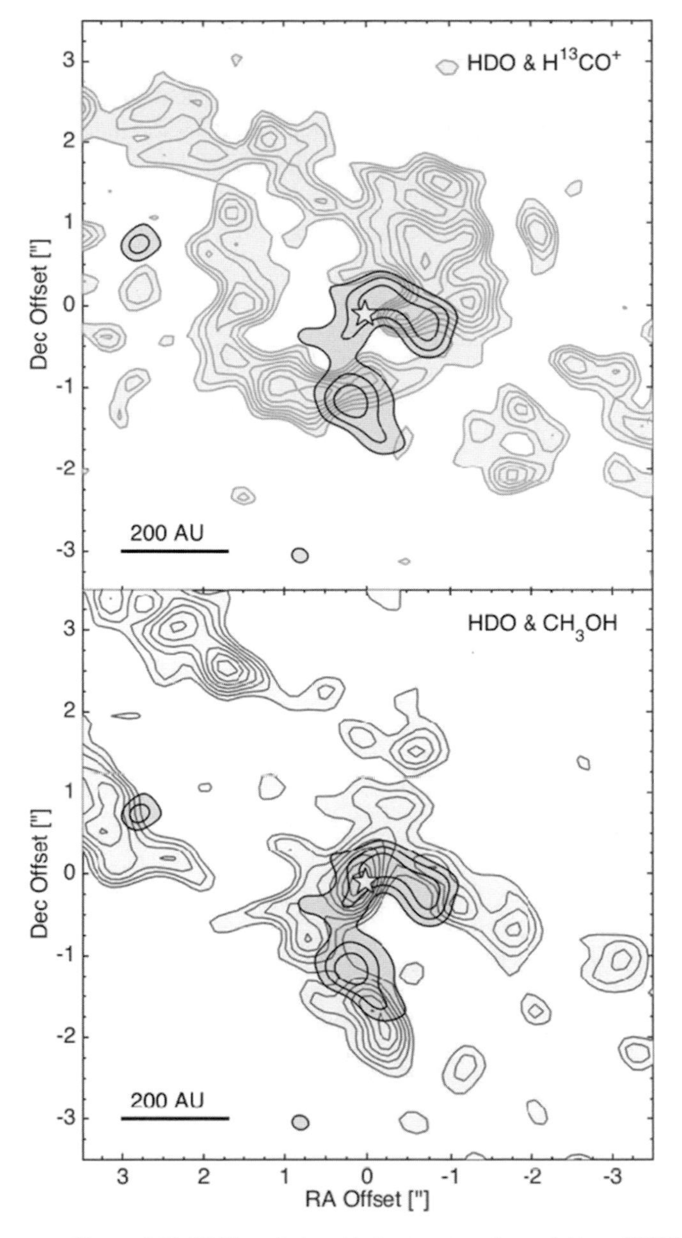

Figure 5.12 HDO emission (darker contours) overlaid on $H^{13}CO^+$ and CH_3OH contours (light grey), respectively (Bjerkeli et al., 2016b).

species and levels of molecular excitation. The previous chapters have illustrated the transition from prestellar to protostellar environments, and distinctions between hot corino chemistry and warm carbon-chain chemistry. As we now look to HMSFRs, we will find both parallels and differences in the emergence and the development of stellar sources and their associated molecular emissions and adsorptions.

PART III

High-Mass Star Formation (HMSF)

6

Two HMSFR Surveys Using APEX and NOEMA

6.1 Introduction

The earliest surveys of HMSFRs targeted UCHII regions as the brightest indicators (e.g. Wood & Churchwell, 1989), and subsequently hot molecular cores were found to be associated with them (e.g. Cesaroni et al., 1992). High-mass protostellar objects (HMPOs) or massive young stellar objects (MYSOs) were delineated in the following decades (e.g. Beuther et al., 2002; Lumsden et al., 2013), as representing even earlier stages of massive star formation, just as Infrared Dark Clouds (IRDCs) were also identified as potential sites of early-life stars (e.g. Zhang et al., 2017). An early, unbiased sampling across the complete dust continuum range on the scale of a molecular cloud complex (that of Cygnus-X) demonstrated the value of multi-wavelength studies that target more than just one specific evolutionary moment of high-mass star formation (Motte et al., 2007). The first of the following two more recent surveys, that of ATLASGAL, distinguishes IR-quiet from IR-bright sources along with HII regions, tracing stages of star formation across ten thousand potential sites. The second survey, CORE, homes in on the kinematic and chemical characteristics of just 20 clearly defined high-mass protostellar locations. The value of a representative survey lies in learning what physical and chemical parameters are shared across a variety of sources. The ATLASGAL survey uses the APEX telescope with sensitivity on a scale of 0.5 pc, sufficient to identify individual protoclusters. The CORE survey uses the NOEMA telescope sensitive to scales of 0.05 pc, sufficient to identify individual cores and small groups of protostars, perhaps even resolving individual high-mass protostars.

6.2 ATLASGAL

The APEX Telescope Large Area Survey of the Galaxy (ATLASGAL) is a large-scale survey in the submillimetre range, covering a significant sector of the inner Galaxy. The Atacama Pathfinder Experiment (APEX) 12 m radio dish, a modified ALMA prototype antenna, operates across far-infrared to radio wavelengths (0.2–1.5 mm). The main objective has been to make a large systematic inventory of dense molecular clumps and to include samples of sources in all of the embedded evolutionary stages associated with high-mass star formation (Schuller et al., 2009; Beuther et al., 2012; Csengeri et al., 2014) and catalogue ~10,000 dense clumps extracted from the emission maps (The ATLAS Compact Source Catalogue; Contreras et al., 2013; Urquhart et al., 2014a). At distances out to 20 kpc from the Solar system, the survey covers ~90 per cent of all dense molecular gas in the inner Galaxy and, with a mass completeness of ~1,000 $M_{\odot}$ beam^{-1} (~700 $M_{\odot}$ pc^{-2}), the Catalogue is taken to include the vast majority of active and potentially active high-mass star-forming clumps (Urquhart et al., 2018). Figure 6.1 shows a mix of 100 bright sources selected to include examples of clumps in all evolutionary stages.

Of the approximately 10,000 distinct massive clumps, analysis suggests ~92 per cent satisfy the size-mass criterion for massive star formation (Kauffmann et al., 2010). The majority (~90 per cent) also already show associations with star formation by hosting embedded 70 μm point sources (Urquhart et al., 2018). Around 15 per cent of these bright infrared sources are associated with compact HII regions (Urquhart et al., 2013a, 2013b, 2014b), and only a small fraction (~15 per cent) appear to be in a cold, quiescent phase (Urquhart et al., 2018). From these early designations, further analyses of the quiescent sources have begun to identify outflows (Feng et al., 2016; Tan et al., 2016; Yang et al., 2018) and likely internal heating sources (Urquhart et al., 2019), and so the evolutionary details continue to emerge. The spatial resolutions available with different instruments are illustrated in Figure 6.2, showing representative ATLASGAL detections of two high-mass sources and one low-mass protostellar source for comparison.

Figure 6.3 shows three snapshots of massive clumps from the sample at different stages of evolution (from the right): an infrared quiet clump, an infrared bright one, and one with an embedded HII region. The three sources in Figure 6.3 were also observed as part of a 128-strong sample in SiO molecular line emission with the APEX telescope, and examples of their corresponding

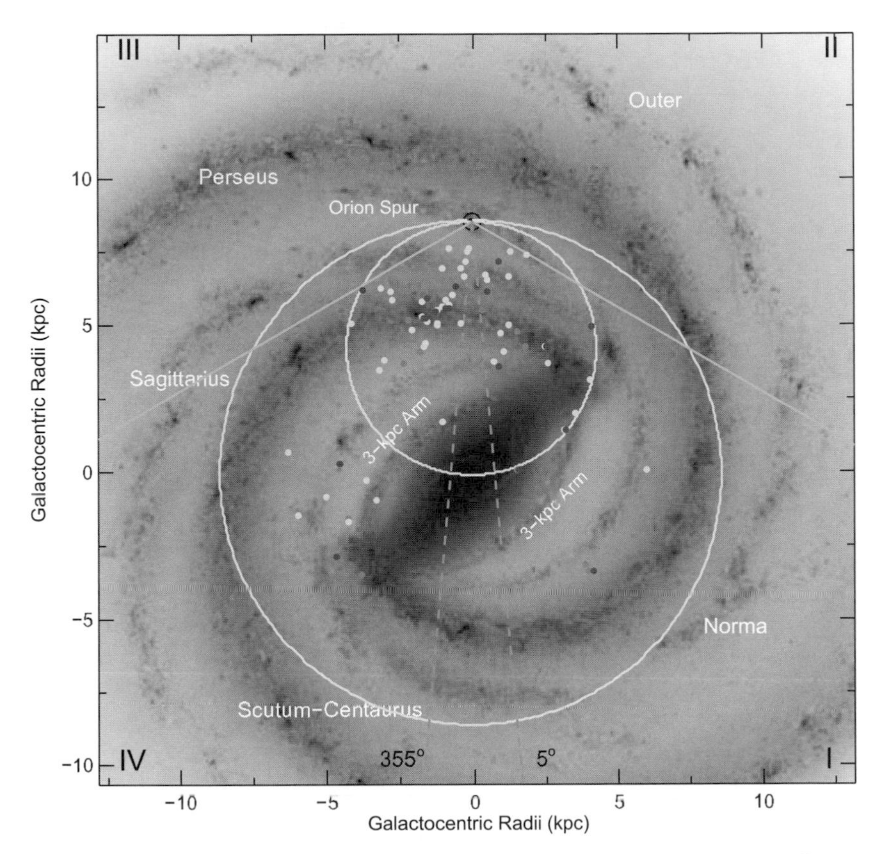

Figure 6.1 Galactic distribution of the ATLASGAL Top100 sample. Dots indicate the HII sources. The background image is a schematic of the Galactic disc as viewed from the Northern Galactic Pole (courtesy of NASA/JPL-Caltech/R. Hurt (SSC/Caltech)) (Konig et al., 2017).

SiO spectra are shown in Figure 6.4. The SiO molecule is a shock tracer, forming inefficiently in cold gas. Silicon, being highly depleted in the ISM, is taken to be a significant dust-grain component in silicate form beneath the more volatile ice layers. As a shock tracer, SiO therefore indicates significant destruction of dust grains. In low-mass YSOs, SiO observations are associated with outflow and jet activity, and typically found to decrease over evolutionary time. The APEX HMSF observations, however, show SiO detections under all embedded clump conditions, reflecting increased excitation in the more evolved sources.

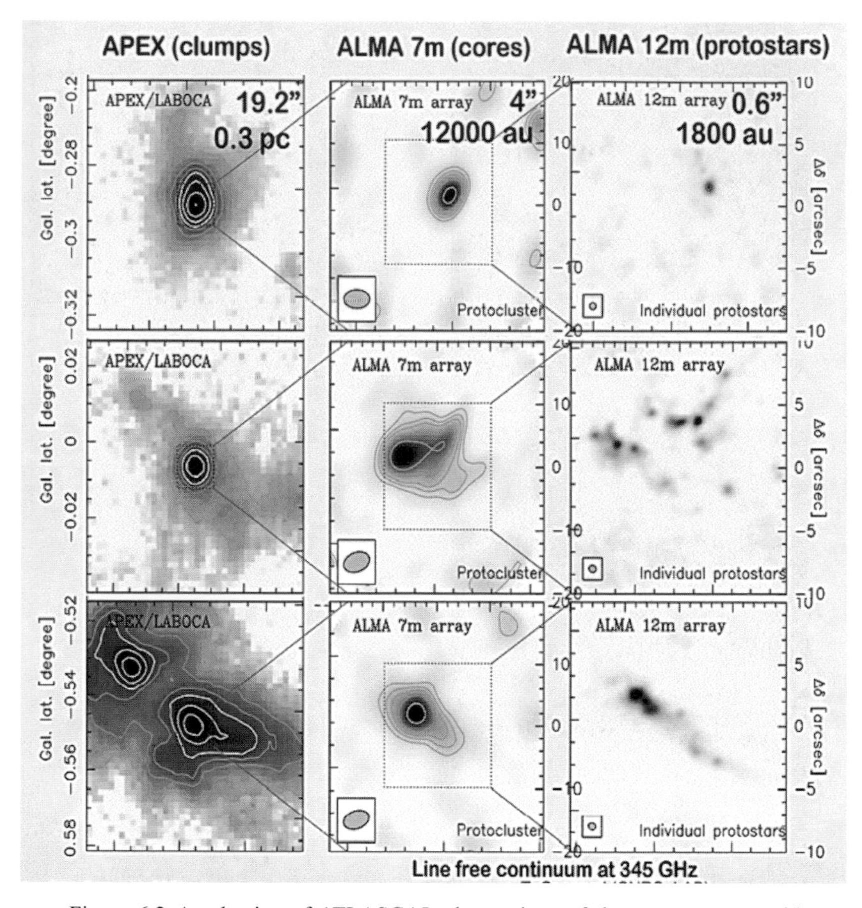

Figure 6.2 A selection of ATLASGAL observations of the same sources with different instruments offering different spatial resolutions. The top row shows a high-mass protostellar source in relative isolation, and the bottom row shows high-mass protostars in association. The middle row shows a low-mass cluster for comparison (Csengeri, private comm. 2021).

6.3 Molecular Fingerprints with Mopra

Beyond the dust continuum and shock tracer SiO, we can also look for molecular line emission fingerprints in the sample, and one spectral survey towards 570 of the ATLASGAL clumps used the Australian 22 m Mopra dish (in the Warrumbungle Mountains north of Sydney) to do so (Urquhart et al., 2019). The sample was selected to include each of the familiar embedded evolutionary stages (quiescent, protostellar, young stellar object (YSO), ionised hydrogen (HII) region, and photodissociation (PDR) region). Table 6.1

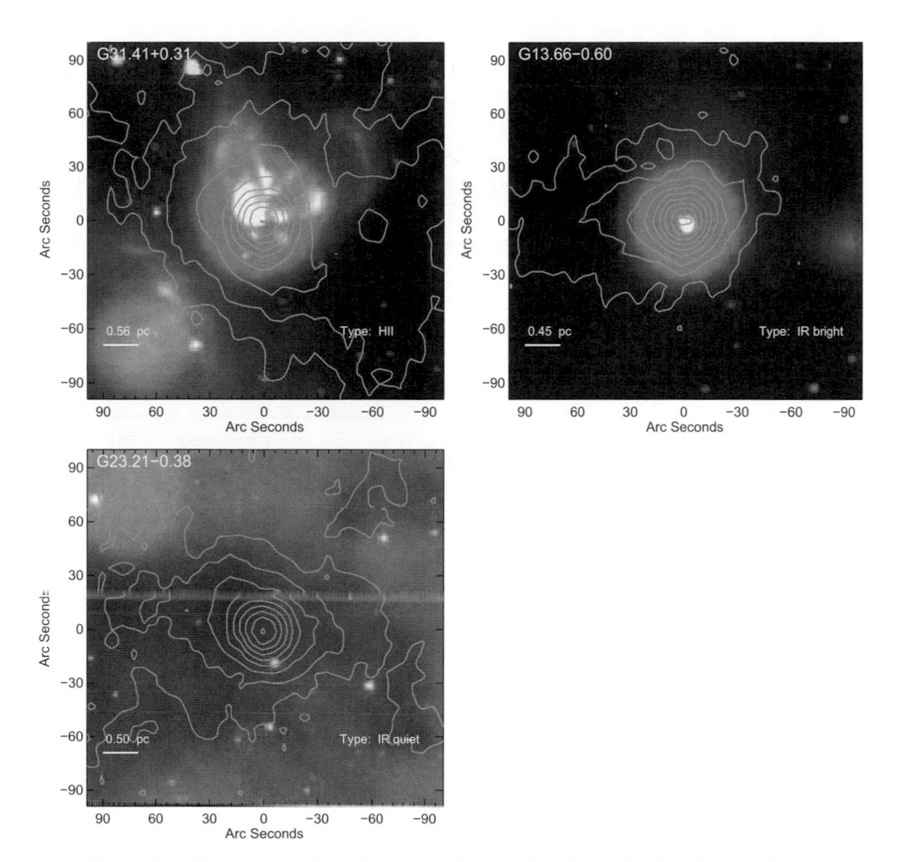

Figure 6.3 Three examples of massive clumps showing an infrared quiet clump, an infrared bright source, and an embedded UCHII region. The colour/greyscale images are Spitzer 3.6 μm, 8 μm, and WISE 22 μm band images. The contours delineate 870 μm ATLASGAL emission (Csengeri et al., 2016).

summarises 26 molecular transitions and radio recombination lines detected in the observed bandpass, with their key characterisations.

The brightest 10 transitions are taken to trace a range of physical conditions, including cold and dense gas (HNC, $H^{13}CO^+$, HCN, $HN^{13}C$, $H^{13}CN$), outflows (HCO^+), early chemistry (C_2H), gas associated with protostars and YSOs (HC_3N and cyclics, such as c-C_3H_2). Cross-comparison with previous observations and modelling typically forms the basis of these characteristic designations. Intensity variations alone can be misleading, with changes in density and/or temperature amplifying differences in excitation energy or critical density, rather than arising from chemical or physical evolutionary history. The 570 ATLASGAL sources are classified according to evolutionary sequence in which 29 are identified as quiescent clumps, 153 as protostellar clumps (70 μm

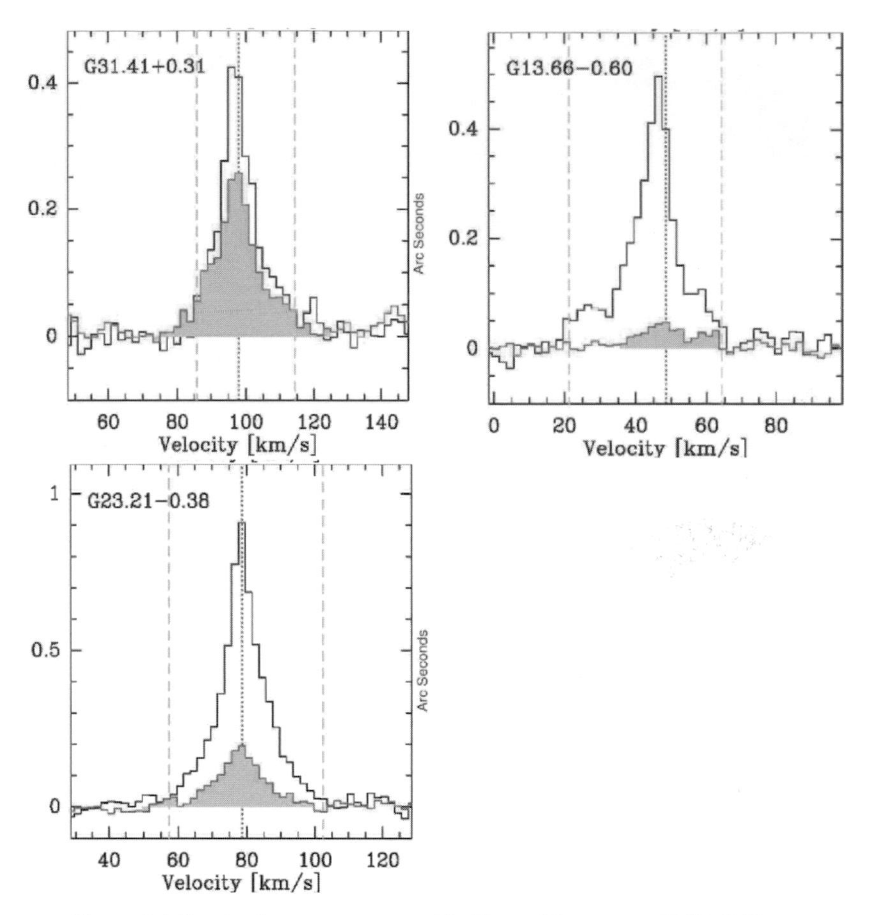

Figure 6.4 For the three sources named in Figure 6.3, spectra of SiO (2-1) transitions (black/higher peak) and SiO (5-4) transitions (red/lower peak), with the shaded area showing the velocity range used to determine integrated emission. The central dotted line shows the systemic velocity of each source (Csengeri et al., 2016).

bright), 128 as YSOs (mid-infrared bright), and 166 as HII regions (bright in mid-infrared and compact in radio continuum). In addition, 48 are now designated PDRs, while the remaining 46 are too complex as yet to resolve. Distances, luminosities, and clump masses for all are broadly known (Urquhart et al., 2018), and integrated line intensities, relative abundances, line ratios, and dust temperatures all contribute to the evolutionary designation.

Figure 6.5 shows the mean values of integrated-intensity ratios for the thirteen most distinct molecular emissions identified. These all show statistically significant differences between the stages categorised and clearly increase as a

Table 6.1 *The 26 transitions in the Mopra 8-GHz bandpass (Urquhart et al. 2019).*

Emission line(s)	ν (GHz)	E_u/k (K)	Log(n_{crit}) (cm^{-3})	Detections	Detection ratio	RMS noise (mK)	T^* (K)	Line width (km s^{-1})	Intensity (km s^{-1} K)	Comments
c–C_3H_2 (2–1)	85.3389	6.4	6.0	446 (9)	0.78	24	0.21	3.3	0.72	Cyclic molecules
HCS^+ (2–1)	85.3479	6.1	4.5	62 (0)	0.11	22	0.13	3.3	0.49	Sulphur chemistry
CH_2CHCN (9–8)	85.4218	-	-	3 (0)	0.01	28	0.11	1.2	0.19	Temperature
CH_3C_2H (5–4)	85.4573	-	-	273 (4)	0.48	26	0.19	2.8	1.34	Temperature
$HOCO^+$ (4_{04}–4_{03})	85.5315	10.3	5.2	3 (0)	0.01	17	0.08	4.5	0.4	Indirect gas-phase CO_2 tracer
H42α	85.6884	-	-	58 (1)	0.10	11	0.16	27.6	1.03	Ionised gas
o-NH_2D (1_{11}–1_{01})	85.9263	20.7	6.6	148 (0)	0.26	24	0.15	2.1	0.89	Deuteration, coldest dense gas
SO_2 (2_2–1_1)	86.0940	19.3	5.2	67 (0)	0.12	26	0.31	5.2	4.64	Sulphur chemistry, shocks
$H^{13}CN$ (1–0)	86.3402	4.1	6.3	403 (13)	0.71	26	0.25	3.5	2.03	Dense gas
HCO (1–0)	86.6708	-	-	36 (0)	0.06	22	0.11	2.8	0.35	Photon-dominated regions
$H^{13}CO^+$ (1–0)	86.7543	4.2	4.8	536 (20)	0.94	25	0.36	3.1	1.2	Dense gas
SiO (2–1)	86.8470	6.2	5.4	218 (9)	0.38	10	0.17	13.2	0.45	Shocks, outflows
$HN^{13}C$ (1–0)	87.0908	4.2	5.5	445 (7)	0.78	26	0.26	3.1	0.82	Dense gas
C_2H (1–0)	87.3169	4.2	6.0	542 (131)	0.95	45	0.68	3.5	6.31	Early chemistry/PDR
HNCO (4–3)	87.9253	10.6	6.0	269 (1)	0.47	25	0.19	3.8	0.82	Hot core & FIR chemistry
HCN (1–0)	88.6318	4.2	5.7	528 (284)	0.93	34	0.98	4.5	9.15	Dense gas
HCO^+ (1–0)	89.1885	4.3	4.8	542 (136)	0.95	39	1.34	4.5	6.26	Kinematics (infall, outflow)
HNC (1–0)	90.6636	4.3	5.1	562 (119)	0.99	33	1.24	4.2	5.54	High column density, cold gas tracer
C_2S (2–1)	90.6864	-	-	3 (0)	0.01	21	0.09	3.8	0.31	Dense gas, evolutionary phase
$^{13}C^{34}S$ (2–1)	90.9260	6.5	5.6	2 (0)	0.00	25	0.11	2.8	0.34	Column density
HC_3N (10–9)	90.9790	24.0	5.0	514 (22)	0.90	26	0.46	3.1	1.64	Hot core
$HC^{13}C_2N$ (10–9)	90.9790	23.9	5.2	3 (0)	0.01	22	0.12	12.7	1.31	Hot core
CH_3CN (5–4)	91.9871	13.2	5.4	142 (5)	0.25	25	0.18	4.5	2.74	Hot core temperature
H41α	92.0345	-	-	60 (0)	0.11	10	0.14	31.8	0.92	Ionised gas
^{13}CS (2–1)	92.4943	6.0	5.5	250 (1)	0.44	23	0.24	3.5	0.95	Dense gas, infall
N_2H^+ (1–0)	93.1738	4.5	4.8	545 (0)	0.96	27	0.66	3.1	8.62	Dense gas, depletion resistant

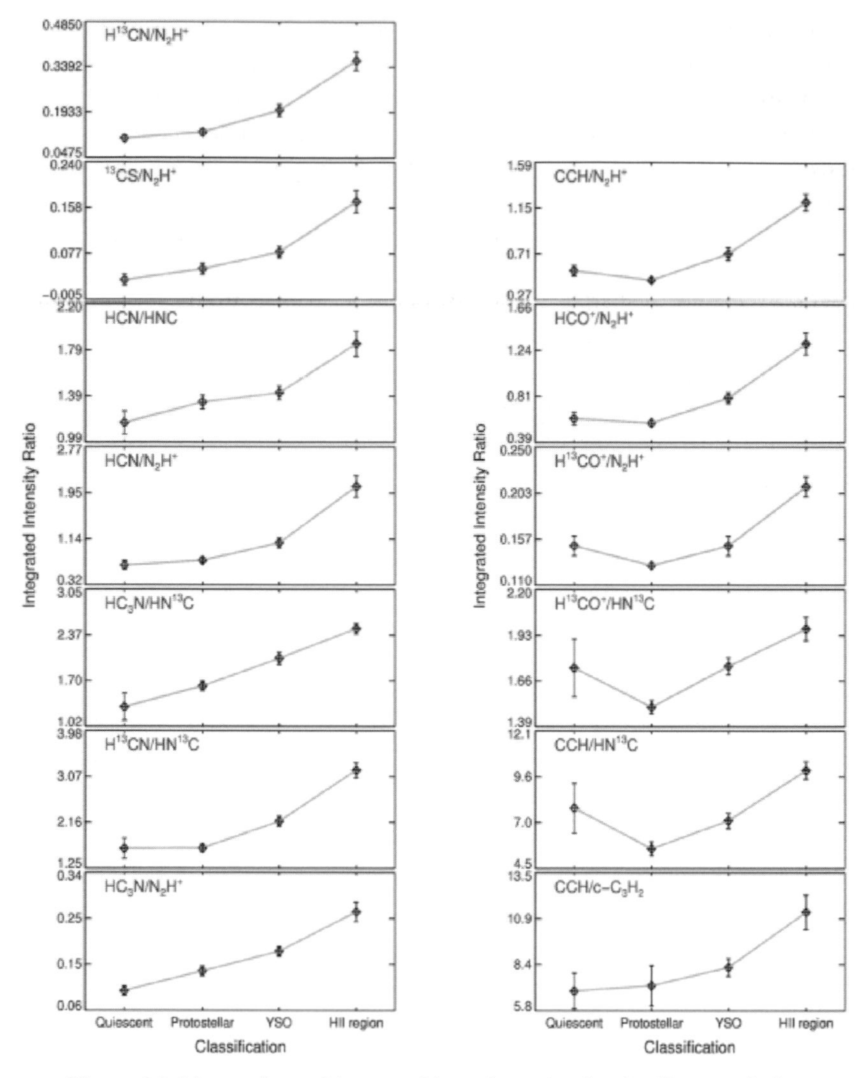

Figure 6.5 Mean values of integrated-intensity ratios for the 13 most distinct emissions identified (Urquhart et al., 2019).

function of evolutionary development. Top to bottom, left to right, is the order of decreasing statistical significance. The plots show that molecular emission from both optically thin and thick line ratios does correlate with dust temperatures as a continuous measure of high-mass star formation, emphasising that the evolutionary distinctions undoubtedly overlap.

Among the conclusions from this detailed study, some species, such as C_2H and $H^{13}CN$, show a steady increase with evolution. However, a small number, including HNCO and SiO, also show detections that peak in the protostellar stage. The abundances of both of these latter molecules are typically enhanced by shocks (slow and fast, respectively) and are, therefore, often associated with outflows. Several other molecules show a significant decline in detection rates at the HII stage, such as N_2H^+ and o-NH_2D, both being sensitive to the far-UV flux, while detections of a molecule such as $HN^{13}C$ appear to differ very little across the evolutionary stages. There are, therefore, useful baseline species with which to compare molecules such as C_2H that are more sensitive to dust temperature, excitation, or far-UV induced changes. While a useful comparator, the insensitivity of these species is most likely a consequence of their widespread distribution in cold gas and high relative abundance and brightness in the submillimetre range, with much remaining even in clumps hosting HII regions. The HCO^+/N_2H^+ ratio proves to be both statistically distinct and to increase with evolutionary stage, this study confirming earlier observations (e.g. Hoq et al., 2013).

While detectable dust temperature in many of the clumps is estimated to be ~25 K, gas temperatures, derived from CH_3C_2H and CH_3CN K-ladders in higher-density cores, are 10 and 20 degrees higher, respectively. The presence of CH_3CH_2 emission in a third of the quiescent sample suggests that some of these clumps may actually be harbouring very young protostellar objects, so deeply embedded that they remain undetectable at 70 μm. Some of them even show SiO emission, as we have noted earlier. This leaves an extremely small number in the sample as genuinely quiescent (~5 per cent), which in turn suggests that high-mass star formation follows efficiently once massive clumps have coalesced.

6.4 NOEMA

As an example of a multiple-source, high-mass star formation survey offering spatial resolutions potentially a factor of 10 better than the APEX telescope, we can consider an IRAM Northern Extended Millimetre Array (NOEMA) survey of fragmentation and disk formation in a sample of 20 regions, known as the CORE program. In this study, kinematic and chemical properties of a relatively homogenous group of HMSF regions have been studied (Beuther et al., 2018; Gieser et al., 2021). The observations are made in both 1.3 mm continuum dust emission and the spectral emission lines of eleven molecular species that we will consider shortly. First, the protostellar sources are chosen for their bolometric luminosity being $>10^4\ L_\odot$, indicating that the stars which are forming are at least 8 $M_\odot$, and for their distances being <6 kpc. An angular

Table 6.2 *The 18 selected sources studied by Gieser et al. (2021) from the CORE sample of Beuther et al. (2018).*

Region	RA	Dec	Distance	Velocity
J(2000)	J(2000)	(kpc)	(kpc)	(km s^{-1})
IRAS 23033	23:05:25.00	+60:08:15.5	4.3	−53.1
IRAS 23151	23:17:21.01	+59:28:47.5	3.3	−54.4
IRAS 23385	23:40:54.40	+61:10:28.0	4.9	−50.2
AFGL 2591	20:29:24.86	+40:11:19.4	3.3	−5.5
CepA HW2	22:56:17.98	+62:01:49.5	0.7	−10.0
G084.9505	20:55:32.47	+44:06:10.1	5.5	−34.6
G094.6028	21:39:58.25	+50:14:20.9	4.0	+29.0
G100.38	22:16:10.35	+52:21:34.7	3.5	−37.6
G108.75	22:58:47.25	+58:45:01.6	4.3	−51.5
G138.2957	03:01:31.32	+60:29:13.2	2.9	−37.5
G139.9091	03:07:24.52	+58:30:48.3	3.2	−40.5
G075.78	20:21:44.03	+37:26:37.7	3.8	−8.0
IRAS 21078	21:09:21.64	+52:22:37.5	1.5	−6.1
NGC7538 IRS9	23:14:01.68	+61:27:19.1	2.7	−57.0
S106	20:27:26.77	+37:22:47.7	1.3	−1.0
S87 IRS1	19:46:20.14	+24:35:29.0	2.2	+22.0
W3 H2O	02:27:04.60	+61:52:24.7	2.0	−48.5
W3 IRS4	02:25:31.22`	+62:06:21.0	2.0	−42.8

resolution ~0.4″ over distances ranging from 0.7 to 5.5 kpc resolves to spatial scales varying between 300 AU and 2,300 AU. Temperature structures have been determined from H_2CO and CH_3CN emission lines, along with column densities for the eleven species. Table 6.2 lists the 18 sources studied by Gieser et al. (2021), and Figure 6.6 shows the original 20-source sample reported by Beuther et al. (2018) in images derived from Spitzer, the Wide Field Infrared Explorer (WISE), and the Submillimetre Common-User Bolometer Array (SCUBA).

6.5 Physical Structure

Within the 18 regions observed, the survey identified 22 individual cores defined by 1.3 mm continuum emission from dust and with radially decreasing temperature profiles. With a linear spatial resolution of order 10^3 AU, the survey cannot resolve potential accretion disks around the protostars, and characterisation through the kinematic analysis of line profiles would require greater spectral resolution (Ahmadi, Kuiper, & Beuther, 2019). However, given the available resolution of dust and gas envelopes around individual

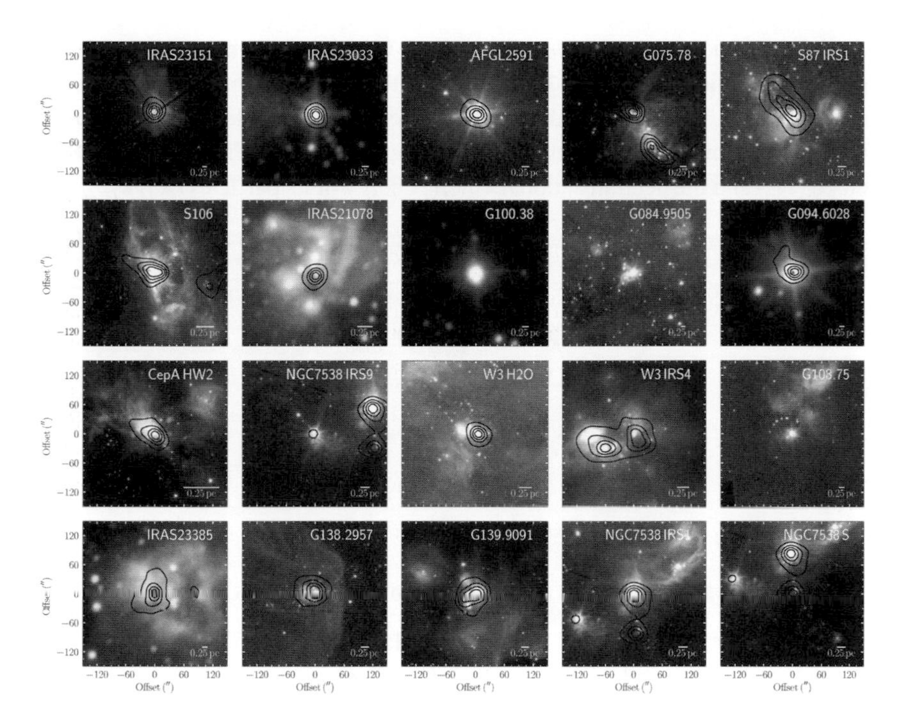

Figure 6.6 An overview of the CORE sample, with a mix of Spitzer 3.6, 4.5, and 8 μm and WISE 3.4, 4.6, and 12 μm images, with SCUBA 850 μm contours superimposed (Beuther et al., 2018).

cores and approximating them as spherically symmetric, temperature profiles (T_r) and radial density profiles (n_r) can be described by power laws:

$$T_r = T_{in}\left(r / r_{in}\right)^{-q}, \text{ and}$$
$$n_r = n_{in}\left(r / r_{in}\right)^{-p},$$

where $T_{in} = T(r_{in})$ and $n_{in} = n(r_{in})$ at an arbitrary radius r_{in}. The core temperature profiles derived from H_2CO and CH_3CN line emission give an average power law index $q = 0.4 \pm 0.1$, which agrees with theoretical prediction (Gieser et al., 2021). The continuum profiles reveal a mean density power law index $p = 2.0 \pm 0.2$, which compares favourably with previous single-dish and interferometric studies, all of which appear between 1.7 and 2.0 from scales of 1 pc down to 1 k AU. The mean core mass across all 22 observed is calculated as 4.1 $M_\odot$, and mean $N(H_2)$ column density ~1.5×10^{21} cm^{-2}.

The relationships between density and temperature in tracing the physical structure of collapsing cores arises from the theoretical understanding briefly sketched here but given more detailed derivations in standard textbooks. Cloud stability is a function of energy balance arising from the cloud's constituent particles, each having mass, position, and velocity, moving in their mutual gravitational field. The time derivative of the cloud's moment of inertia at equilibrium (Γ) is zero, and the virial theorem expresses that relationship for the equilibrium condition between the self-gravitating potential energy of the cloud (Ω_G) and the net translational kinetic energy of its particles (K_T) as

$$1/2\Gamma = \Omega_G + 2K_T = 0.$$

For a given cloud mass (M), radius (R), and velocity dispersion (σ):

$$K_T = 1/2M\sigma^2, \text{and}$$
$$\Omega_G = -GM^2/R.$$

Substituting these into the virial theorem, we obtain

$$\sigma^2 \sim GM/R,$$

and from this a virial mass is definable as

$$M_{vir} \sim \sigma^2 R/G.$$

Exceed this mass, and a cloud is likely to collapse. Less than this, and the cloud is likely to disperse unless confined by external pressures. The proportionality of virial mass to velocity dispersion accounts for cloud dispersion by stellar winds during formation.

For simplicity, the first approximation of free-fall time for a collapsing isothermal, spherically symmetric cloud ($M > M_{vir}$) is given by

$$t_{ff} = (3\pi/32G\rho_0)^{1/2},$$

where ρ_0 is the initial density. With free-fall time having an inverse dependence on density, the high-density regions collapse more rapidly, and the density distribution becomes more centrally peaked with a density profile roughly proportional to r^{-2}. The simplicity of this breaks down because gas compression generates heat and, where density (and therefore optical thickness) increases sufficiently to prevent the radiative escape of excess heat, isothermal assumptions no longer apply. The maximum density beyond which radiative cooling no longer matches compressional luminosity is essentially an extension of Stefan's radiation law and therefore proportional to T^4.

More complex modelling applies especially in the case of the particularly uncertain circumstances within high-mass star formation regions. Theoretical proposals include a monolithic collapse of turbulent cores (McKee & Tan, 2002, 2003); protostellar collisions and coalescence in dense clusters (Bonnell et al., 1998; Bonnell & Bate, 2002); and competitive accretion in clusters (Bonnell et al., 2001; Smith et al., 2009; Hartmann, 2002; Murray & Chang,

2012). For theoreticians, the density and temperature structure are important parameters for the initial cloud and the developing clumps and cores that follow. Early models (Shu, 1977; Shu et al., 1987) that trace the gravitational collapse of an isothermal sphere find the density index $p = 2$ in the outer envelope and $p = 1.5$ in the inner region where gas is free-falling onto the central position. There are later refinements, for example for the isothermal case (McLaughlin & Pudritz, 1996, 1997), or assuming some turbulence (McKee & Tan, 2002, 2003), and others that model a variety of power law index values (e.g. Bonnell et al., 1998; Murray & Chang, 2012). Hydrodynamic simulations have also explored steep and shallow density profiles (Chen et al., 2020; Girichidis et al., 2011) and have found that the steeper ones engender centrally concentrated clusters, while shallower ones promote hierarchical fragmentation. More-uniform profiles have been shown to generate large fractions of low-mass stars. In general, all agree it is likely that the initial density profile of the cloud and the ongoing turbulence dominate the determination of star-formation outcome.

While additional forces such as magnetic fields and mechanical turbulence undoubtedly can slow down or even prevent core collapse, as is particularly evident in the diffuse ISM (Ballesteros-Paredes et al., 2007; Burge et al., 2016), CORE results do suggest that thermal fragmentation in clouds having density $>10^4$ cm^{-3} dominates (Beuther et al., 2018), a conclusion consistent with that of other surveys (e.g. Palau et al., 2015).

6.6 Chemical Structure

In considering core collapse timescales, we can turn to the 11 molecular species detected in the CORE survey within a 4 GHz spectral bandwidth, all of which take their own time to evolve. The species observed are ^{13}CO, $C^{18}O$, SO, OCS, SO_2, DCN, H_2CO, HNCO, HC_3N, CH_3OH, and CH_3CN. Column densities for these species are derived, and some show a distribution having a clear column-density peak (^{13}CO, $C^{18}O$, SO, DCN, H_2CO, and HNCO). Other species show a clear double-peaked distribution indicating a separation between core and offset positions (OCS, HC_3N, CH_3OH, and CH_3CN). In these cases, column density is enhanced by a factor of 10–100 towards the core positions. Generally, across all species detected, the highest column densities are found towards core positions, and the lowest column densities towards the offset positions. It is noticeable that it is the larger molecules that show a clear difference between core and non-core position, while for simpler species (such as ^{13}CO and DCN) it is less obvious. This could be partly a consequence of critical densities since CO is easily excited and therefore ubiquitous. But it may also be that the emission from high-column-density COMs is associated exclusively with the cores. If we had no information from any other study, it could still be said that the influence of likely energetic processes (outflows, shocks, disk accretion, protostellar radiation) on higher-density core gas and dust seems to correlate significantly with the larger COMs abundances observed at those locations.

The CORE researchers make use of their observations and deductions in an iterative chemical-fitting model, MUSCLE ('MUlti Stage CLoud codE') (Feng et al., 2016) to constrain some mean physical properties and chemical ages during the evolution from the conjectured earliest IRDC phase through to the late UCHII phase of high-mass star formation. The derived power laws provide radial density and temperature profiles reduced to a grid of radial cells and linked to a time-dependent gas–grain chemical model based on the ALCHEMIC code (Semenov et al., 2010) in which the chemical rate file links a network of ~1,000 reactions. Cosmic-ray particles (CRP) and CR-induced FUV radiation are adopted as the only external ionising sources. The UV dissociation and ionisation photorates correspond to the spectral shape of the interstellar FUV radiation field. In addition to pure gas-phase chemical processes, the chemical network includes gas-grain interactions, assuming molecules stick to grain surfaces at low temperatures with 100 per cent probability (except for H_2, which is taken as completely non-sticking). Ices are taken to be released back to the gas phase by thermal, CRP-, and CRP-induced UV photodesorption. Grain re-charging is modelled by dissociative recombination and radiative neutralisation of ions on grains (electrons sticking to grains), while chemisorption of surface molecules is not considered. The UV photodesorption yield for ices is $\sim 10^{-3}$ (e.g. Oberg et al., 2009a, b). The synthesis of complex molecules is included using a set of surface reactions, together with desorption energies and photodissociation reactions of ices (Garrod & Herbst, 2006; Semenov & Wiebe, 2011).

The net result of the modelling is that estimates of the chemical age are derived, ranging from 20,000 to 100,000 years, with a 60,000-year mean across the sample. Multiple cores within a single region do show that there can be an age gradient for the more widely separated cores (e.g. IRAS 23033 and G108.75), while closer-grouped cores show similar chemical ages (e.g. towards IRAS 21078 and W3 H2O), both of which support a sequential star-formation scenario. From the perspective of astrochemical evolution, we have already sought out correlations in earlier chapters between species as a function of temperature, column density, and location, as being significant. These give clues as to potential formation and destruction pathways in the gas–grain chemistries.

The CORE study finds high correlation coefficients (>0.8) between the following pairs for their column densities relative to that of $C^{18}O$, an optically thin stand-in for $N(H_2)$ detected towards all positions: HC_3N-SO; HC_3N-OCS; CH_3CN-SO; CH_3CN-OCS; CH_3CN-H_2CO; CH_3CN-HNCO; CH_3CN-HC_3N; and CH_3CN-CH_3OH. This strong positive correlation between methyl cyanide, CH_3CN, and other nitrogen and sulphur species follows directly from the fact that all of these species are formed from efficient gas-phase reactions characteristic of dense molecular cloud conditions and, observationally, the line-integrated ratios of sulphur- and nitrogen-bearing species clearly show correlations with dust temperature (and thus warming gas) in a sample of HMSF clumps (Urquhart det al., 2019).

The particular correlation between CH_3OH and CH_3CN observed in these high-mass sources is also observed towards low-mass star-forming regions (Bergner et al., 2017; Belloche et al., 2020) as we discussed in the LMSF context. However, there are longstanding questions arising from, and about, this rather ubiquitous molecule, CH_3CN, which may be instructive, and we will consider these separately in subsequent chapters. For the moment, given no chemical link between CH_3OH and CH_3CN, we might argue that this correlation is a temperature effect resulting from chemically unrelated species evaporating off icy grain mantles following from the range of possible desorption mechanisms noted in Chapter 1 (Belloche et al., 2020). The question nevertheless arises, can we deduce a timescale for the kinematics of these high-mass cores that is compatible with the chemical evolution of these observed species in the gas?

6.7 Comparing Timescales

The free-fall timescale for a spherical cloud to collapse under gravity alone is a function of its initial density:

$$t_{ff} = [3\pi / 32G\rho(H_2)]^{1/2},$$

where G is the gravitational constant and ρ (H_2) is the initial mass density. Turbulence and the mixing of dense cloud gas and dust will limit the maximum mass of a star that is able to form because small structures will preferentially coalesce (e.g. Elmegreen, 2000). Statistically there is a time constraint on star formation, the so-called crossing timescale, for any given mass, expressible as

$$\tau_{cross} = R / v,$$

where R is the clump radius and v the velocity dispersion.

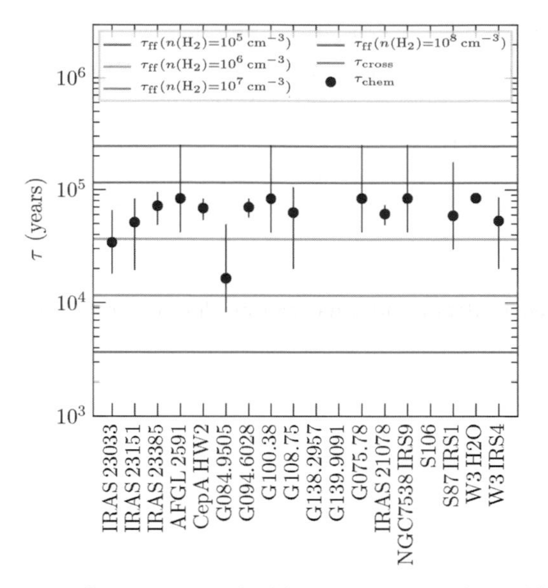

Figure 6.7 For each of the 18 cores, comparisons of free-fall timescale (τ_{ff}), crossing timescale (τ_{cross}), and chemical timescale (τ_{chem}) are shown. The five horizontal lines from bottom to top mark τ_{ff} for n(H_2) = 10^8 cm^{-3}, 10^7 cm^{-3}, 10^6 cm^{-3}, and 10^5 cm^{-3}, with τ_{cross} as the top line. The black circles mark τ_{chem} values (Gieser et al., 2021).

The CORE study compares free-fall timescales for a range of $n(H_2)$ values between 10^5 cm^{-3} and 10^8 cm^{-3} and crossing timescales (assuming R = 1 pc, v = 4 km s^{-1}, and chemical ages for each observed core derived with MUSCLE). Figure 6.7 shows the comparisons of free-fall timescale (τ_{ff}), crossing timescale (τ_{cross}), and chemical timescale (τ_{chem}). The derived chemical ages are in agreement with initial clump densities in the 10^5–10^6 cm^{-3} range, which corresponds to an expected IRDC value >10^5 cm^{-3} (e.g. Carey et al., 1998).

The free-fall timescales for gas densities of 10^8 to 10^5 cm^{-3} span ~3 × 10^3 yr to 10^5 yr, respectively. The crossing time estimate at ~2 × 10^5 yr is an upper limit, but in agreement with the scenario that suggests that star formation occurs within a few crossing times (Elmegreen, 2000). Chemical timescales, as we noted earlier, average ~6 × 10^4 yr. In conclusion, the chemical evolution of the observed COMs is readily accomplished within the apparent HMSF timescale.

6.8 Summary

The value of a representative survey of HMSFRs, such as the very large-scale ATLASGAL study or even the much smaller CORE study, lies in learning what physical and chemical parameters are shared across a variety of sources.

The results of statistically large samples of detected, or non-detected, sources such as that of ATLASGAL give us very secure data from which to generalise about the typical star formation process. The results of smaller but still multi-location studies such as CORE give us greater specific details, albeit from a self-selecting sample, which may or may not be typical but that we can certainly say are common, at least until future wider surveys prove us wrong. In the following three chapters, we go to three of the most well-known HMSF sites to see in even greater detail what, in terms of molecular line emission complexity, has been observed in detail and what can be derived from that about the physical conditions and the ongoing activity in the region.

7
Sagittarius B2(N)

7.1 Introduction

Sagittarius (Sgr) B2 is the most massive star-forming region in the Galaxy, with a total bolometric luminosity 10,000,000 times that of the Sun. As a complex of GMCs and HII regions, it is located about 8 kpc away from the Solar system, little more than 100 pc from the Galactic Centre (Boehle et al., 2016; Anderson et al., 2020). With a diameter ~40 pc, it has a mass $\sim 10^7$ $M_\odot$ (Lis & Goldsmith 1990). It has been characterised as a mini-starburst region, containing as it does more than 70 high-mass stars and much evidence of ongoing high-mass star formation, including methanol (CH_3OH) and water maser emission, HII regions, and more than 200 compact millimetre sources. SiO emission maps obtained with, for example, the IRAM 30 m telescope show turbulent kinematics on a vast scale, indicative of not only star-formation but large-scale intercloud collisional shocks (Armijos-Abendaño et al., 2020b). Given the proximity to the supermassive black hole Sgr A* at the Galactic Centre, the Sgr B2 clouds are strongly irradiated by an UV flux 100 times greater than the average local field (Clark et al., 2013), and with an impinging cosmic ray ionisation rate two or three orders of magnitude higher than the standard (Le Petit et al., 2016). The B2 location is shown in Figure 7.1 in a Spitzer image built from 3.6 μm, 8 μm, and 24 μm continuum emission (Anderson et al., 2020).

7.2 Sgr B2 (N), (M), and (S)

Sgr B2 is described as having three principal features. The first is that of the high-mass protoclusters Sgr B2(N), Sgr B2(M), and Sgr B2(S) having gas densities $\sim 10^7$ cm^{-3} and temperatures >200 K (Etxaluze et al., 2013). The second is

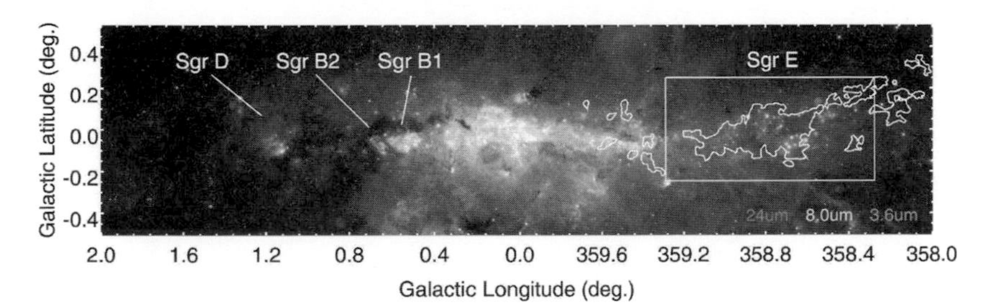

Figure 7.1 *Spitzer* (MIPSGAL, GLIMPSE) view of the Galactic Centre in continuum emission, with massive star-formation regions labelled (Anderson et al., 2020). The contours around Sgr E are ^{13}CO emission (Schuller et al., 2017).

a moderately dense surrounding envelope (~10^5 cm^{-3}) extending ~ 2.5 × 5.0 pc, with temperatures ~100 K (Schmiedeke et al., 2016). The third delineator is a lower-density envelope or halo cloud (~10^3 cm^{-3}) extending nearly 40 pc further out from the dense cores, with its outer margins subject to raised temperatures due to the impinging UV field from the Galactic Centre (Huttemeister et al., 1995).

The dense inner region is filled with the multiple hot cores and compact- and ultracompact-HII regions associated with high-mass star formation (Gaume & Claussen, 1990), and it is here that many molecular species have been found. For example, 56 different species towards Sgr B2 (N) and 46 towards Sgr B2(M) have been detected with the IRAM 30 m telescope (Belloche et al., 2013). A large population of high-mass protostellar sources are actually found along the length of the B2 cloud, not just in the three cores (Ginsburg et al., 2018; Ginsburg & Kruijssen, 2018). Sgr B2(N) itself has a mass ~10^5 $M_\odot$ on a scale of 0.2–0.4 pc, and the compact HII 'K' region was resolved into subcomponents K1 to K6 during the 1990s (Gaume & Claussen, 1990; Gaume et al., 1995; de Pree et al., 1996). K1, K2, and K3, are particularly compact (<0.6″) and spatially coincident with the maser clusters associated with the B2(N) core (Kobayashi et al., 1989). They were also the location in which early hot core chemistries were observed by interferometric observations of NH_3 and CH_3CN line emission (Vogel et al., 1987; Gaume & Claussen, 1990) as well as dust continuum emission at millimetre wavelengths (Carlstrom & Vogel, 1989; Lis et al., 1993; Kuan et al., 1996). Further line emission has continued to reinforce observations of the hot core molecular complexity (Goldsmith et al., 1987; Kuan & Snyder, 1994; Kuan et al., 1996; Liu & Snyder, 1999; Hollis et al., 2003; Belloche et al., 2008; Qin et al., 2011; Belloche et al., 2013, 2014).

It is worth reflecting on the 8 kpc distance of the Sgr B2 complex for a moment. There are multiple intervening molecular clouds associated with

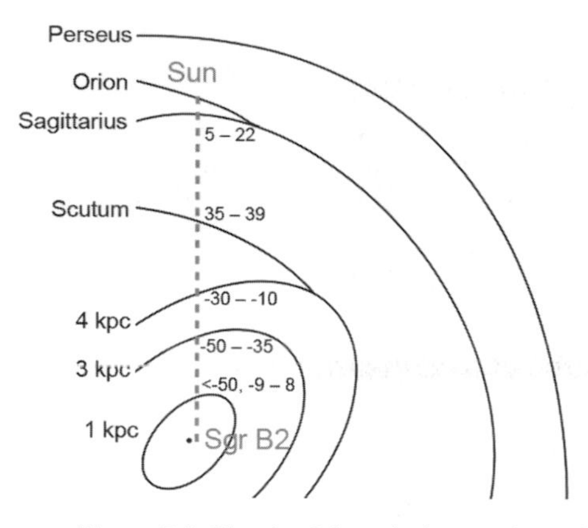

Figure 7.2 Sketch of the spiral arms along the line of sight to Sgr B2, with velocity ranges of the intervening diffuse and translucent clouds given in km s^{-1} (Thiel et al., 2019).

those spiral arms that are closer to the Galactic Centre, and Figure 7.2 shows a sketch of that line of sight (Thiel et al., 2019).

Of the 15 main velocity components along the line of sight, two identify the Sgr B2 envelope at 63 km s^{-1} and 80 km s^{-1}. Of two distinctive categories of absorption with significant c-C_3H_2 column density differences, one (v_{LSR} < –13 km s^{-1}) identifies Galactic Centre clouds plus clouds belonging to the 3 kpc and 4 kpc arms. The other (–13 km s^{-1} < v_{LSR} <42 km s^{-1}) includes the Scutum and Sagittarius clouds. Opacity maps in a variety of molecules (c-C3H2, $H^{13}CO^+$, ^{13}CO, HNC, CS, and their isotopologues, plus SiO, SO, and CH3OH) reveal generally homogenous structures, although SO and SiO show greater complexity of structure within small clumps (5–8″).

7.3 Sgr B2 (N)

The dense molecular core Sgr B2 (N) contains several H II regions (Gaume et al., 1995) as well as several hot molecular cores (Bonfand et al., 2017; Sánchez-Monge et al., 2017). Its continuum emission in the millimetre wavelength range consists of free–free radiation and thermal dust emission (e.g. Liu & Snyder 1999). ALMA observations of Sgr B2(N) in SO_2 and SiO at an angular resolution ~2″ have also specifically confirmed a bipolar outflow (collimation ~60°) and a rotating core (with Keplerian ring-like structure, radius ~6,000 AU), with the outflow driving force located at K2 (Higuchi et al., 2015). The outflow is characterised as young (5 × 10^3 yr) and massive (~2,000 $M_\odot$). Figure 7.3 shows the ALMA continuum map against which the

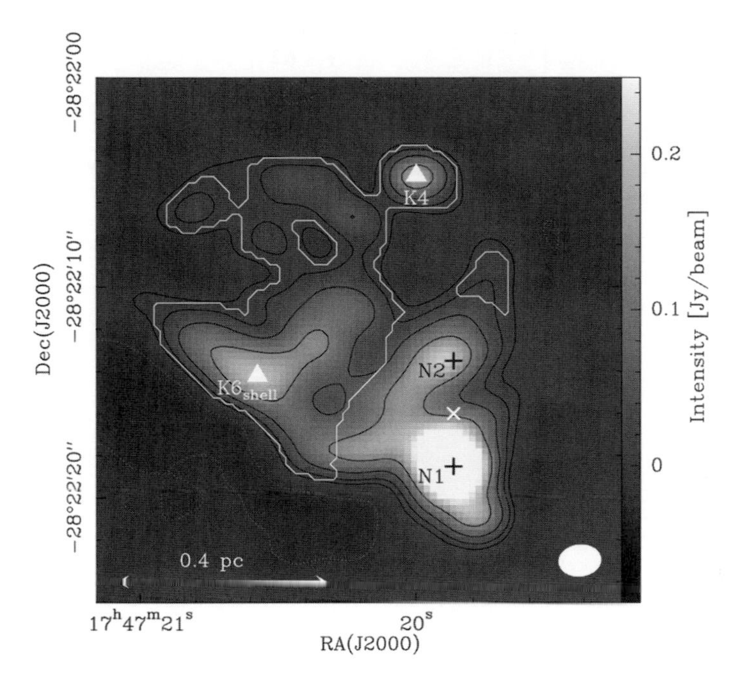

Figure 7.3 ALMA continuum map of Sgr B2 (N) at 85 GHz against which the intervening line-of-sight clouds were detected in c-C_3H_2 absorption lines. The black crosses identify hot cores N1 and N2, and the white triangles the UCHII regions K4 and K6. The white cross identifies the focus of the ALMA observation and the white ellipse the synthesised beam size (Thiel et al., 2019).

line-of-sight absorption detections were made (Thiel et al., 2019), with hot core and UCHII locations marked.

7.4 COMs in Sgr B2 (N)

At least a third of all COMs observed anywhere in the ISM have been detected towards Sgr B2 , and Sgr B2(N) is one of its major sites (Belloche et al., 2019). B2(N) itself is fragmented into multiple dense and compact objects. In addition to the two sites marked in Figure 7.3, three further hot cores (N3, N4, and N5) have been detected in 3 mm emission (Belloche et el., 2016) as well as 11 star-forming cores within a 0.4 pc region centred on N1 (Ginsburg et al., 2018). Figure 7.4 shows some representative COMs-to-CH_3OH abundance ratios towards the hot cores N2–N5 (Bonfand et al., 2019).

The COMs detected towards Sgr B2 hot cores include glycoaldehyde (CH_2OHCHO), propenal (C_2H_3CHO), propanal (C_2H_5CHO), cycloproprenone

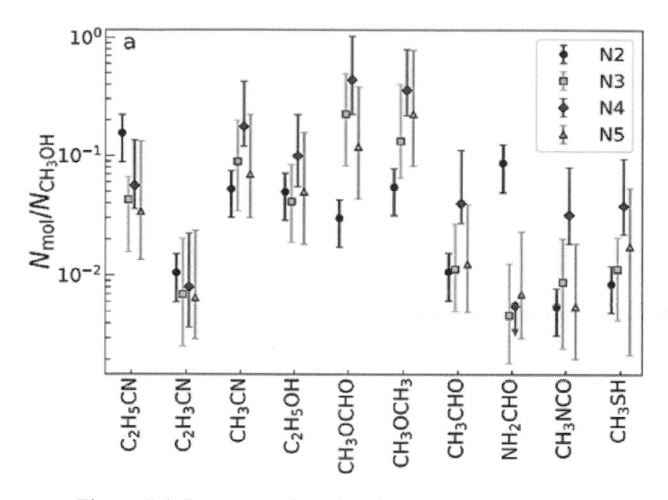

Figure 7.4 Representative abundance ratios of COMs against CH_3OH detected towards Sgr B2 (N2–N5) (Bonfand et al., 2019).

(c-CH_2C_3O), acetamide (CH_3CONH_2), cyanoformaldehyde (CNCHO), ethanimine (CH_3CHNH), and cyanomethanimine (HNCHCN) (Hollis et al., 2004a, b; 2006a, b; Remijan et al., 2008; Loomis et al., 2013; Zaleski et al., 2013). However, it is also the case that many COMs are identified with the cooler and less dense surrounding envelope, some of which are familiar from our discussions in earlier chapters, such as CH_3OH and CH_3CN, as well as the less familiar methyl acetylene (CH_3CCH), formamide (NH_2CHO), acetaldehyde (CH_3CHO), and cyanodiacetylene (HC_5N), observed with the Australian 22 m Mopra telescope (Jones et al., 2008, 2011).

What has been possible is to use large-scale mapping of the spatial distribution of several COMs around Sgr B2 (e.g. with the Arizona Radio Observatory 12 m telescope) to distinguish the 'extended' from the 'compact' emission (Li et al., 2020). Glycoaldehyde (CH_2OHCHO), methyl formate (CH_3OCHO), formic acid (HCOOH), ethanol (C_2H_5OH), and methylamine (CH_3NH_2) are extended over several arcminutes around Sgr B2(N) and Sgr B2(M), while dimethyl ether (CH_3OCH_3) and ethyl cyanide (C_2H_5CN) show compact distributions close to the B2(N) and B2(M) centres. While it seems clear that during the cold early stages of star formation many atoms and molecules adsorb onto grain surfaces, where they thermally hop across the surface and react with each other or react more slowly within bulk ice via radical chemistry energetically driven by UV photons (Herbst & van Dishoeck, 2009), there are a range of desorption possibilities to consider as a cloud collapses and warmer conditions evolve. Desorption may be thermal, when dust temperature is high enough, or non-thermal, through UV or CR-induced photodesorption (Oberg et al., 2010; Cernicharo et al., 2012; Vastel et al., 2014). It may also be reactive desorption driven by exothermic surface reactions (Garrod et al., 2007; Vasyunin & Herbst, 2013). Shocks can also strip ice mantles from grain surfaces (Requena-Torres et al., 2006; Burkhardt et al., 2019), and there is a variety of complex induced reaction mechanisms that could contribute (Ruaud et al., 2015; Chang & Herbst, 2016; Shingledecker et al., 2018; Garrod, 2019; Jin & Garrod, 2020). While some or all of these mechanisms incorporated into gas–grain models can successfully reproduce many COMs in both cold cloud and hot core conditions, anomalies remain.

One puzzle is the observed abundances of CH_3OCH_3 and C_2H_5CN in the extended region around B2, which are low compared with model expectations. One group of modellers have explored a range of possible factors in both static physical models and evolving ones (Wang et al., 2021). They include the influences of chain reaction mechanisms, shock impacts, X-ray bursts (from Sgr A* flares), as well as enhanced reactive desorption and low surface diffusion barriers. None of their static physical models can reproduce observed CH_3OCH_3 and C_2H_5CN abundances. On the other hand, most evolving models (cold phase plus warm-up phase) do reproduce many COMs abundances but only where the best-fit temperature used (30–60 K) seems well above the dust temperature ~20 K that is observed. The very best-fit for all the COMs modelled, including the two puzzling observations, occur at a gas temperature ~27 K when additionally subjected to a short-duration X-ray burst lasting 10^2 yr.

An X-ray flare released by the supermassive black hole Sgr A* at the Galactic Centre has been observed and is assumed to arise from accretion of gas onto the black hole surface (Baganoff et al., 2001; Marrone et al., 2008; Rea et al., 2013). The dynamics suggest this probably originated around 100 years ago (Churazov et al., 2017) and could be one of a multiple succession of short-duration flares. An X-ray echo from Sgr B2 has been observed (Terrier et al., 2010; Nobukawa et al., 2011), supporting the scenario and the enhancement of two other organic molecules, CH_3OH and H_2CO, by X-ray irradiation simulated previously (Liu et al., 2020). On the assumption that a short-duration X-ray burst might engender a CR ionisation rate ~10^{-13} s^{-1} at an early stage of the warm-up phase when dust and gas are still close to 20 K, the modelling shows reactive desorption as the key mechanism in COMs production. Under this scenario, the radicals on grain surfaces responsible for COMs production become plentiful for all the COMs in the study except CH_3OCH_3 and C_2H_5CN. While this may account for chemical differentiation in the extended region, the observed differentiation would endure for only a few hundred years, which perhaps suggests short-term X-ray flares associated with the accretion activity of Sgr A* may be occurring quite often.

7.5 Complex Isocyanides in Sgr B2 (N)

The astrochemistry of complex cyano (-CN) and, more particularly, isocyano (-NC) functional group molecules has not been widely explored. However, exceptionally in the Sgr B2 (N) case, the ALMA-EMoCA (Exploring Molecular Complexity) spectral line survey has detected half a dozen alkyl cyanides (-CN) and isocyanides (-NC) and searched for cyano- and isocyano-polyynes (Willis et al., 2020). Figure 7.5 shows the integrated intensity maps for four bright species, and we can note that ALMA resolves the B2(N) source into two hot cores, N1 and N2. Column densities for these species are calculated assuming LTE, a valid assumption given the high particle densities ($>10^7$ cm^{-3}) (Belloche et al., 2016; Bonfand et al., 2019). For each molecule a synthetic spectrum is produced with variables including line opacities and the finite resolution of the observations in the radiative transfer calculations. Five free parameters are employed: the intensity of emission (assumed to be Gaussian), the column density, the temperature, the line width, and the velocity offset with respect to the assumed systemic velocity of the source. Each of these are adjusted to obtain a best-fit with the observed spectrum.

The tentative detections of CH_3NC and HCCNC towards Sgr B2(N2) are among the first identifications of these molecules. The derived abundance ratios for isomeric pairs include $CH_3NC/CH_3CN \sim 5 \times 10^{-3}$, and for HCCNC/ $HC_3N \sim 1.5 \times 10^{-3}$ (Willis et al., 2020). The observing team also calculated

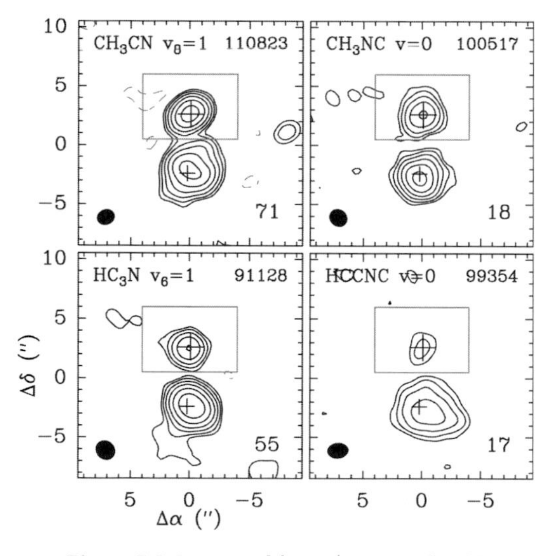

Figure 7.5 Integrated intensity maps for CH_3CN, CH_3NC, HC_3N, and HCCNC, around the two hot core sources Sgr B2 (N1) (lower location) and (N2) (upper) (Willis et al., 2020).

upper limits for the abundances of C_2H_5NC, C_2H_3NC, HNC_3, and HC_3NH^+. In the chemistries of these species described in what follows, the cosmic-ray ionisation rate has a complex effect, with some ratios increasing with ionisation (e.g. CH_3NC/CH_3CN) but not others. In fact, it is modelling with variable CR-ionisation rates, rather than consistently high ones, that best matches the observational ratios. The models of Willis et al. (2020) use a kinetics code that simulates gas–grain–ice chemistry (Garrod, 2013), and we can look at the formation and destruction networks assumed for some of these species as follows.

CH_3NC (Methyl Isocyanide)

This species is thought to form, primarily, through the gas-phase radiative association of CH_3^+ and HCN, producing two isomers, CH_3CNH^+ and CH_3NCH^+, in a ratio dominated by the first after internally energised unimolecular isomerisation (DeFrees et al., 1985). Subsequent electron recombination is thought to produce about 40 per cent saturated products, CH_3CN and CH_3NC respectively, with the remaining 60 per cent of isomers fragmenting (Loison et al., 2014; Plessis et al., 2012). Destruction reactions at $T < 100$ K are dominated by ion–molecule exchange, particularly with abundant C^+, H^+, and H_3^+, while above 100 K, reaction with atomic hydrogen on grain surfaces is more likely:

$$CH_3NC + H \rightarrow CH_3 + HCN.$$

C_2H_5NC (Ethyl Isocyanide)

Given the possibility, if not likelihood, that large isocyanide radicals form on the surfaces of grains and within the ice mantles via reactions such as

$$CH_2 + CH_2NC \rightarrow CH_2CH_2NC, \text{ or}$$
$$H + C_2H_3NC \rightarrow CH_2CH_2NC,$$

the first reaction is likely to dominate, having minimal activation energy compared to the second. An alternative route involving the CH_3 radical is also possible but will be less efficient because heavier radicals are less mobile as reactants either on the surface or within ice layers. The resulting large isocyanide radicals can then undergo hydrogenation to make the saturated product:

$$CH_2CH_2NC + H \rightarrow C_2H_5NC.$$

Standard destruction mechanisms are likely once the molecule enters the gas phase, including photodissociation and CR-induced (photo)dissociation, although multiple ion–molecule destruction routes are likely to dominate.

HC_3N (Cyanoacetylene), HCCNC (Isocyanoacetylene)

At low temperatures, cyanoacetylene forms efficiently through the gas-phase reaction

$$CH_2CN + C \rightarrow HC_3N + H,$$

while at higher temperatures

$$C_3H_3 + N \rightarrow HC_3N + H_2$$

dominates in the modelling. The more familiar combination of C_2H and HCN is also a production route, although a less efficient one. Equally, there are grain-surface reaction routes to cyanoacetylene, such as the hydrogenation of C_3N, but none are particularly efficient, not least because they compete with the grain-surface destruction route by further hydrogenation to C_2H_2CN, as well as myriad gas-phase ion–molecule dissociations.

The chemistry of the HCCNC isomer is less complex. There are two primary gas-phase routes:

$$H + HCNCC \rightarrow HCCNC + H, \text{ and}$$

$C_3H_2N^+ + e^- \rightarrow HCCNC + H$, the first favoured at higher temperatures, the second at lower.

There are no known formation routes to isocyanoacetylene on grain surfaces. It can be destroyed, just as for HC_3N, by further hydrogenation and by the many ion–molecule gas-phase routes.

C_2H_3NC (Vinyl Isocyanide)

In the gas phase, C_2H_3NC forms in one of the primary recombination channels of $C_2H_6NC^+$ (protonated ethyl isocyanide):

$$C_2H_6NC^+ + e^- \rightarrow C_2H_3NC + H_2 + H.$$

This links the ethyl and vinyl isocyanide chemistries, with 40 per cent of recombinations assumed to go through this channel, in line with experimental results (Vigren et al., 2012).

On grains, it forms through the hydrogenation of the C_2H_2NC radical, which itself is hydrogenated on grains from an HCCNC precursor. We have noted that HCCNC forms through two primary routes in the gas phase, the dominant one at lower temperatures being the dissociative recombination of $C_3H_2N^+$. This involves the rearrangement of the carbon backbone atoms from –CCCN to –CCNC. The more favoured reaction at higher temperatures involves hydrogenation of HCNCC, which itself comes from dissociative recombination of $C_3H_2N^+$, the gas-phase product of C_3NH^+ with H_2 following that of C_3N^+, also with H_2.

7.6 Summary

The canonical HMSFR Sgr B2 is perhaps the richest source of molecules detected in the Galaxy to date, not least in the number of COMs recorded. The variable and higher-than-standard cosmic ionisation rate in this region close to the Galactic centre has a complex effect on COMs chemistries, offering both an unusual test bed for chemical evolution theory, while not being conditions representative of more widely observed HMSF cores. The particular case of cyanides and isocyanides stands out, and modelling that uses an enhanced but extinction-dependent CR ionisation rate brings best agreement between model results and observations. Nonetheless, the modelled column densities of some species are much lower than observed, which suggests either key formation routes are still missing in the model or the physical structure profile of the region is inaccurate. In the next chapter we will look at more tightly constrained observational conditions in which to examine such hot core chemistries close to young high-mass stars.

8
G29.96-0.02 in W43

8.1 Introduction

Given the evidence that high-mass stars form in sequential clusters, the differing evolutionary stages alongside one another can present us with locations and events such as those where UCHII regions clearly sit adjacent to hot molecular cores. In the case of W43 in Aquila, we will look at the clustering in more detail and at those molecular emissions that offer direct observational evidence for the interaction of the two distinctive evolutionary structures. In this and the next chapter we will see more of the detailed consequences of that interaction on the observed chemistry and the kind of computational modelling that has been explored.

8.2 Westerhout 43

The massive GMC complex Westerhout 43 (W43) imaged in the infrared in Figure 8.1 lies close to the tangential point of the Scutum-Centaurus spiral arm, almost 6 kpc from Earth, close to where the arm meets the Galactic central long-bar. It spans a huge scale, stretching some 150–200 pc across, a site of much turbulence and exceptional stellar activity. Supersonic cloud–cloud collisions account for the dense gas formation and a star formation rate that represents 5 to 10 per cent of the entire Galactic output.

Figure 8.2 shows a composite Spitzer image of the three main star formation locations within W43 – sub-sources named Main, G30.5, and South. The composite consists of 3.6 μm, 8 μm, and 24 μm distributions, tracing thermal emission from stars, polycyclic aromatic hydrocarbon (PAH) features, and hot dust grains excited by the high-mass stars, respectively. The 8 μm emission shows diffuse distribution covering the whole GMC complex, while the

Figure 8.1 The W43 molecular cloud complex in the infrared (NASA/Spitzer/JPL).

24 μm emission is bright towards W43 Main and W43 South, foci of multiple HII regions. Best estimates based on maser sources put these at ~5.5 ± 0.4 kpc distance from the Solar system.

Observations of CO in W43 across a ~200 pc width show a broad velocity dispersion of 20–30 km s^{-1}, with velocity separation between the three features (Main, South, and G30.5) being too large for each cloud to be gravitationally bound. The highly turbulent conditions are driven by a converging gas flow from the Scutum Arm and the long bar, supersonic cloud–cloud collisions that generate dense gas and local starburst activity. Figure 8.3 shows a schematic of the GMC complex and Galactic bar interaction.

8.3 The W43 Sub-sources

W43 Main and W43 South are both concentrated star formation regions on scales ~10 pc, while G30.5 consists of five star-formation sites more widely dispersed. Figure 8.4 shows two of the three locations in close-up, with the

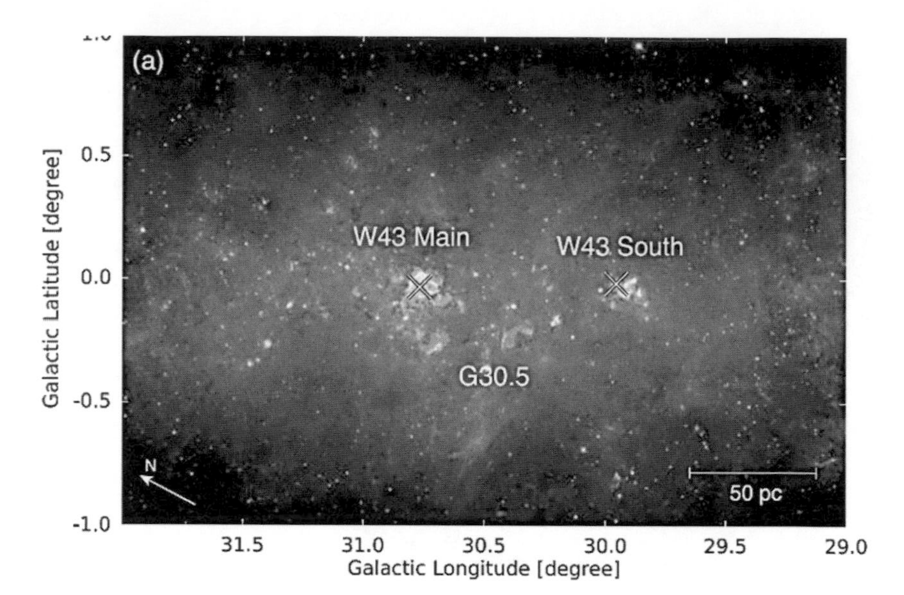

Figure 8.2 Spitzer three-color composite image of the W43 GMC complex from combined Spitzer/IRAC 3.6 μm, Spitzer/IRAC 8 μm (Benjamin et al., 2003), and Spitzer/MIPS 24 μm (Carey et al., 2009) distributions. The X marks indicate the loci of peak integrated emission (Kohno et al., 2021).

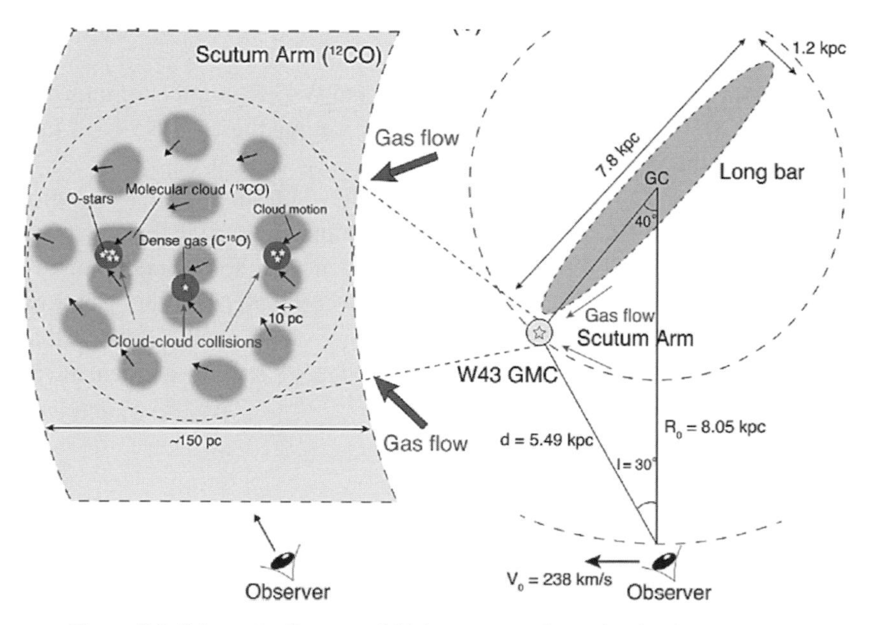

Figure 8.3 Schematic diagram of high-mass star formation in the W43 GMC complex (Kohno et al., 2021).

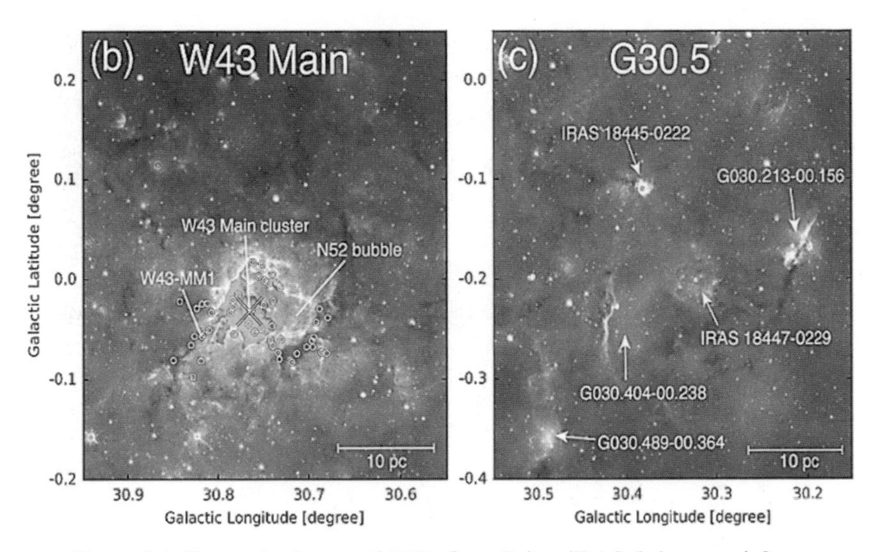

Figure 8.4 Composite image of W43 from Spitzer/IRAC 3.6 μm and 8 μm (Benjamin et al., 2003), and Spitzer/MIPS 24 μm (Carey et al., 2009) (Kohno et al., 2021).

background 8 μm emission mostly from warm dust and the bright 24 μm identifying ionised hydrogen. There are over 50 protocluster candidates in W43 Main, some of which are indicated by small white circles in Figure 8.4.

The central cross in the W43 Main image marks the position of an OB cluster, and the white circles the protocluster candidates identified by sub-millimeter dust continuum observations (Motte et al., 2003). W43 Main is the most active high-mass star-forming region in the W43 GMC complex (e.g. Nguyen-Luong et al., 2013, 2017; Cortes et al., 2019). A ring-like structure in the 8 μm emission corresponds to an identifiable infrared bubble, N52 (Churchwell et al., 2006). The total infrared luminosity and ultraviolet luminosity are estimated to be ~10^6–10^7 $L_\odot$ (e.g. Hattori et al., 2016; Hanaoka et al., 2019), equivalent to ~14 O5V or ~40 O7V stars. ALMA observations have also identified 131 individual massive dust clumps specifically at the W43-MM1 location shown in Figure 8.4 (Motte et al., 2018b). The total luminosities and estimated ages of Main and South are listed in Table 8.1.

Focusing on the multiple sources within W43 South, otherwise known as G29.96-0.02, Figure 8.5 first shows a close-up equivalent to Figure 8.4, with 10 radio continuum sources corresponding to the OB stars listed in Table 8.2.

The total luminosity of this cluster is estimated to be 2–6 × 10^6 $L_\odot$ (Beltran et al., 2013; Lin et al., 2016) and the stellar age determined from the evolutionary development of the UCHII region as ~0.1 Myr (Watson & Hanson, 1997). The brightest infrared source is #1, identified with the UCHII/hot core locus

Table 8.1 *Properties of W43 Main and W43 South (Kohno et al., 2021).*

	Total Luminosity	O-type Stars	Cluster Age
W43 Main	7–10 × 10^6 $L_⊙$	~14–50	1–6 Myr
W43 South	2–6 × 10^6 $L_⊙$	6	~0.1 Myr

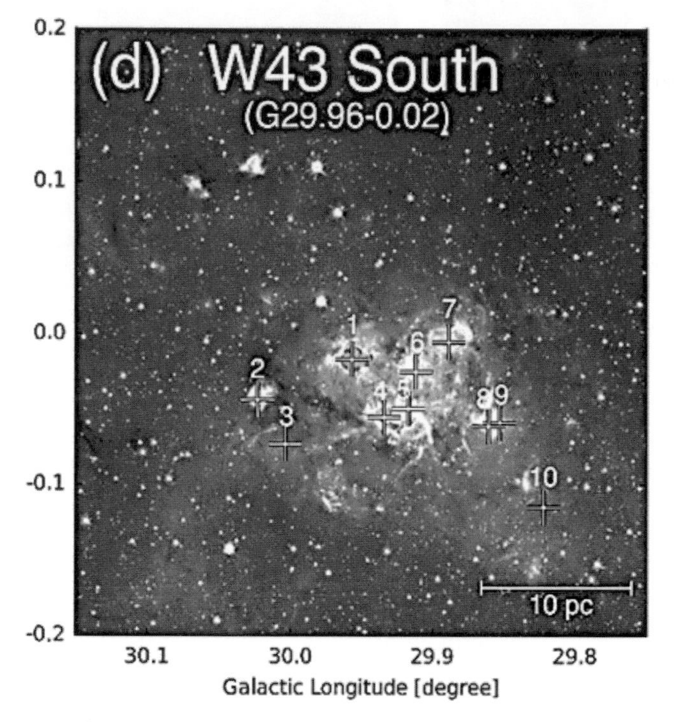

Figure 8.5 Close-up image of W43 South in the same frequencies as Figure 8.4. The crosses indicate radio continuum sources corresponding to 10 identified late O-type and early B-type stars (Kohno et al., 2021).

(also known as IRAS 18434-0242), a very well-studied object (e.g. Pratap et al., 1994; Hoffman et al., 2003; Roshi et al., 2017) which we will look at in detail in the next section.

8.4 G29.96-0.02 #1

The cometary UCHII feature in Figure 8.6 was identified first over 30 years ago in an early classic survey of ultracompact ionised sources (Wood & Churchwell, 1989). A near-infrared spectral classification of the likely ionising

Table 8.2 *OB Stars in W43 South (from Beltran et al., 2013)*

Star	Longitude	Latitude	Spectral Type
#1	29°.957	−0°.0170	O6
#2	30°.023	−0°.0438	B0
#3	30°.004	−0°.0730	B0
#4	29°.934	−0°.0555	O5
#5	29°.917	−0°.0497	O6.5
#6	29°.912	−0°.0252	O9.5
#7	29°.889	−0°.0056	O8.5
#8	29°.860	−0°.0614	O9.5
#9	29°.853	−0°.0591	B0
#10	29°.822	−0°.1150	B0.05

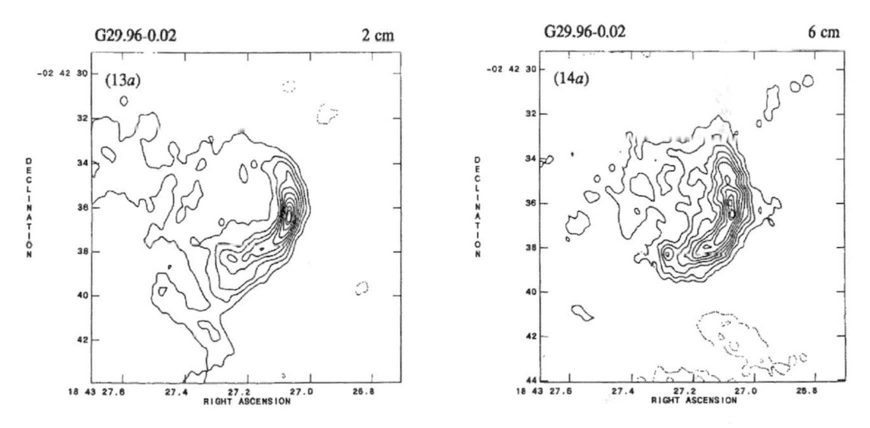

Figure 8.6 Continuum emission maps (2 cm and 6 cm) made with the VLA (Wood & Churchwell, 1989).

O5-O8 star was then made, the first such for any UCHII region (Watson & Hanson, 1997). Two additional sources towards the rim of the UCHII region and an enhanced density of reddened sources indicative of an embedded cluster followed (Pratap et al., 1999). The hot core detection was clarified through NH_3 emission, shown in Figure 8.7 (Cesaroni et al., 1994, 1998) and as being co-spatial with CH_3CN (Olmi et al., 2003), these two molecular species being the prime indicators in the early days of hot core observations (to which we will return in the next chapter). Delineation of an infrared dark cloud was followed by identification of multiple massive, pre-protostellar, low-temperature cores (Pillai et al., 2011), several radio continuum sources, and many YSOs in different stages of evolution (Beltran et al., 2013). The molecular gas distribution was studied using isotopologue ^{13}CO and $C^{18}O$ emission detected with the IRAM 30 m telescope (Carlhoff et al., 2013), and observation

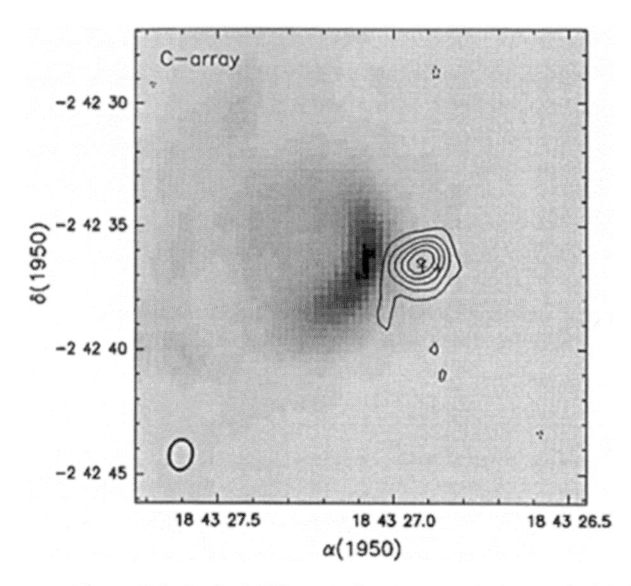

Figure 8.7 Excited NH_3 emission (contours) observed with the VLA, with associated water masers (crosses), adjacent to the 1.3 continuum of the UCHII region (greyscale) (Cesaroni et al., 1994).

of diffuse X-ray emission suggested irradiation from massive stars escaping their natal envelope clouds (Townsley et al., 2014).

The succession of observations and detections, both tentative and more certain, indicates how research on such remote and obscured sources progresses as the capability of the observing instruments evolves. For example, one can go back to 1970 for the first tentative maser detection (Turner, 1970) with just an 18′ resolution, before significant receiver improvements followed in the mid-1970s (Knapp & Morris, 1976; Genzel & Downes, 1977). Masers have since become highly resolved tracers of energetic and kinetic conditions within star formation regions.

While NH_3 and CH_3CN line emission proves particularly useful in determining warm, dense envelope and core gas temperatures in HMSFRs, the deuterated species NH_2D, in contrast, has proved most useful in tracing the cold, high-density pre-cluster gas. For example, there are two massive filaments, one of them an infrared dark cloud (IRDC) seen in extinction about 2′ east of the G29.96 UCHII region. Using the VLA and PdBI, dust and line observations reveal fragmentation into the multiple massive cores strung out in filamentary structures shown in Figure 8.8. Most of these cores are cold and dense (<20 K and >10^5 cm^{-3}), highly deuterated (NH_2D/NH_3 >6%), and at an early stage of potential star formation, with just one protostar evident in each filament through maser and outflow activity (Pillai et al., 2011).

Returning to the UCHII–hot core interfacial relationship, as we have noted, with high-mass star lifetimes short compared with those of their low-mass cousins, observable high-mass star locations are relatively rare and their typical distances from the Solar system remote. Their associated hot cores offer an early evolutionary indicator prior to UCHII formation, and single-dish emission line surveys first revealed high abundances of saturated species and temperatures exceeding 100 K from observations in the mid-1990s (Macdonald et al., 1996; Schilke et al., 1997; Hatchell et al.,1998). Except for the Orion-KL hot core (which, as we will see in the following chapter, is far from being a typical example), the kiloparsec distances require high-spatial-resolution studies if we are to disentangle structure, multiple heating sources, and chemical complexities. It is only in recent years that sub-arcsecond submillimetre instrumentation and dust continuum observations have begun to characterise prototypical regions such as G29.96.

Figure 8.9 shows the UCHII on the left in centimetre continuum emission traced by the dashed contours. On the right, the greyscale shows submillimetre continuum emission from warm dust. The continuous line contours trace NH_3 emission defining the hot core object; the triangles, circle, and pentagon mark

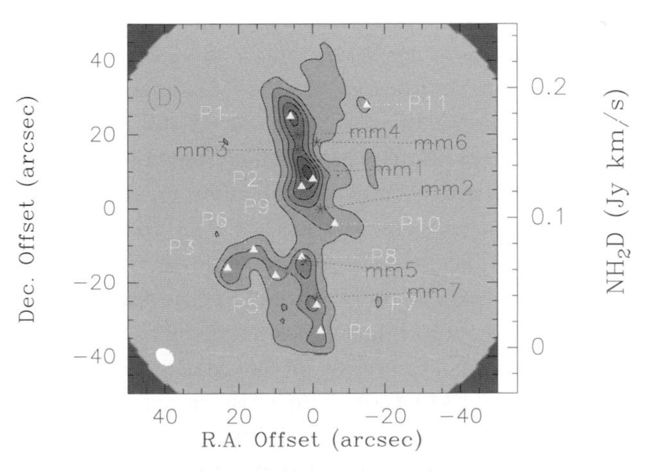

Figure 8.8 NH_2D line emission (contours) in G29.96 East, observed with the PdBI. The NH_2D cores are marked as stars, and the 1.3 continuum cores (mm1-7) marked as triangles (Pillai et al., 2011).

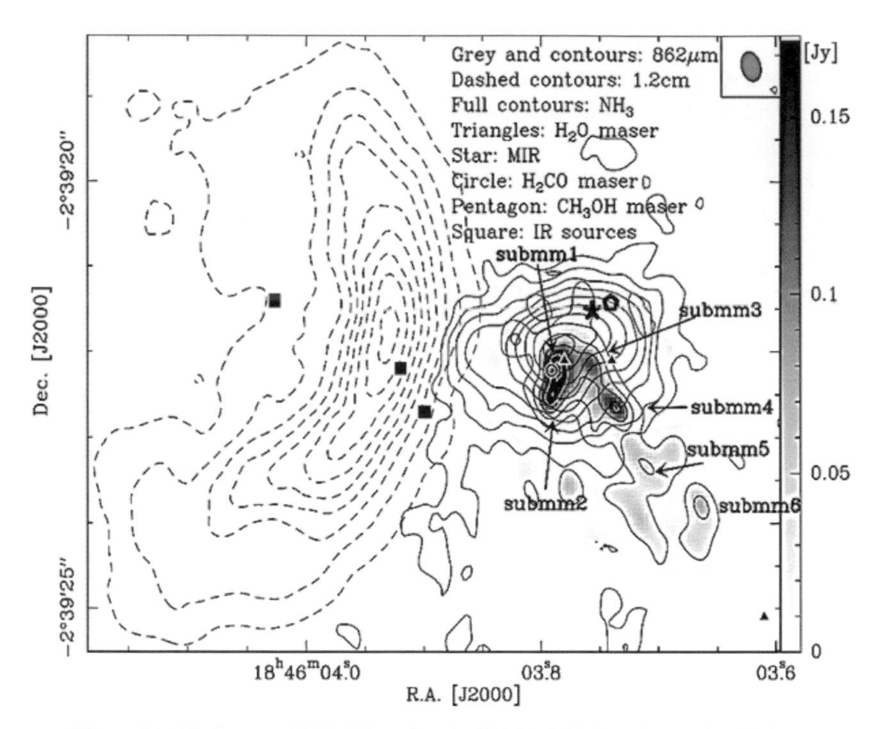

Figure 8.9 The hot core/UCHII interface in G29.96-0.02 (Beuther et al., 2007).

maser sites; the squares mark infrared sources; and the star symbol, which is not an actual star location, identifies the peak of mid-IR-range emission. This submillimetre map was obtained using the SMA with a sub-arcsecond spatial resolution corresponding to linear scales ~1,800 AU. This image resolves the hot molecular core source into the sub-sources labelled, with the central core containing four submillimetre continuum peaks which resemble an Orion Trapezium-like multiple system at a very early evolutionary stage, but clearly more advanced than the filamentary cores just described in G29.96 East. Within the hot core-UCHII region of central G29.96, three sites of massive star formation are identifiable – the UCHII region itself, the mid-infrared source, and the submillimetre continuum sources, the implication being of sequential high-mass star formation within the same evolving protocluster.

A decade later and we have an ALMA continuum emission map, shown in Figure 8.10, revealing the same six submillimetre peaks in the central core, but with ALMA's resolution of 0.2″ being just under 1,000 AU. It identifies each bright core (all thought to be up to 30 $M_\odot$) aligned in a filament-like structure. The brightest is Core A, which shows both the richest associated COMs emission and the most powerful bipolar outflow traced in SiO.

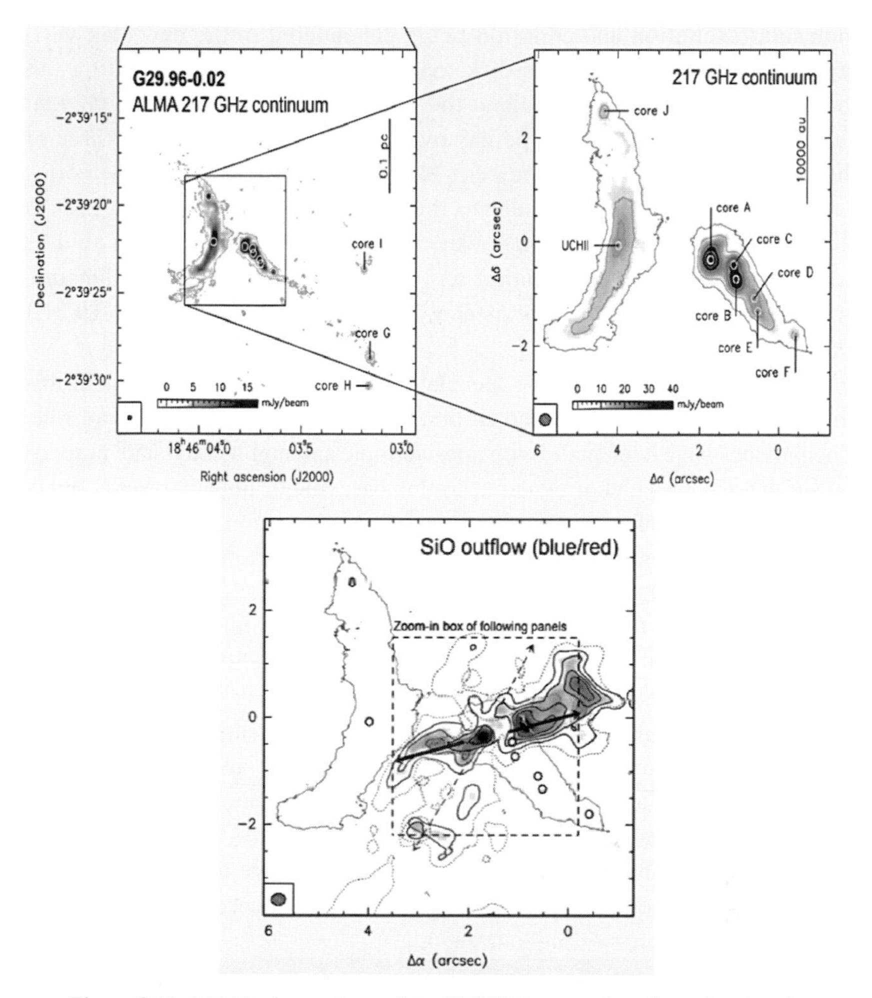

Figure 8.10 ALMA observations of the UCHII–hot core interface, showing the six O-stars and the SiO outflow red-shifted south-east, blue shifted north-west (Sánchez-Monge, 2019).

While disks and outflows are thought to be common to the formation process of all stars, whatever their mass, few clear examples of disks around O-stars (>20 $M_{\odot}$, >100 $L_{\odot}$) have been observed. The observations of a decade ago found little correlation between molecular line peak emission and the submillimetre continuum peaks. While high-column-density regions generate many optically thick spectral lines, restricting our ability to analyse and interpret, the differences were also taken to indicate possible chemical evolution and temperature effects. Furthermore, with molecular excitation and emission being engendered in the expected variety of circumstances (such as disk, outflow, or envelope), separating out the contributions is not possible at the available resolution. In the G29.96 case, a velocity gradient for the gas over a spatial scale spanning three of the central core protostellar sources (~4,800 AU) was shown to flow north-east to south-west, perpendicular to the molecular outflow seen in SiO. A Keplerian accretion disk is not, however, indicated – just as likely would be a larger-scale toroid of material or an envelope rotating and/or collapsing, or perhaps the complex consequence of outflow influence and/or UCHII expansion.

Putting more detail into these speculations, the ALMA observations of this site suggest that an energetic and collimated outflow plays the dominant role. Comparisons have been made with non-isotropic and non-steady-state numerical simulations in which irregular accretion has a major impact, progressively changing the disk plane and hence the outflow orientation. The G29.96-0.02 case has most recently, therefore, been taken to be a truncated disk around an O-type protostar, caught in the act of disruption by anisotropic accretion. Figure 8.11 shows the result of numerical simulations mimicking the shifting outflow axis that would accompany anisotropic accretion and showing the outflow cavity, with a simulated gas density distribution.

Since the first SiO outflow observations (Gibb et al., 2004), a second outflow has been detected in what appears a perpendicular direction. In addition, the isotopologue $C^{34}S$ offers a distinctive morphology, being weak towards the hot core but strong in its surroundings, particularly close to the UCHII interface. The reduced abundance in the hot core could be explained through early desorption from grains, followed by gas-phase reaction with OH

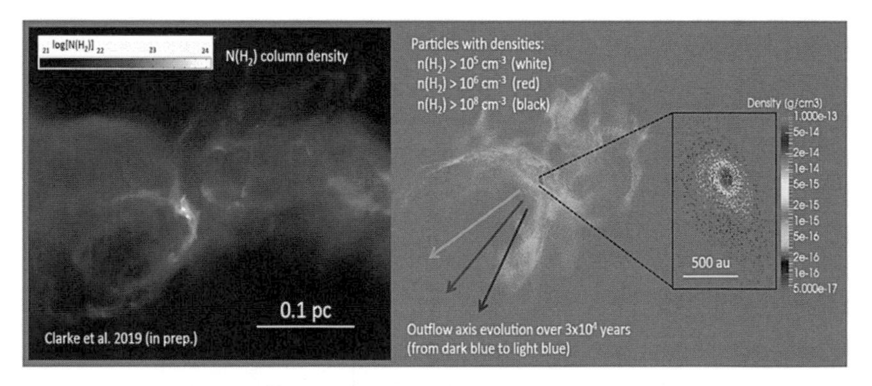

Figure 8.11 On the right, numerical simulations of the shifting outflow axis associated with anisotropic accretion. On the left, a simulated gas density distribution illustrating the outflow cavity (Sánchez-Monge, 2019).

(the dissociative product of desorbed H_2O). The sulphur/OH reaction engenders SO and then, at a later stage, SO_2.

8.5 Star Formation Efficiency (SFE)

If we pull back again for a moment to the larger-scale view of W43, we can use the isotopologues of CO to redefine the spatial distribution and velocity structures of molecular gas in the GMC complex. Figures 8.12a, b, and c show low-density gas traced by ^{12}CO across a 150 pc × 100 pc region. The velocity distribution range is 20–30 km s^{-1}. All three isotopologues trace dense gas, but $C^{18}O$ is detectable only towards the densest regions of the three star formation sites.

Estimates of star formation efficiency (SFE) in GMCs can be approximated as $M_* / M_* + M_{cloud}$ (stellar mass and cloud mass, respectively). W43 turns out to have a low (~4%) SFE compared with some other complexes (such as W51 and M17), but the high local concentrations of bright $C^{18}O$, shown in Figure 8.13, suggest much more star formation is to come (Kohno et al., 2021).

Taking this data back to the G29.96-0.02 scale, Figure 8.14 shows both temperature and column-density maps derived from ^{12}CO (Paron, Areal, & Ortega, 2018). In the case of $^{13}CO/C^{18}O$ column-density ratios across the interface between UCHII and hot core, the ratio increases in line with expectations closer to the ionised gas, since selective far-UV irradiation dissociates $C^{18}O$ preferentially. Comparing the spatial distribution of CO isotopologue ratios and total gas densities in maps such as Figure 8.15, along with maps of the ionised gas distribution, indicates the penetration of far-UV irradiation into the dense molecular gas.

The evidence from this kind of detailed mapping is that higher values of $^{13}CO/C^{18}O$ appear in the north-eastern sector of the UCHII region (~8), while towards the south-west, coincident with a dark, dense cloud component, are the lowest (~3). This suggests ionising radiation is escaping north-east and photodissociating the envelope gas, while to the west and south-west the radiation stalls as it enters the dense cloud (traced by dust emission at 850 μm) and heats the hot core (Beuther et al., 2007). The pattern of far-UV penetration is echoed to a reasonable degree in the distribution of other molecular species. However, G29.96-0.02 as a high-mass star-forming region is complex, as we would expect, and studies do show that the molecular gas may have very different physical conditions both spatially across the cloud and along our line of sight. The source remains a prime target for ever better-resolved observations.

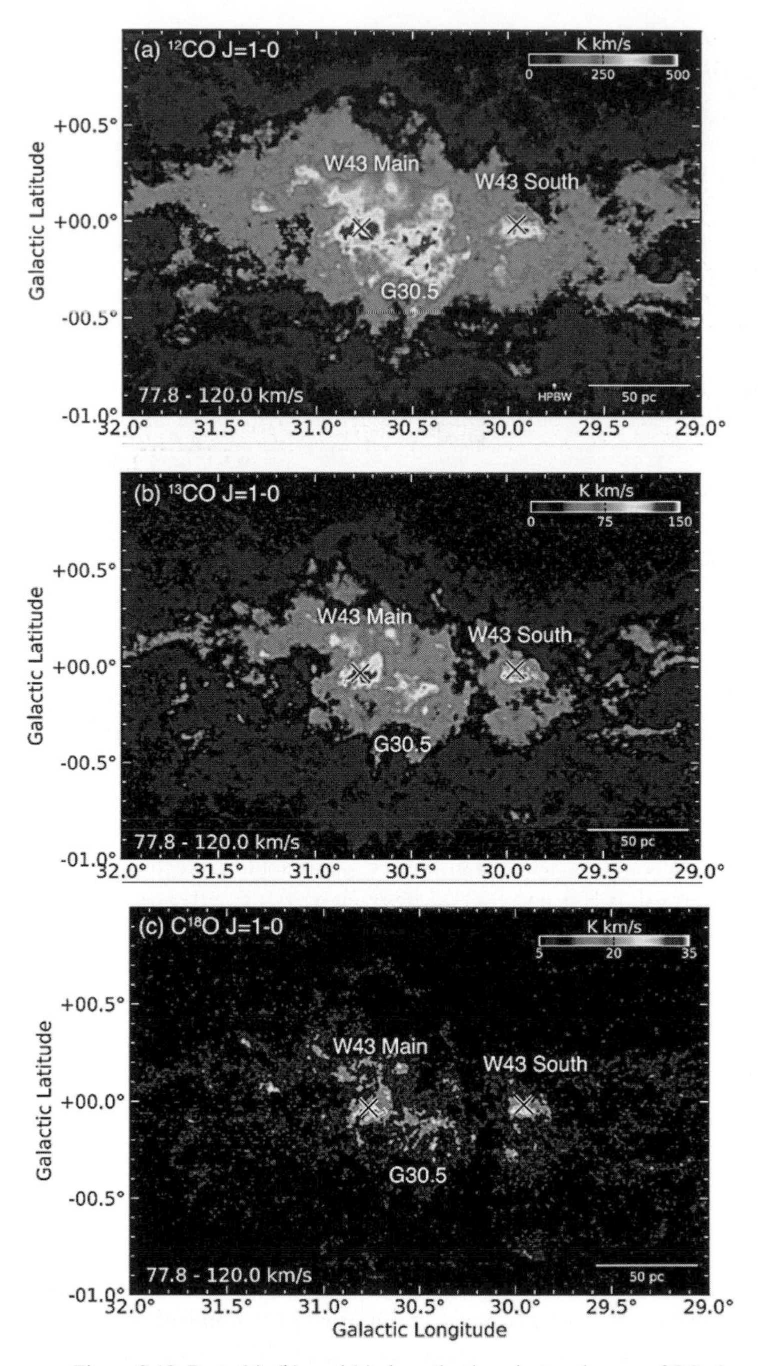

Figure 8.12 Parts (a), (b), and (c) show the three isotopologues of CO observed to trace gas of increasing column density in the sequence (Kohno et al., 2021).

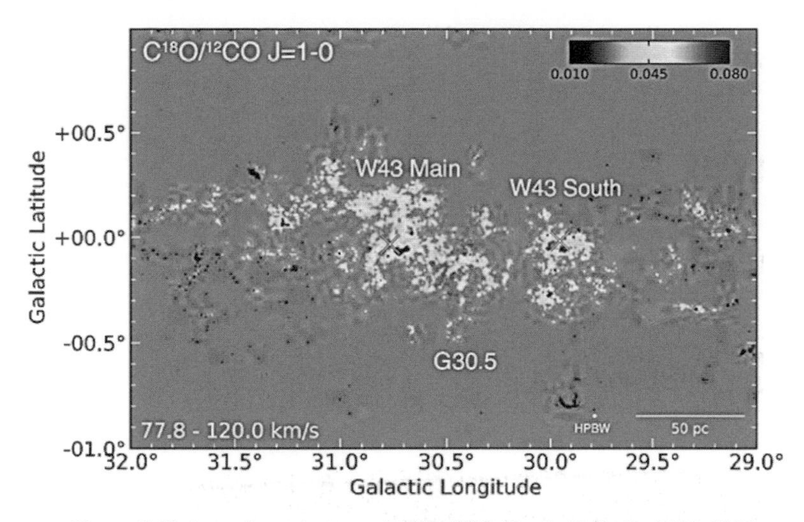

Figure 8.13 Intensity ratio map of $C^{18}O/^{12}CO$ (J = 1–0) for the W43 GMC complex (Kohno et al., 2021).

8.6 G29.96 COMs

Just over 20 years ago, the G29.96 hot core was first mapped in a variety of molecules, including NH_3, HCO^+, CS, CH_3CN, HNCO, and $HCOOCH_3$ (Cesaroni et al., 1998; Pratap et al., 1999; Maxia et al., 2001; Olmi et al., 2003; Beuther et al., 2007). The east–west velocity gradient was measured in NH_3, CH_3CN, and $HN^{13}C$ and interpreted as a rotation (Cesaroni et al., 1998; Olmi et al., 2003, Beuther et al., 2007). In addition, the south-east to north-west outflow was mapped in H_2S and SiO (Gibb et al., 2004; Beuther et al., 2007) and, as we have noted, masers (H_2O, OH, H_2CO, CH_3OH) were also detected (Hofner & Churchwell, 1996; Walsh et al., 1998; Minier et al., 2000, 2002; Hoffman et al., 2003; Breen & Ellingsen, 2011; Caswell et al., 2013; Breen et al., 2015, 2016; Chen et al., 2017; Surcis et al., 2019), with a representative maser distribution shown in Figure 8.16.

From the many spectral lines and identifiable species, Figure 8.17 shows the spatial distribution of some of those molecules in relation to the UCHII region and the submm continuum peaks (Beuther et al., 2007).

The abundance of CH_3OH lines identifies a temperature gradient peaking centrally ~300 K, decreasing towards the core margins. The detection of $HCOOCH_3$ (methyl formate), C_2H_5CN (ethyl cyanide), and HNCO (isocyanic acid), and a tentative identification of $CH_2(OH)CHO$ (glycolaldehyde) across half a dozen massive star-forming hot cores, including G29.96, have prompted detailed chemical modelling of the formation of these COMs. Table 8.3 shows

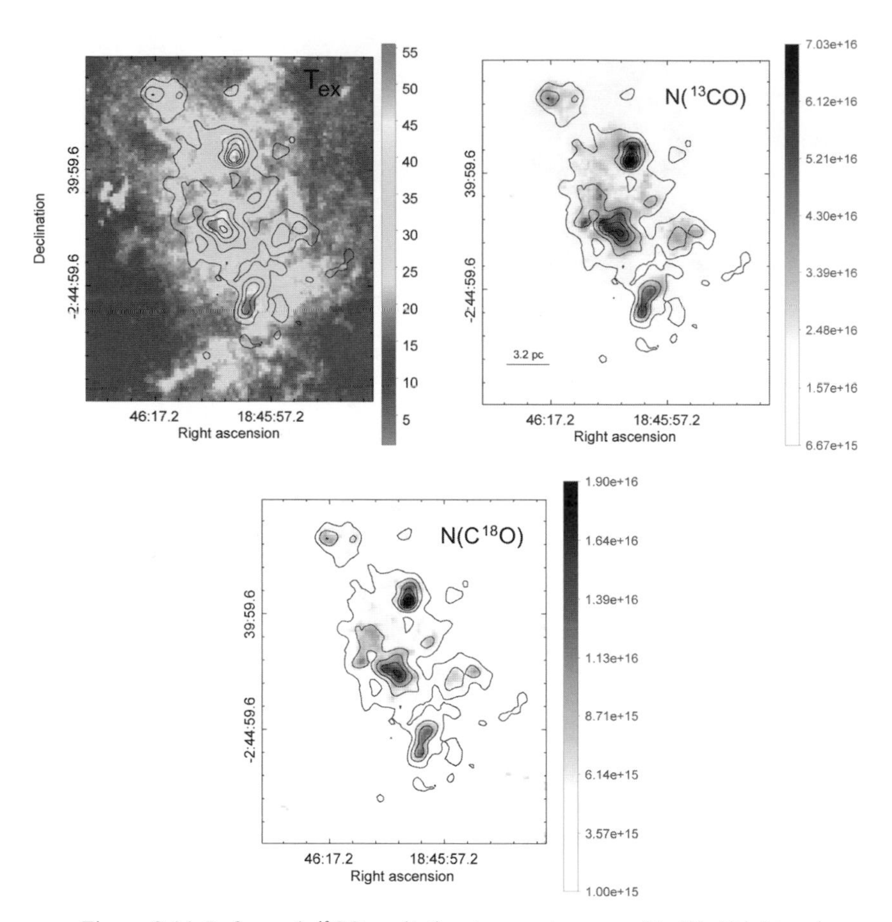

Figure 8.14 Left panel: ^{13}CO excitation temperature map (T_{ex} K); 'Right and lower panels': corresponding ^{13}CO and' $C^{18}O$ column density map (cm^{-2}) (Paron, Areal, & Ortega, 2018).

representative comparisons between modelled and observed column densities for three of the species. The correspondence across the sample of evolved hot cores (≥ 20 k yr) and between model and observations is consistent, within a factor of just a few for G29.96.

8.7 Summary

In the W43 sub-cores and G29.96 in particular, HMSF in clusters is evident, and the impact of disk winds and outflows on the observable chemistry clear. Modelling of the hot core COMs abundances matches observations for many key species observed in both this source and other Galactic

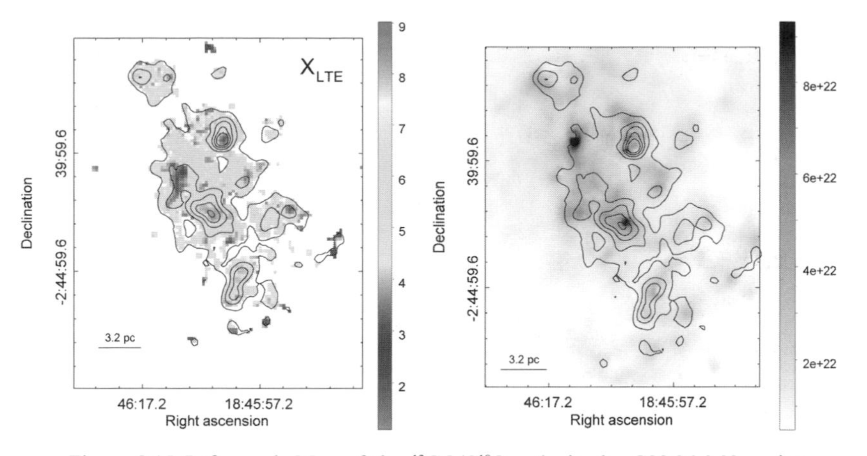

Figure 8.15 Left panel: Map of the $^{13}CO/C^{18}O$ ratio in the G29.96-0.02 region obtained from the ^{12}CO, ^{13}CO, and $C^{18}O$ J=3–2 line under the LTE assumption. Contours of integrated $C^{18}O$ J=3–2 emission are included for reference. Right panel: H_2 column density (cm^{-2}) derived from Herschel (160, 250, and 350 mm) and SCUBA 850 mm data, with the same reference contours (Paron, Areal, & Ortega, 2018).

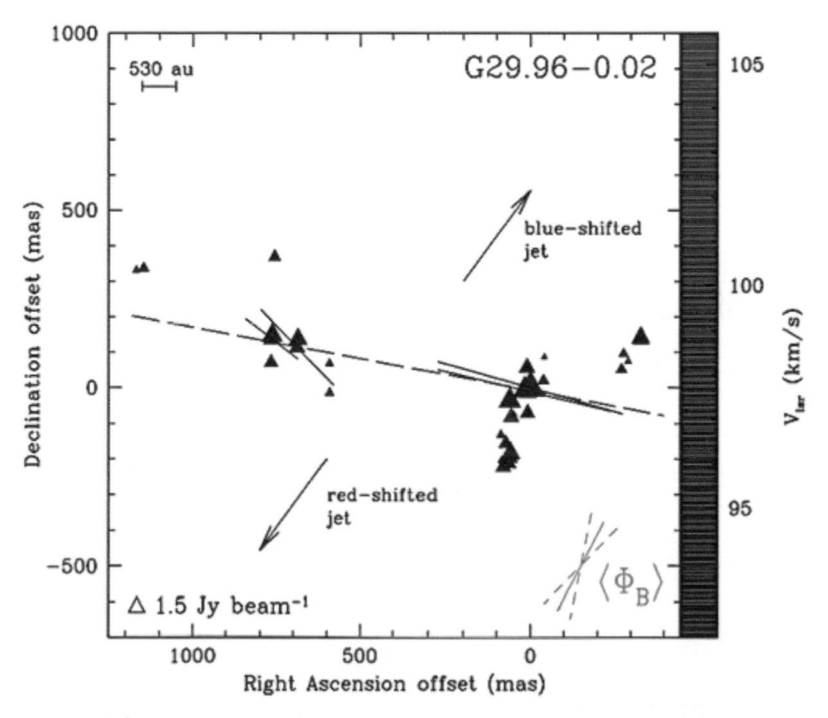

Figure 8.16 Some CH_3OH maser features (triangles) in relation to the red- and blue-shifted outflow directions indicated. The orientation of the magnetic field Φ_B is also shown, closely aligned to the outflows, with uncertainties indicated by the dashed lines (Surcis et al., 2019).

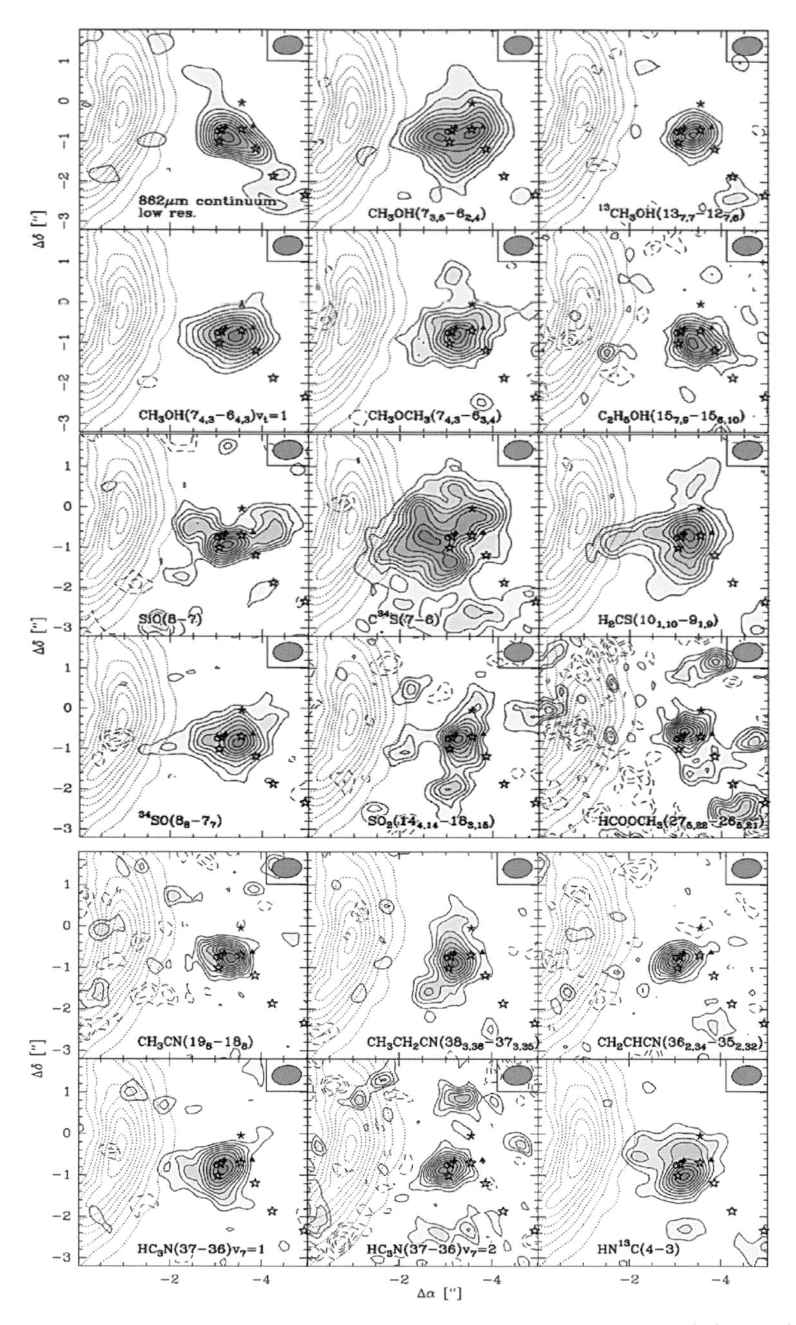

Figure 8.17 Selected COMs observed towards the hot core in relation to the UCHII region (dotted contours) and the submm continuum sources marked with stars (Beuther et al., 2007).

Table 8.3 *Modelled versus observed column densities for three COMs across a variety of hot cores, including G29.96 (Calcutt et al., 2014).*

Hot Core	T_{ex} (K)	N_{model} (cm^{-2})	N_{obsv} (cm^{-2})
		CH_3CN	
G31	450	9×10^{15}	5×10^{16} (300 K)
G29	410	5×10^{15}	2×10^{16} (300 K)
G19	450	4×10^{15}	1×10^{16} (300 K)
G24a1	310	4×10^{15}	4×10^{16} (300 K)
G24a2	450	6×10^{15}	3×10^{16} (300 K)
		$HCOOCH_3$	
G31	130	4×10^{17}	4×10^{17} (150 K)
G29	300	5×10^{16}	2×10^{16} (300 K)
G24a1	300	9×10^{16}	3×10^{16} (300 K)
G24a2	300	1×10^{17}	4×10^{16} (300 K)
		$CH_2(OH)CHO$	
G31	130	3×10^{16}	2×10^{16} (150 K)
G29	300	8×10^{15}	5×10^{15} (300 K)
G19	300	9×10^{15}	3×10^{15} (300 K)
G24a1	300	1×10^{16}	5×10^{15} (300 K)
G24a2	300	1×10^{16}	6×10^{15} (300 K)

sources. The interaction between an HII region and an associated hot, dense core is exemplified in G29.96, despite the evident complexity of physical conditions in the surrounding region. As in all of our studies made through the lens of molecular emission, we are able to probe the physical conditions from the chemical tracers. In Chapter 9 we look to sources in Orion that contributed much to the development of COMs formation and destruction theory, but which have since been shown to be anything but conventional in the hot core sense.

9
Orion BN/KL

9.1 Introduction

The Orion Molecular Cloud (OMC) is a complex of components totalling several 10^5 $M_\odot$, having its closest features at about 400 pc distant from Earth and its furthest ones another 50 pc beyond that. Both its proximity and the rich diversity of sources have made this, not surprisingly, the most studied star formation region historically. The three principal component giant clouds are designated Orion A, Orion B, and Monoceros R2. Each of the three are accumulations of several 10^3 $M_\odot$ of gas and dust. Between 10 and 20 degrees below the Galactic plane, all three link to planar material by long, slender gas and dust filaments, one of which (named the S. Orioni filament) is evident in Figure 9.1, panel (a) of which shows a segment of a Galactic map in CO emission centred on the OMC. Panel (b) shows a corresponding schematic of the same region.

The OMC is thought to have had its origin in a super-cloud that fragmented 60 million years ago. Part of that fragmentation was a huge compression bubble pushing away a massive ring (named Lindblad's ring), which broke up to form the OMC among other GMCs about 20,000,000 years ago. In more recent astronomical times, it is estimated that up to 100 OB stars likely formed in Orion during the past 12,000,000 years, each with lifetimes of just a few million years, and of which perhaps 20 ended in supernova explosions. The combined kinetic energy that was released ($>10^{52}$ ergs) energised a large X-ray-emitting gas expansion now evident in the massive shell of gas and dust known as the Orion-Eridanus super-bubble. Barnard's Loop that wraps around the OMC is a crescent of Hα emission extending 40° west into Eridanus, part of the super-bubble, shown in Figure 9.2.

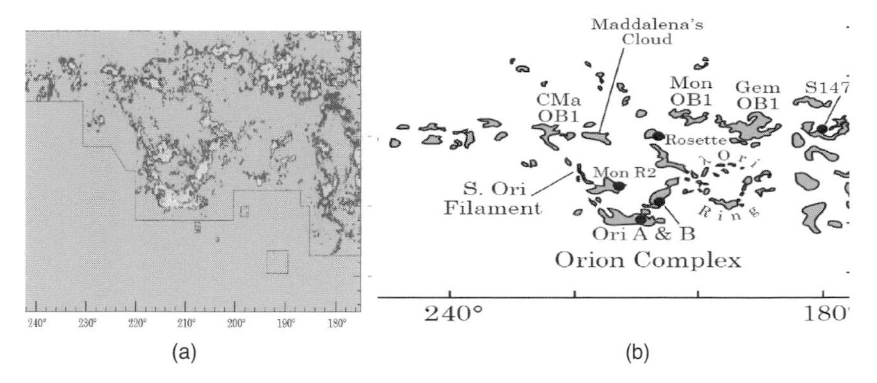

Figure 9.1 A Galactic CO map of the Orion region with a corresponding schematic (Dame, Hartmann & Thaddeus, 2001).

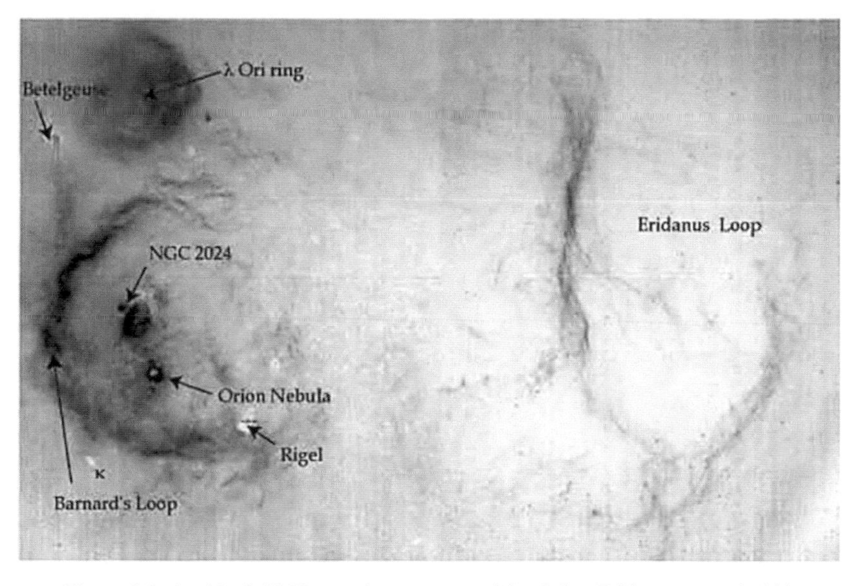

Figure 9.2 A wide field Hα continuum map of the Orion-Eridanus super-bubble (Bally, 2008).

9.2 OMC

Orion A and Orion B are just two of the three massive molecular cloud complexes in the wider OMC. Figure 9.3 shows two dust emission maps. Image (a) shows the spatial relationship between Orion A and Orion B, the latter above the former in both panels. The second image (b) records the dust temperature range (15–30 K) in greyscale, based on 2MASS near IR dust extinction.

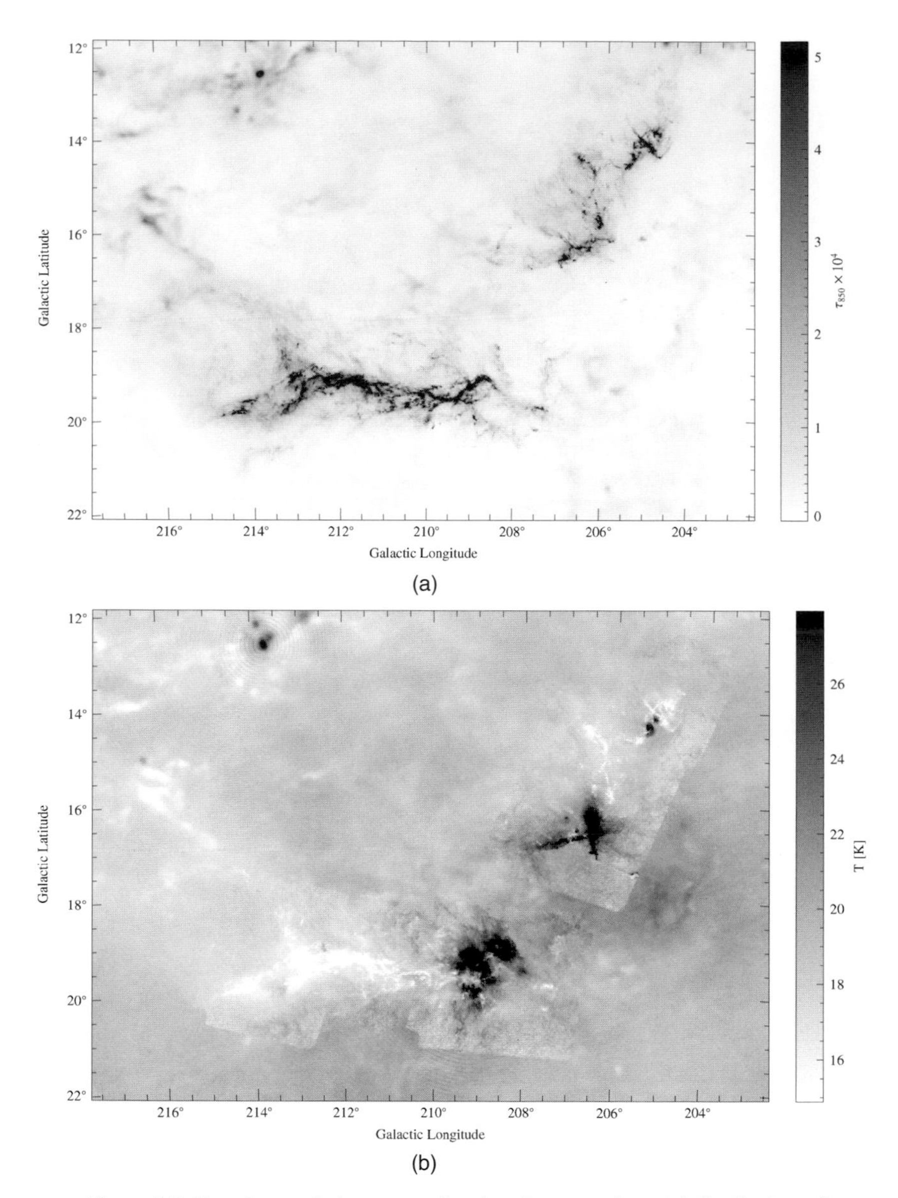

Figure 9.3 Two dust emission maps, showing the general spatial distribution of Orion A and Orion B (a) and the dust temperature range (15–30 K) in greyscale (b) (Herschel/Planck/RAS/Lombardi et al., 2014).

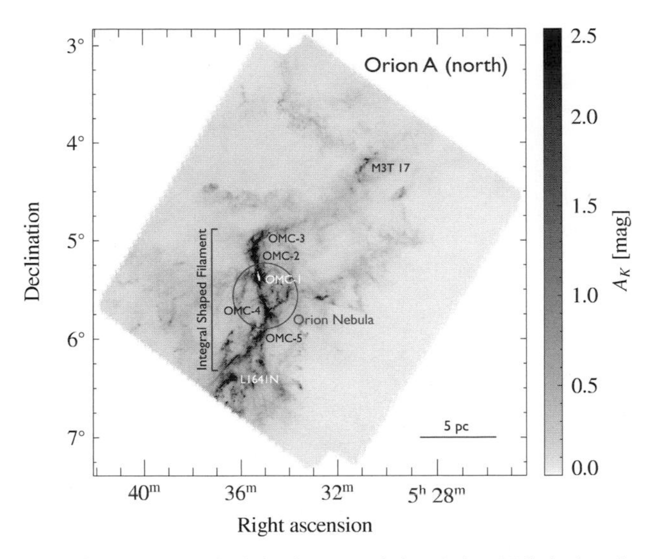

Figure 9.4 Optical-depth map of the Orion Nebula location in Orion A at a resolution of 18 arcsec (Lombardi et al., 2014).

Of the distinctive locations in the Orion A complex, OMC-1 shown in Figure 9.4 lies behind the Orion Nebula (M42), the latter containing many well-studied features including the Horsehead nebula, the Orion Bar, and OMC 1-5. The Orion Nebula has the highest number of densely packed young stars in the OMC, including the brilliant Trapezium stars, but there are several other OB associations of distinctive age that are strongly indicative of sequential clustered formation. Estimates put Orion B1*a* at 12,000,000 years, Orion B1*b* at 6,000,000 years, Orion B1*c* between 2,000,000 and 5,000,000 years, and the Trapezium stars at 2,000,000 years. While it is the hot bright OB stars having masses >10 $M_{\odot}$ that dominate the large-scale morphology of reflection nebulae such as M42, there are also huge numbers of lower-mass stars scattered throughout.

9.3 BN/KL

OMC-1 contains a cluster of luminous infrared, young, high-mass stars, having a total luminosity of 10^5 $L_{\odot}$ and a combined mass of 200 $M_{\odot}$ across a 0.1 pc scale. Emerging towards Earth along the line of sight from the heart of OMC-1 is an explosive outflow of gas and dust called Orion KL (Kleinmann–Low), with its principal feature BN (Becklin–Neugebauer) shown in Figure 9.5 in relation to the Trapezium stars as distant neighbours that lie at the centre of the Orion Nebula.

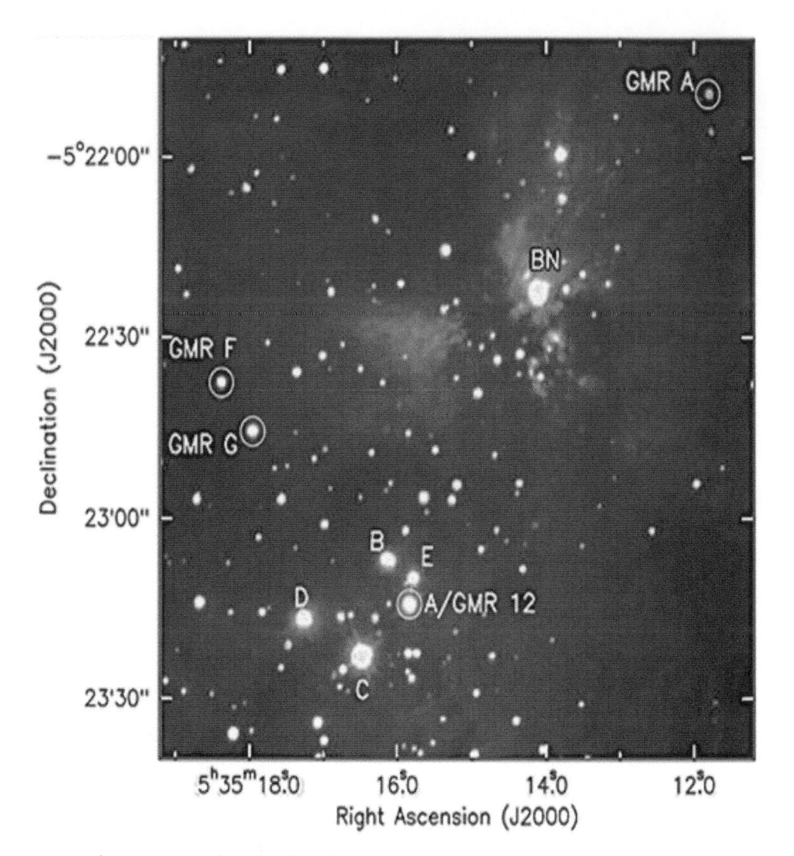

Figure 9.5 Orion BN/KL in relation to the Trapezium cluster (A-E) in infrared emission (ESO/VLT/ Menten et al., 2007).

Filaments of molecular gas known as Orion's fingers appear to be expanding from a common origin in BN/KL at velocities ranging from tens to hundreds of kilometres per second. Figure 9.6 shows a vibrationally excited H_2 emission image of the ejecta in which boxes and arrows mark the sites whose proper motions have been measured, all pointing back to an apparent explosion centre at RA 05 h 35 m 14.4 s, DEC –0.5° 22′ 28″. Figure 9.7 shows an equivalent, equally dramatic, outflow image in FeII and H_2, with 'fingers' much in evidence.

The ejection is thought to have been powered by the dynamical decay of a multiple high-mass star group about 500 years ago, energetically triggered perhaps by the coming together of a massive binary pair. There are good reasons to suppose that this is just the latest in a succession of cycles of relaxation and ejection in a compact complex star-formation region on timescales of tens of thousands of years (Rivera-Ortiz et al., 2020). Three of the ejecta

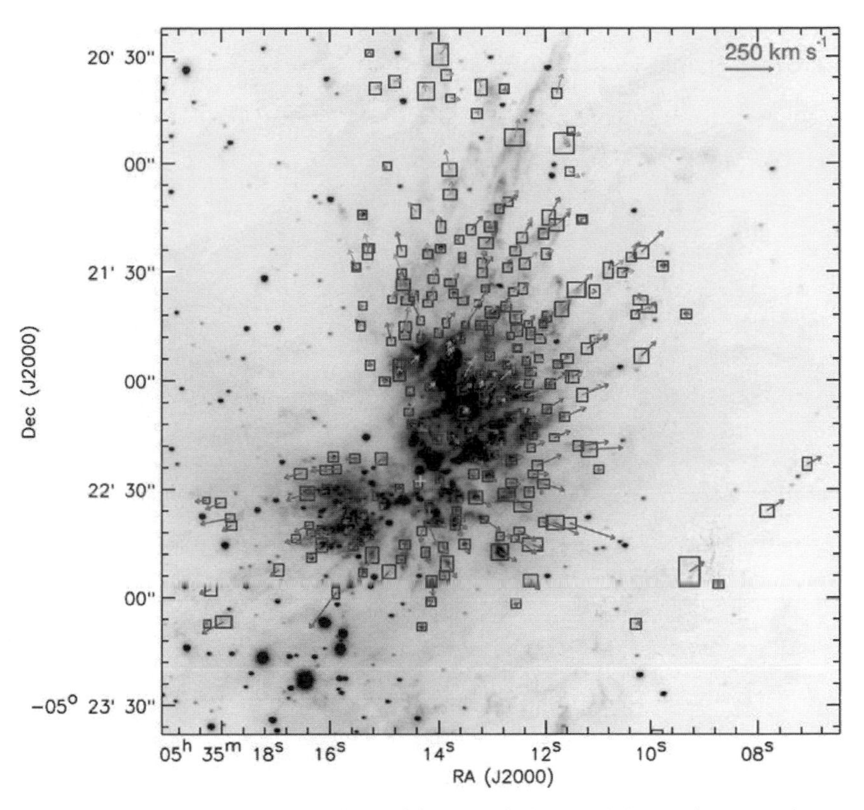

Figure 9.6 BN/KL outflow imaged in H_2 emission. Each box-and-arrow pair denotes a proper-motion measurement (Bally et al., 2011).

objects, designated *I*, *n*, and the already named BN, are self-luminous and assumed to be protostellar, all within a 10″ (0.02 pc) location at the heart of Orion KL. It seems likely that these three are just part of the disintegration of a stellar cluster, with location *I* showing clear evidence of a bipolar outflow. Figure 9.8 shows each of these and other sites of interest marked against a variety of molecular line emission contours that delineate components that collectively constitute Orion KL. EC marks the point of origin as the conjectured 'explosion centre' (Zapata et al., 2009, 2020) – BN is top right, and below it is a fourth source, IRc6. South-east is IRc7 and, east of that, continuum source SMA-1, south-west of which is source *n* and north of which is source *I*.

Although not spatially resolved by single-dish observations, the components of Orion KL are differentiated using high-resolution spectroscopy that identifies their significantly different line widths and central velocities. In addition, interferometric observations have exploited molecular tracers. We can identify four distinct structural components that have been examined with

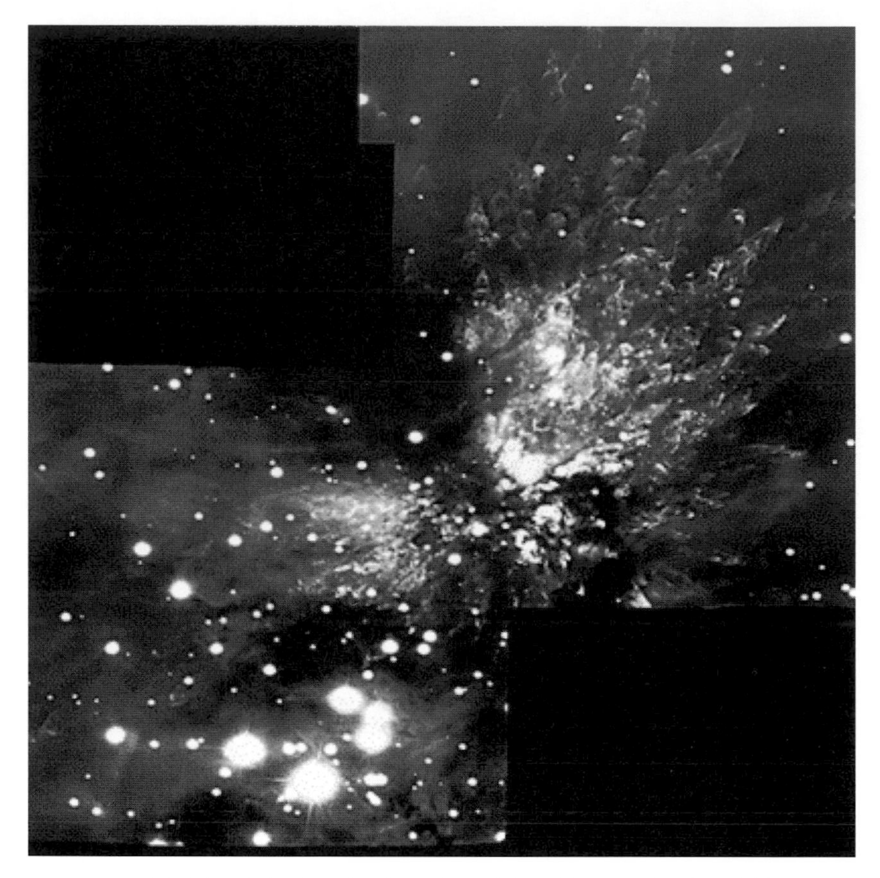

Figure 9.7 Orion BN/KL outflow in 1.64 μm FeII and 2.12 μm H_2. (Gemini South GSAOI) (casa.colorado.edu/ Bally).

successive instrumentation for at least the past three decades: the Orion Hot Core, the Compact Ridge, the Plateau, and the Extended Ridge, two of which are labelled in Figure 9.9.

9.4 The Four Features

The Hot Core was first identified forty years ago as a compact source of hot ammonia emission embedded in a more extended ridge of dense material. Within a few years, using an early incarnation of the VLA telescope with a 2″ resolution, the core was defined as a heart-shaped region about 15″ square (0.03 pc^2). A second conspicuous region, the Compact Ridge, was then observed abutting the Hot Core, showing its own distinctive

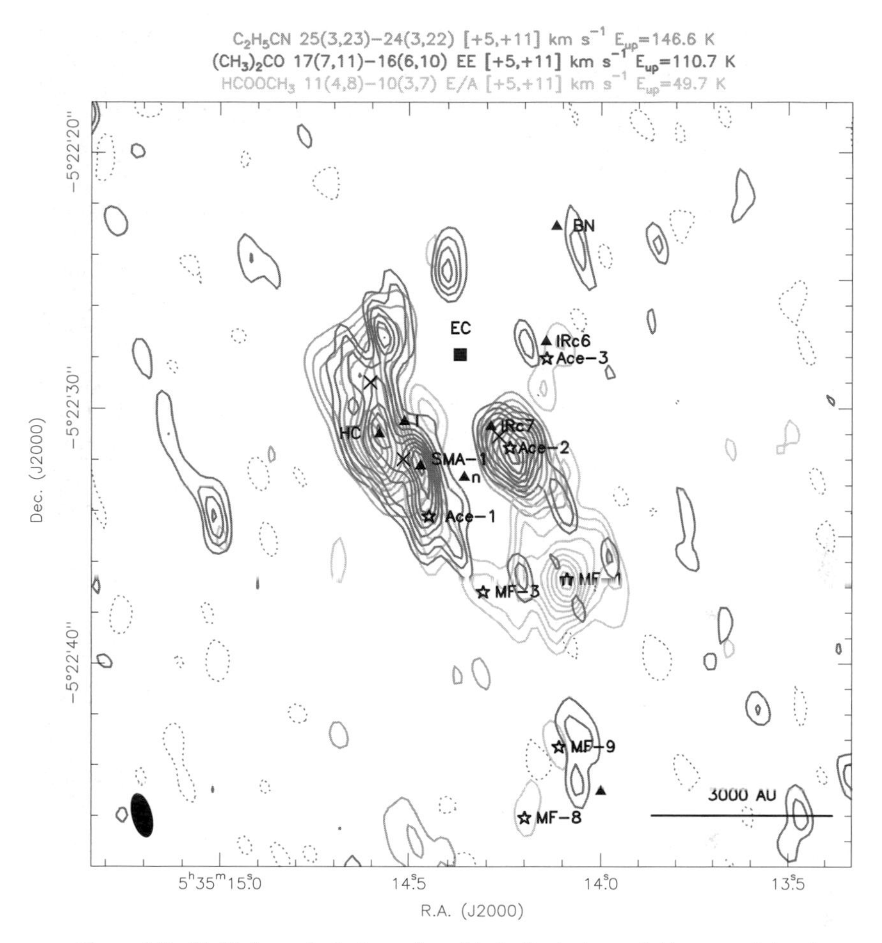

Figure 9.8 Multiple marked sites referred to in the text, overlaid on a variety of molecular emission contours delineating Orion KL (from Peng et al., 2013).

complex gas-phase chemistry. While the Orion KL Hot Core was the first such high-temperature (150 K), compact (<0.05 pc), dense (10^7 cm^{-3}) source with a rich chemical complexity to be identified, and many others have followed, whether it contains one or more embedded YSOs still remains uncertain. It might therefore not turn out to be the classic form of hot core directly associated with a single internal high-mass star formation source. What is certain is that interferometric detections of CH_3CN and NH_3 reveal a nitrogen-rich gas phase, a clumpy structure on scales of 1–2″ (<800 AU), and a temperature distribution ranging over several hundreds of degrees (e.g. Crockett et al., 2014).

The Compact Ridge, on the other hand, consists of clumps of dense, quiescent gas of density ~10^6 cm^{-3} seeming to be externally heated and extending

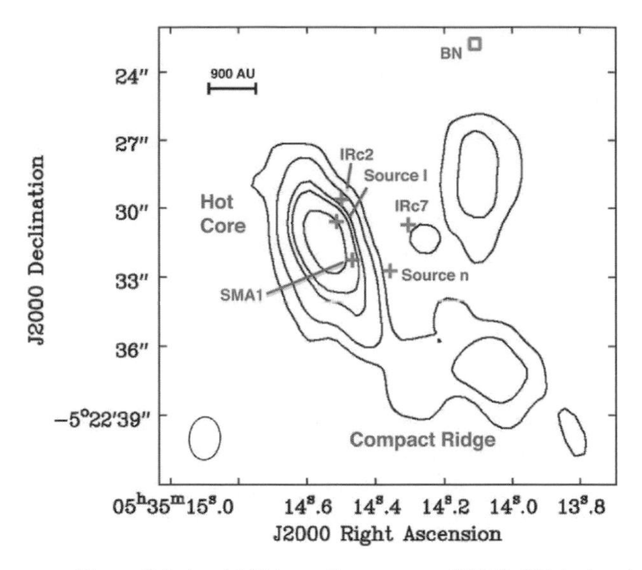

Figure 9.9 An ALMA continuum map (230.9 GHz) showing the locations of hot core, compact ridge, and BN/KL, along with additional localised sources (Crockett et al., 2014).

12″ (5 k AU) to the south-west of the Hot Core. The temperature is lower (80–150 K) than the Hot Core and, as quiescent gas, the emission line widths are also much narrower. The Compact Ridge gas phase is rich in oxygen-bearing molecules, a chemical distinction from the Hot Core which challenges our understanding and to which we will return shortly.

The third location, the so-called Plateau, is a mixture of outflows, shocks, and interactions with ambient cloud. Emission line widths are characteristic of turbulence, and two dominant outflows are evident. Of those, one is a low-velocity flow (LVF) oriented north-east to south-west and probably originating from radio continuum source *I*, itself most likely an embedded massive protostar. The high-velocity outflow (HVF) is more spatially extended (>30″) and oriented perpendicular to the LVF. The HVF may be powered by the dynamical decay of a multi-star system, perhaps that of *I*, *n*, and BN together. The more collimated part of the LVF has been traced by interferometric observations of SiO (Plambeck et al., 2009).

A sub-mm source (SMA-1 in Figures 9.8 and 9.9) possibly associated with an embedded protostar, even perhaps driving the HVF, has been observed in vibrationally excited CH_3OH, HC_3N, and SO_2 (Beuther & Nissen, 2008). With radio continuum source *I*, plus *n* and BN as strong infrared sources, all three are undoubtedly YSO sites. Two objects, likely identifying protostellar emergence (also marked in Figure 9.9), are the infrared clumps IRc2 and IRc7.

Finally, the Extended Ridge, as its name suggests, is a widespread quiescent region warmed to no more than 60 K and rich in unsaturated carbon-rich species normally indicative of cold, dense conditions, in which exothermic ion–molecule reactions with low activation energies dominate.

9.5 Chemical Differentiations

In simple summary, the chemical complexity as a first approximation identifies the following: (i) the Hot Core is rich in complex, saturated N-bearing species, such as CH_3CH_2CN; (ii) the Compact Ridge is rich in organic, saturated O-bearing species, such as $HCOOCH_3$ and CH_3OCH_3; (iii) the Plateau molecular outflows show typical shock products such as SiO and SO; and (iv) the Extended Ridge is rich in simple, cold, dense cloud species such as CS and HCN.

While evidence suggests that the Orion KL Hot Core has no internal high-mass protostellar heat source, as hot core–like gas and dust subjected to the energetic impact of material associated with the proposed explosive event ~500 years ago, the COMs we observe are those we might anticipate. The Hot Core chemical composition would be shock-driven, with shorter-lived higher temperatures than the conventional more slowly evolving case in which grains are gradually heated sufficient to evaporate COMs over an extended period (Herbst & van Dishoeck, 2009). What kind of shock is the question: low-velocity (~50 km s^{-1}) continuous shocks (C-shocks) are strong enough to disrupt grain mantles by sputtering and probably more efficiently than grain–grain collisions or even J-shocks (jump-shock) (Flower & Pineau des Forets, 1995). Fast shocks driven by the explosive event in the KL region are likely to dissociate most molecules and initiate high-temperature, fast gas reaction networks that underlie the chemistry actually observed.

Of note is that CH_3CN traces a clumpy distribution of N-bearing species in the Hot Core (Wang et al., 2010), echoing apparent variations in gas and dust densities noticed in earlier studies, for example in the dust continuum (Beuther et al., 2004) and early NH_3 detections (Migenes et al., 1989). Early explanations for the nitrogen–oxygen distinctions between Hot Core and Compact Ridge included different conjectured ice compositions (the Hot Core as NH_3-rich, the Compact Ridge as CH_3OH-rich) taken to precede gas-phase processing, with the differences in ice composition reflecting different thermal histories (Charnley et al., 1992; Caselli et al., 1993). The simple influence of high cloud-core temperatures on NH_3 gas-phase chemical modelling is also suggestive (Rogers & Charnley, 2001). For example, above about 200 K, oxygen atoms and OH radicals mostly convert to H_2O, removing oxygen from the available pool of reactants. High abundances of HCN can form quickly under these conditions, with CH_3CN as subsequent product to follow, provided the NH_3 supply from ices is substantial. If high-temperature, dusty conditions favour CH_3CN formation, its lesser but non-negligible presence in the Compact Ridge (given no satisfactory grain surface formation route) might be accounted for as a low-velocity outflow of material from the Hot Core region (Blake et al., 1987; Millar et al., 1991).

Another source recorded in panel (a) of Figure 9.10 is that of the Hot Core South. Does this mark a transition between Hot Core and Compact Ridge conditions? Where the principal Hot Core continuum is pinpointed by NH_3 and $^{13}CH_3CN$ peak emission, HDO and $^{13}CH_3OH$ transitions seem specifically concentrated on a clump 1″ south of it. The abundance ratios of methyl cyanide to methanol decrease in moving north to south from Hot Core to Compact Ridge, and then between the two, the HDO emission peak marks Hot Core South. There are also interesting differences between Hot Core South and the Compact Ridge, even though both are rich in O-bearing species. In combination, these sources offer among the strongest interstellar detections of the more-complex O-bearing organic molecules, including a range of alcohols (CH_3OH, CH_3CH_2OH, $OHCH_2CH_2OH$, and CH_3OCH_2OH), ethers (CH_3OCH_3, CH_2OCH_2, and $CH_3CH_2OCH_3$), ketones (CH_3COCH_3), aldehydes ($OHCH_2CHO$), esters ($HCOOCH_3$, $HCOOCH_2CH_3$, and CH_3COOCH_3), and carboxylic acids (HCOOH and CH_3COOH) (Tercero et al., 2018). The spatial differentiation of these functional groups in Orion KL has become increasingly clear in recent years, so that we may now say that, while species with the hydroxyl group bound to a carbon atom (-C-O-H) peak at Hot Core South, those with an ether linkage (C-O-C) are detected in both Hot Core South and the Compact Ridge but peaking in the latter. Given the north–south trend already noted in the CH_3CN/CH_3OH abundance ratio, we may envisage the shock impact of a recent explosive event striking the principal Hot Core location first and being transmitted southwards as evidenced by the gas-phase chemistry, first to the Hot Core South and thereafter to the Compact Ridge, in the manner represented schematically in panel (a) of Figure 9.10.

The progressive impact is both temporal and spatial. If radical–radical formation reactions on grain surfaces occurred under quiescent conditions, essentially before the impact event, then we would expect the Compact Ridge gas-phase peak to be where we find it, following relatively gentle evaporation of ices. Equally, with the Hot Core South subjected to greater exposure from the oncoming effects to which the Hot Core itself has already been subjected (shocks, higher temperatures, increased gas–grain and grain–grain collisions),

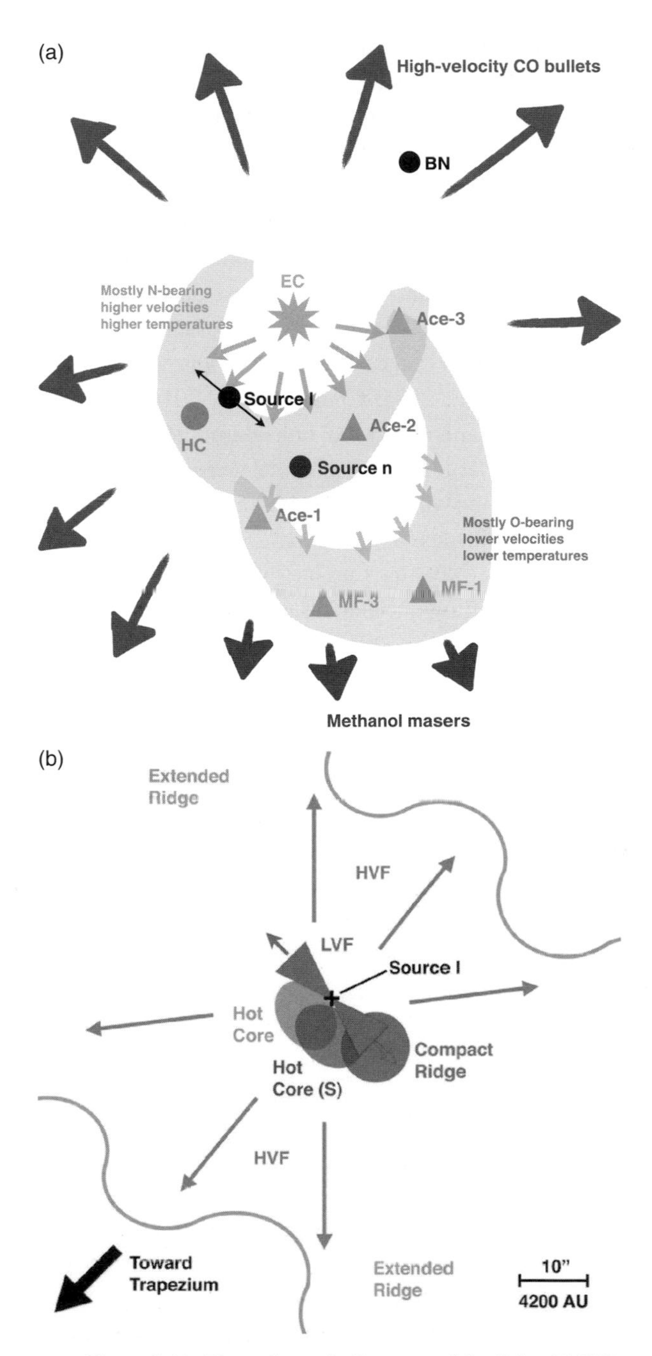

Figure 9.10 Two schematic diagrams of the Orion BN/KL structure and apparent dynamics. The plateau feature is unmarked but associated with the LVF and HVF outflows as discussed in the text (Peng et al., 2013; Crockett et al., 2014).

subsequent acceleration of grain-surface processing to produce hydroxy products before thermal or collisional release might also account for the observations. We are being simplistic, of course, because we have no data on initial conditions in any of these sources, and there is a range of conditions to which each may be subjected, whether UV photolysis of ices, cosmic ray grain interactions, grain–grain collisions, infrared heating, and so on. It may be that all the COMs species observed have a predominantly grain-surface formation route, such as between mobile radicals and strongly bound ones (Charnley & Rodgers, 2005; Garrod et al., 2008; Goumans et al., 2008; Belloche et al., 2009; Ioppolo et al., 2011; Occhiogrosso et al., 2014), or additions and abstractions of simple atoms (Fedoseev et al., 2015; Chuang et al., 2016). Equally there are possible gas-phase mechanisms to be considered (Neill et al., 2012; Balucani et al., 2015; Taquet et al., 2016; Jimenez-Serra et al., 2016), in support of which much interesting work on reaction routes and gas–grain modelling has also been done (Charnley & Rodgers, 2005; Herbst & van Dishoeck, 2009; Simons et al., 2020). Not knowing but having the opportunity to seek out an answer to these challenging questions is the lucky research place where we find ourselves.

9.6 Summary

Orion KL offers much evidence for the sequential chemical processing of shocked molecular cloud material and reminds us of just how violent the dynamic processes associated with HMSFRs can be. In pursuit of understanding the star formation process in general, and high-mass star formation in particular, the following six chapters focus on the molecular astronomy of ionised hydrogen (HII) regions and photodissociation regions (PDRs), which are so central in plotting the evolutionary path of star formation.

PART IV

Ionisation

10
Two HII Surveys Using JVLA and ALMA

10.1 Introduction

Our understanding of star formation comes from studies of conditions and processes both before and after nuclear ignition. The examples discussed in the preceding chapters have given attention to a breadth of evolutionary developments, from the prestellar coalescence of molecular gas and dust through to protostellar accretion, along with jet, wind, and outflow activity in their contiguous, if not simultaneous, action. The impacts of individual stellar processes after ignition, beyond time zero (ZAMS) on the Main Sequence, are as much on a young star's immediate surroundings as on itself and, particularly in the high-mass star case, the individual becomes a significant component in a complex network of multiple stellar evolutionary events. Both compact ionised (HII) sources and photon-dominated regions (PDRs) are two of the pre- and post-protostellar ignition phenomena impacting surrounding molecular clouds that have been much studied over the last five decades. We will look at the ionised hydrogen regions first.

10.2 HII with the Jansky-VLA

We have referred already in earlier chapters to the radio emission associated with ionised hydrogen. Dust in the molecular and ionised gas that surrounds a newly formed high-mass star invariably absorbs virtually all of the stellar radiation, so even these very luminous stars are invisible to us at visible wavelengths. The warm dust does, however, re-radiate most of the stellar luminosity subsequently in the far-infrared (FIR) and, of the many point sources detected by the Infrared Astronomical Satellite (IRAS) launched by NASA in 1983 as the first infrared space telescope, two-thirds were subsequently reconfirmed as associated with compact-HII regions by the first interferometric surveys at the end of that decade. At radio wavelengths, transparent to a star's molecular gas and dust cocoon, single-dish observations originally identified only unresolved continuum (HII) sources, although these were closely linked with OH and H_2O masers, indicative of shocks and dynamic motions arising from high-mass star formation. The first high-resolution observations (<1″) capable of resolving some of the structure of compact-HII sources were made with the US National Radio Astronomy Observatory's (NRAO) Very Large Array (VLA), a survey in which 75 ultracompact (UCHII) sources were identified (Wood & Churchwell, 1989). Studies then categorised the evolving sequence of ionised hydrogen regions engendered by high-energy UV photon fluxes as expanding through hypercompact, ultracompact, compact, classical, and diffuse stages (Churchwell, 2002). The ultracompact (UCHII) phase defines an HII region having a small diameter (<0.1 pc), a high electron density (>10^4 cm^{-3}), and high emission measures (>10^7 pc cm^{-6}), surrounding the youngest and most massive O and B stars (Wood & Churchwell, 1989). The even earlier identifiable evolutionary phase, the hypercompact (HCHII) region, is defined as having a diameter <0.05 pc, an electron density >10^5 cm^{-3}, and an emission measure >10^8 pc cm^{-6}, as well as radio recombination line widths >40 km s^{-1} (e.g. Kurtz, 2005). Of course, the evolutionary process is continuous, so the delineations are convenient rather than definitive, but nonetheless useful in categorising emerging characteristics.

While the physical properties of HII regions, such as emission measure (EM), electron density (n_e), and Lyman-continuum flux, can be estimated from observed angular size and flux density at a given frequency as have been measured over a 30-year period (e.g. Wood & Churchwell, 1989; Murphy et al., 2010; Urquhart et al., 2013a; Kalcheva et al., 2018; Medina et al., 2019), the use of single-frequency observations can be misleading. Firstly, the ionised gas may be optically thick at the observed frequency (e.g. Cesaroni et al., 2015). Secondly, apparent angular size is frequency dependent (e.g. Yang et al., 2019). Therefore, acquiring a spectral energy distribution (SED) over a wide frequency range that covers both optically thick and thin emission is essential (e.g. Murphy et al., 2010). One multi-band survey of 114 HCHII candidates was conducted in 2019 using the 27 antennae (each 25 m in diameter) in the C-configuration of the newly upgraded and renamed Jansky-VLA in New Mexico (Yang et al., 2021).

The research combined archival with the new high-frequency observations to measure SEDs between 1 and 26 GHz for each source and modelled the free–free emission within ionisation-bounded HII regions assumed, as a first approximation, to be spherical. While the structure of these regions is not known, the assumption of spherical geometry is adopted in the knowledge that n_e and an object's diameter represent averaged properties over the entire emitting gas at multiple bands. Emission measure (EM) values for each region are calculated (EM = n_e^2 × diameter), with mean errors in both flux density at each frequency and the distance measurements of ~10 per cent (giving typical errors in n_e ~20 per cent, in diameter ~10 per cent, and in EM ~40 per cent). The distribution of derived physical properties of 116 HII regions is shown in Figure 10.1 Across the row, 'a, b, and c' show the distribution against three parameters: diameter, electron density, and emission measure, respectively (with the smallest diameter distribution magnified in the insert). The physical sizes peak at 0.02 pc, while 57 per cent of sources (66/116) have diameters <0.1 pc. This is consistent with the majority of these

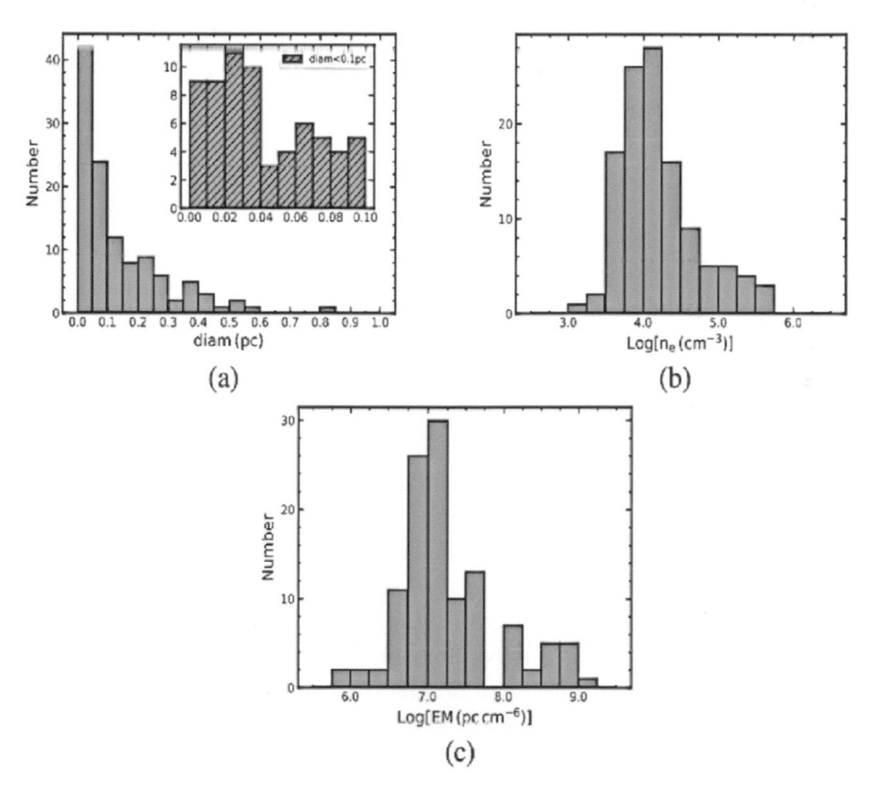

Figure 10.1 Distribution of derived physical properties of 116 young HII regions (Yang et al., 2021).

being classified as UCHII regions or smaller. Nine sources have a diameter <0.01 pc with a mean ~0.06 pc (~1,000 AU). This scale is comparable to, and would be coincident with, radio jets and jet candidates from massive young stellar objects (e.g. Purser et al., 2016).

Figure 10.1b shows the distribution of electron density peaking at 10^4 cm^{-3}, with 60 per cent (70/116) of sources having densities $>10^4$ cm^{-3}. With their mean diameter ~0.06 pc, these are small-scale, high-density objects. As for the EM distribution in 'c', the peak distribution is at 10^7 pc cm^{-6}, with two groups (less or greater than 10^8 pc cm^{-6}), the latter indicating very early stellar evolution indeed.

The bulk of sources have low peak and mean turnover frequencies (ν_t), implying that they are optically thin. For an optically thin HII region in photoionisation equilibrium, the Lyman continuum ionising flux (N_{Ly}) emitted by a single embedded massive star can be calculated from the radio continuum flux and the heliocentric distance to the source (Sanchez-Monge et al., 2013). The typical error in flux intensity derived from the observations is ~40 per cent, given distance and integrated flux measurement uncertainties. Figure 10.2 shows the turnover frequency range, the Lyman continuum flux, and the fraction of UV photons estimated to be absorbed by dust (f_d). Among 67 HII regions in the sample with confirmed dust emission (at 24 μm and 70 μm from previous studies), 43 per cent (29/67) of those with diameters <0.1 pc have a mean f_d ~0.79, and 57 per cent (38/67) of those with diameters >0.1 pc have a mean f_d ~0.58. These results suggest that the dust absorption fraction tends to be more significant for the more compact and presumably younger HII regions, compared to their larger and more evolved counterparts, with the fraction of ionising photons absorbed by dust decreasing with time (Yang et al., 2021; Arthur et al., 2004).

Twenty of the survey sources are identified as optically thick, having turnover frequencies larger than 5 GHz, and 14 of them satisfy the expected criteria for HCHII character. The others satisfy the size criterion for HCHII regions, but their electron densities are low enough to suggest a transitional stage towards UCHII status. Figure 10.3 reiterates the distribution of diameter, electron density, and emission measure but for 18 of the 20 optically thick sources and plotted in three-parameter space. The plot location in which HCHII regions are expected to reside (n_e $>10^5$ cm^{-3} and diameter <0.05 pc) is marked, as is the evident evolutionary trend from HCHII towards UCHII stages. Table 10.1 summarises the characterisation of the three HII stages.

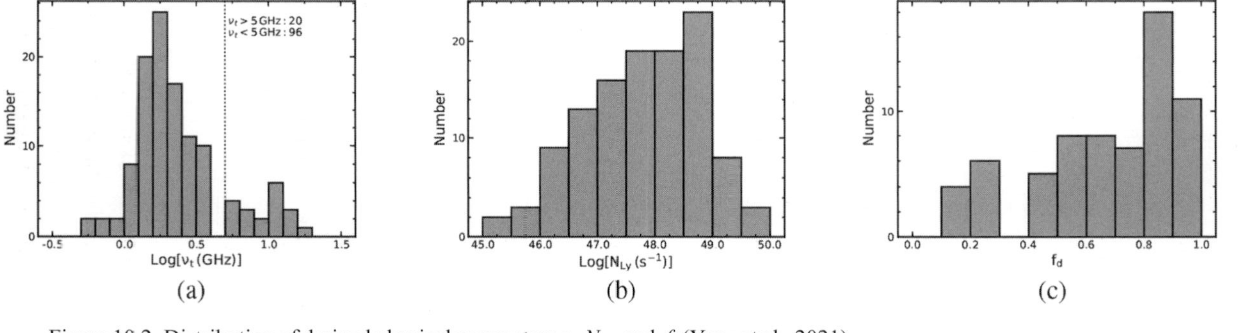

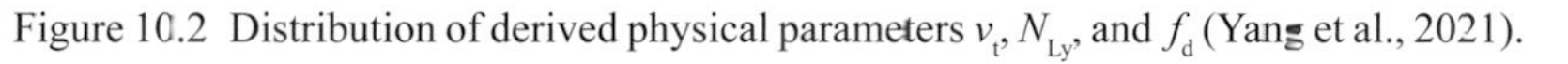

Figure 10.2 Distribution of derived physical parameters ν_t, N_{Ly}, and f_d (Yang et al., 2021).

Table 10.1 *The three classifications of pure HCHII, pure UCHII, and transitioning HCHII-to-UCHII (Yang et al., 2021).*

Parameters	HCHII	HCHII -> UCHII	UCHII
Size (pc)	≤ 0.05	~ 0.05–0.1	≤ 0.1
n_e (cm^{-3})	$\geq 10^5$	$\sim 10^4$–10^5	$\geq 10^4$
EM (pc cm^{-6})	$\geq 10^8$	$\sim 10^7$–10^8	$\geq 10^7$
RRL ΔV (km s^{-1})	≥ 40	~ 25–40	< 40

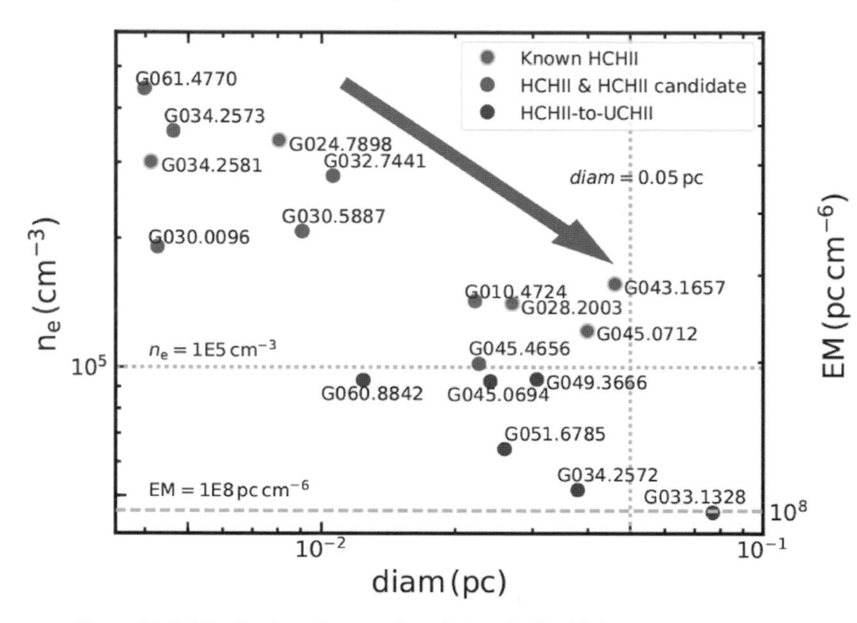

Figure 10.3 Distribution of properties of 18 optically thick HCHII regions. In the absence of colour coding, the six below the horizontal dotted line are HCHII to UCHII, all those above show an equal scattering of known HCHII (five) or candidate HCHII (seven) (Yang et al., 2021).

10.3 A Second Survey: ATOMS-ALMA

High-mass stars are thought to spend at least 10 per cent of their lifetime after nuclear ignition still embedded within their natal molecular clouds. Observationally, the hyper- or ultra-compact HII stage is the fourth emergent stage often used to characterise high-mass star formation (after massive dense core, high-mass protostellar object, and hot molecular core). UCHII lifetimes are taken to be of order 10^5 yr, and the lifetime of a HCHII region ~10^4 yr. The HMSFR examples of Chapters 6 to 9 have shown how massive dense

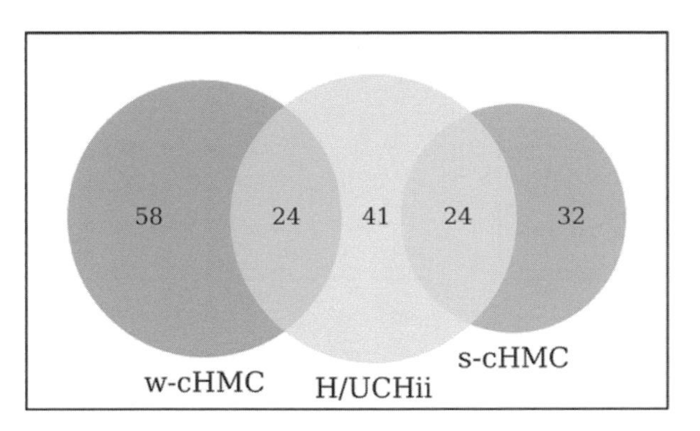

Figure 10.4 The distribution of 138 COMs-containing and 89 HCHII/UCHII cores (Liu et al., 2021).

cores coalesce within infrared dark clouds (IRDCs) and how observations show the early embedded stages of high-mass star and cluster formation (e.g. Sanhueza et al., 2019). The deep, high-resolution ALMA 3 mm observations of the ATOMS survey (ALMA Three-millimetre Observations of Massive Star-forming regions, Liu et al., 2021) identifies 453 compact dense cores and catalogued three groups of note within the sample: 89 cores that cocoon HCHII/UCHII sources observed in H40α; 32 hot molecular cores (HMCs) showing >20 COMs emission lines and closely associated with HII sources; and 58 HMC candidates not associated with HII sources, as shown graphically in Figure 10.4 (Liu et al., 2021). If hot cores reflect the molecular gas of protostellar cores that have been heated by an emerging high-mass star, and that they are frequently observed to be closely coincident with emergent HII gas locations, then the lifetimes of these two evolutionary stages must be broadly comparable.

The strong hot core candidates (showing >20 COMs lines) are designated 's-cHMC'; the weaker hot core candidates (showing between 5 and 20 COMs lines) are designated 'w-cHMC'. In each category there happen also to be 24 closely associated with ionised HCHII/UCHII cores as well. The masses of 'pure' HMCs are found to be $\sim 10^2\ M_\odot$, whereas those of HMCs associated with HII cores are $\sim 10^4\ M_\odot$.

10.4 UCHII or HCHII

Figure 10.5 shows luminosity–mass ratios and dust temperatures for the categories identified (plus that of cores showing neither HMC nor HII signatures). As we might expect, the HII sources show higher luminosity to

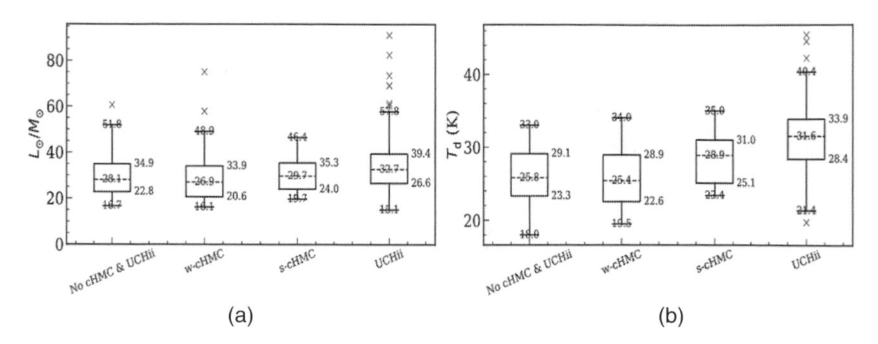

Figure 10.5 Box and whisker plots showing the luminosity/mass ratio (a) and the dust temperature (b) for four groupings taken from the ATOMS-ALMA clumps surveyed (Liu et al., 2021)

mass of gas and higher dust temperature. Equally, as we might expect, the stronger HMC COMs sources are also characterised by higher dust temperatures. Plenty of statistical analysis goes into managing the observational reductions and deriving results in a survey such as ATOMS-ALMA. Among other interesting correlations deduced in this survey is the core size versus line width of the H40α radio recombination line (RRL) relationship for the HCHII/UCHII cores. There are two types of sizes, one measured from 3 mm continuum and the other from H40α emission. Figure 10.6(a) shows that these two methods are comparable until core size exceeds ~0.1 pc, at which point increasing amounts of cold dust emission may interfere, leading to an overestimate of the ionised region. Of the 89 smaller cores for which the RRL data defines sizes, the median size is 0.057 pc. Figure 10.6(b) then shows how RRL emission line widths ($\Delta V_{H40\alpha}$) appear as a function of distance to the source, with little apparent distance dependence. This study finds line widths ranging from ~15 to 55 km s^{-1}, with a mean value of 28 km s^{-1}. This compares with previous estimates for UCHII regions ~10 to 40 km s^{-1} and for HCHII regions at least 40 to 60 km s^{-1} (Hoare et al., 2007; Rivera-Soto et al., 2020). Radio recombination line widths generally are a function of thermal, dynamical, and/or pressure broadening mechanisms in combination (Keto, Zhang & Kurtz, 2008; Nguyen-Luong et al., 2017; Rivera-Soto et al., 2020). Thermal broadening is in part a function of electron temperature. By line width measure, 90 per cent in the sample are probably UCHIIs and 10 per cent probably HCHIIs. A line width ~19 km s^{-1} corresponds to an electron temperature ~8,000 K, typical of ionised plasma (Osterbrock & Ferland, 2006). However, for the 90 per cent with a mean line width of 28 km s^{-1}, the ionised plasma must also be subject to non-thermal mechanisms (dynamical and pressure).

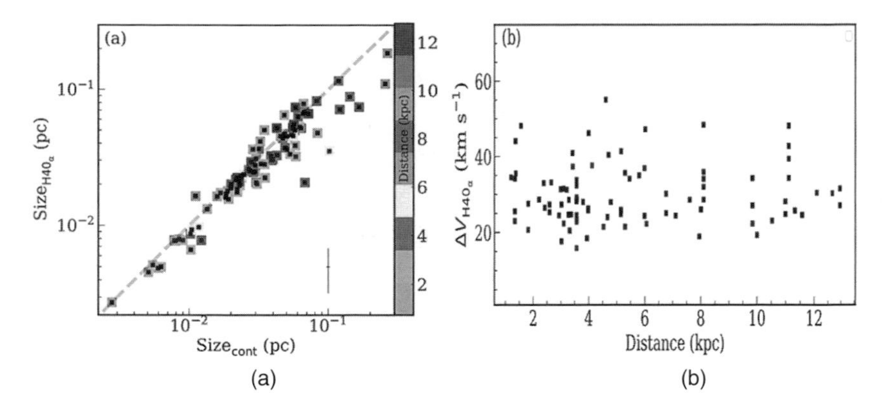

Figure 10.6 (a) comparison of the two core size measures: 3 mm continuum and H40α RRLs; (b) distribution of H40α line width as a function of source distance (Liu et al., 2021).

10.5 Summary

The ATOMS-ALMA survey finds that half of the HCHII/UCHII cores are candidates for associated hot molecular cores. However, this is taken to be an underestimate due to the relatively poor linear resolution and line sensitivity with distance. In extracting the locations of compact cores from the ATOMS 3 mm interferometric survey data towards HMSF cores, the survey does provide a template for how kinematics and dynamics in the vicinity of newly formed OB protostars and HII regions at an early stage of evolution can be studied. Translating that into understanding the dynamics of infall, outflow, and rotational motions, as well as the feedback roles of outflows, stellar winds, and HII regions, are aspects we will consider in the following two chapters.

11

G24.78+0.08 in Scutum

11.1 Introduction

The stellar mass for the onset of an HII region depends critically on the geometry and mass accretion rate on to the star but is predicted to be in the range of 10–30 $M_{\odot}$ (Hosokawa et al., 2010). Simulations also suggest that high-mass star formation at least partially follows a similar pattern to that of low-mass star formation in so far as disk-outflow systems balance mass accretion against ejection. That is, until the point at which photoionisation and radiation forces engender sufficiently broad cavities for protostellar outflow and to eventually quench accretion (Moscadelli et al., 2021). There are some high-mass YSOs that exhibit irregular accretion and luminosity bursts, which results in 'flickering' HCHII or UCHII emission on timescales of just tens of years (e.g. Galvan-Madrid et al., 2008). However, much more usually the high angular resolution (~0.1″) and sensitivities of an interferometer such as the Jansky VLA show continuous ionised winds and radio jets (e.g. Purser et al., 2016), with the three-dimensional velocity distributions of masers, which trace shocked gas near a YSO, generally being consistent with the predictions of disk winds (e.g. Moscadelli et al., 2019). Typically, radio winds and jets that are shock-ionised represent a stage in massive star formation prior to the onset of photoionisation of ambient cloud and the emergence of HCHII regions. The following example is one in which these characterisations are evident and to which we can also look for evidence of interactions between expanding ionised hydrogen regions and neighbouring hot molecular cores.

11.2 G24.78+0.08

Twenty years ago, an early generation of interferometric instruments identified four high-mass protostellar molecular cores embedded in what had been observed as a 0.5 pc scale molecular cloud in NH_3 emission (Codella et al.,

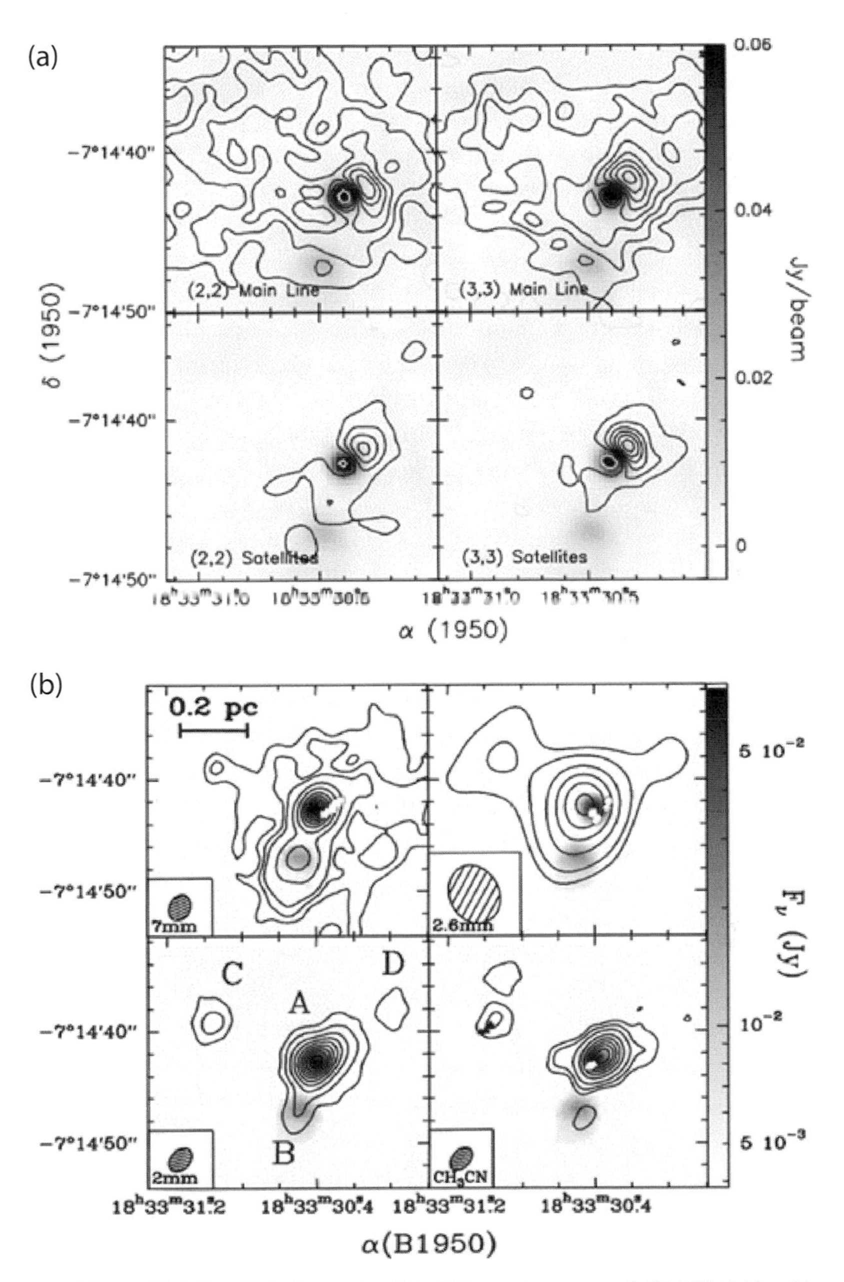

Figure 11.1 Panel(a) shows the first NH_3 contour map of G24.78+0.08 with 1.3 mm continuum in greyscale (Cordella et al., 1997). In panel(b), overlaid also on 1.3 cm continuum (greyscale), early maps of 7 mm (top left), 2.6 mm (top right), 2 mm (bottom left), and CH_3CN (8-7) (bottom right) emission observed with the VLA, PdBI, and NMA, respectively. The four cores are marked as white circles on the 2.6 mm map (Furuya et al., 2002).

1997) in the massive star-forming complex G24.78+0.08. Figure 11.1 shows the four cores in 1.3 mm dust continuum, along with emission maps at 7 mm, 2.6 mm, and 2 mm, as well as CH_3CN (8-7), obtained with the Very Large Array (VLA), the Plateau de Bure Interferometer (PdBI), and the Nobeyama Millimetre Array (NMA), respectively.

When these observations were made, sources A and D had already been identified as compact HII sites (Forster & Caswell., 1987). Source C is here also seen in the mm continuum and CH_3CN lines which, along with H_2O masers previously seen to the north-east, confirm a compact molecular core and potential protostellar source. The fourth detection at 2 and 2.6 mm of a continuum peak, D, to the north-west of the UCHII region A, appeared to trace a dusty core, but one not detected in any molecular lines, suggesting molecular depletion in high-density but low-temperature conditions (Furuya et al., 2002).

Two bipolar outflows were detected in ^{12}CO (1-0) emission lines centred on cores A and C, with one of them likely originating from the YSO that ionises source A. The other would originate in a more deeply embedded YSO in C. Both outflows, shown in Figure 11.2, show closely similar maximum velocities of

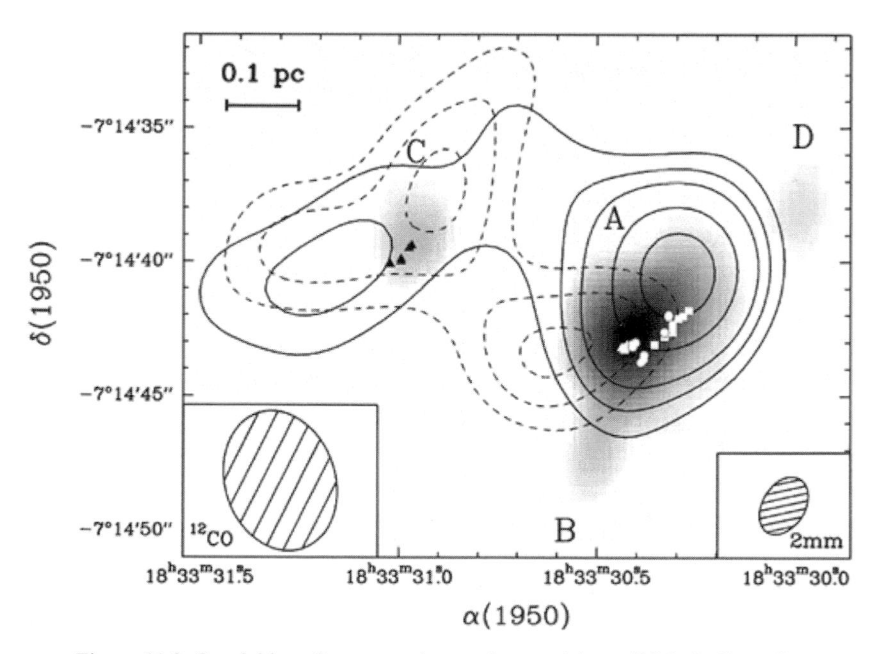

Figure 11.2 Overlaid on 2 mm continuum (greyscale) are ^{12}CO (1–0) outflow emission maps (contours) (Furuya et al., 2002).

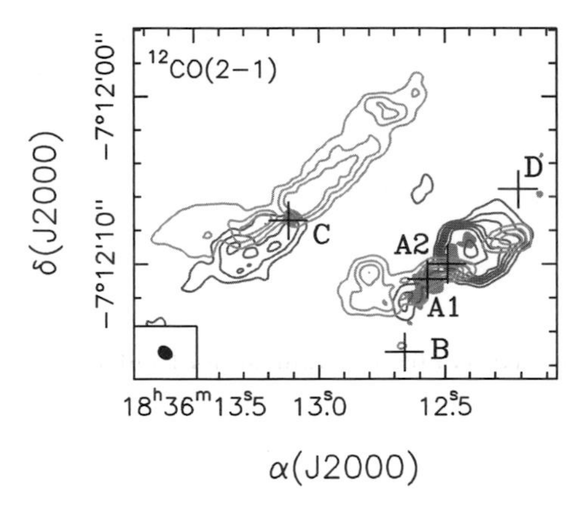

Figure 11.3 Outflow lobes in G24.78 traced in ^{12}CO emission with the IRAM 30 m dish in Pico Veleta, Spain. In greyscale contours, the red-shifted is north-west of C and south-east of A1; the blue-shifted is south-east of C and north-west of A2 (Beltran et al., 2011).

20 km s^{-1} relative to the systemic velocity, and from the size of the lobes (~0.45 pc) a kinematical timescale is estimated to be ~2×10^4 yr, with outflow masses ~10 $M_\odot$, mechanical luminosities ~10 $L_\odot$, and mass loss rates ~5×10^{-4} $M_\odot$ yr^{-1} (Furuya et al., 2002). Uncertainty over the optical thickness of the ^{12}CO lines and the projected dimensions of the flow meant that these estimates were likely to be lower limits, but they were typical of high-mass stars known at that time (Churchwell, 1997). A more detailed follow-up ^{12}CO (2-1) map in Figure 11.3 shows IRAM 30 m detections (at <1″ resolution) of red- and blue-shifted lobes on both G24 C and A, the latter divided into two sub-sources A1 and A2 (Beltran et al., 2011).

Of the four molecular cores in G24.78, Figure 11.4 shows emission lines in N_2H^+, $C^{34}S$, and CH_3OH detected with the Hat Creek Observatory's BIMA (Berkeley-Illinois-Maryland Association) array in California which operated for 20 years until 2005 (Beltran et al., 2005). At higher resolution, continuum and CH_3CN emission towards core A resolve into two sub-cores, A1 and A2. Figure 11.5 homes in on the A-cores detected in dust continuum by both BIMA and the Plateau de Bure Interferometer (PdBI) in the French Alps, showing the A1 and A2 YSO sources marked as crosses (Beltran et al., 2005). Although the surroundings of G24 A1 are quite complex, with multiple cores, bipolar outflows, and the nearby compact HII region G24 B (Codella et al., 2013), the A1 HCHII region seems little affected as indicated

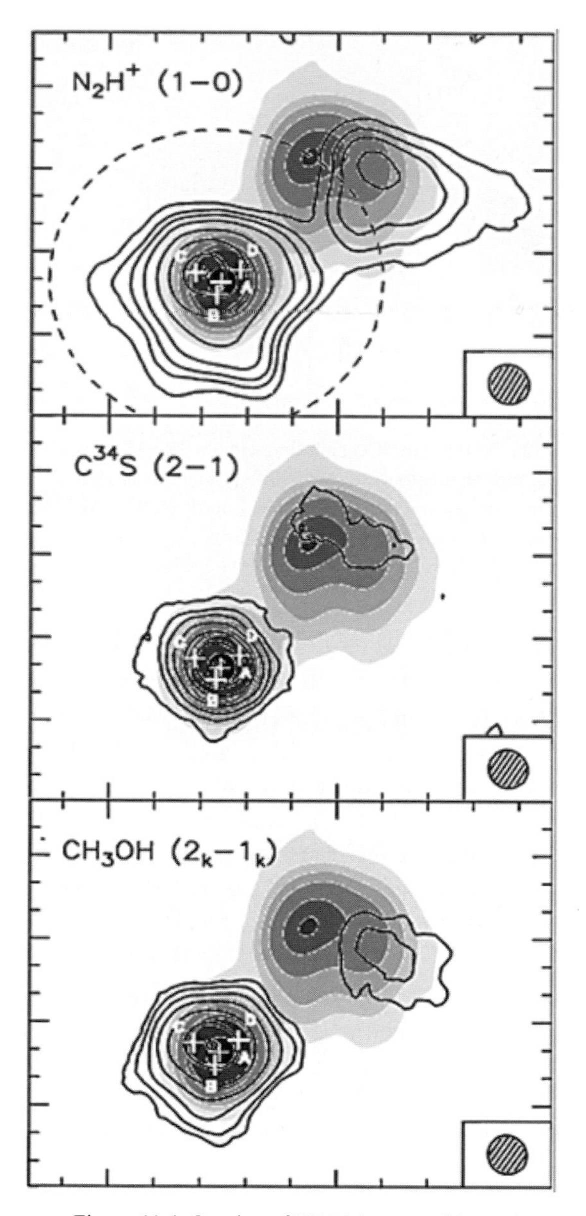

Figure 11.4 Overlay of BIMA integrated intensity maps of the N_2H^+ (1–0), $C^{34}S$ (2–1), and CH_3OH (2_k–1_k) emission on BIMA 3.16 mm continuum emission (greyscale) towards G24.78+0.08 (Beltran et al., 2005).

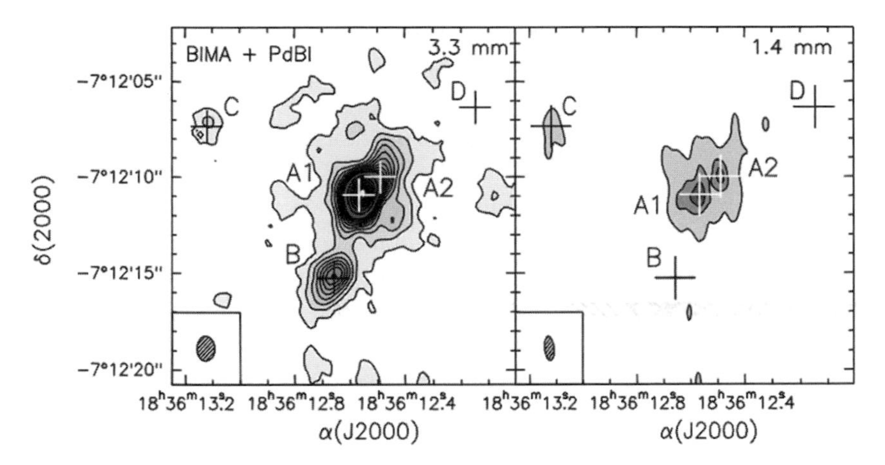

Figure 11.5 Combined BIMA plus PdBI map of the 3.3 mm continuum emission (*left*), and PdBI map of the 1.4 mm continuum emission (*right*) continuum emission towards the A core of G24.78+0.08 (Beltran et al., 2005). The crosses mark the position of five YSOs identified in the region (Furuya et al., 2002).

by high-resolution observations (<670 AU) (e.g. Moscadelli et al., 2007; Beltran et al., 2007).

11.3 RRL Emission

The formation of massive stars (>8 $M_\odot$, >10^4 $L_\odot$), is, of necessity, characterised by high mass-accretion rates (e.g. Tan et al., 2014) and strong radiation feedback from the star even before zero-main sequence age (ZAMS) is reached. The feedback includes both radiation pressure and photoionisation of the circumstellar gas. Hypercompact ionised hydrogen regions (HCHII) represent the earliest identifiable phase in the expanding ionisation process around a newly formed high-mass star, and observation has shown that these sources emit radio recombination lines (RRLs) at line widths much broader than expected from pure thermal and pressure turbulence (Cesaroni et al., 2019; Sewilo et al., 2004). The implication is that in HCHII regions the ionised gas is subject to systematic large-scale motions and, across a variety of sources, evidence of gas infall, expansion, and rotation have all been observed (Sollins et al., 2005; Klaassen et al., 2018; Moscadelli et al., 2018). New and upgraded instruments such as the Jansky-VLA, e-MERLIN, NOEMA, and ALMA allow us to detect and image the RRL emission and resolve the free–free continuum emission from the ionised gas in many of these objects (e.g. Klaassen et al., 2018). They also allow us to quantify mass infall, along with outer molecular and inner ionisation disk dimensions, bringing confirmation that molecular disks around young high-mass stars will become internally ionised while also continuing to accrete mass.

Continuum and hydrogen recombination lines towards G24 A1 have been successfully modelled as an expanding ionisation shell, assuming a shell thickness of almost 1,000 AU, expanding at ~16 km s^{-1}, powered by a stellar Lyman continuum of 5.3×10^{47} photon s^{-1} corresponding to an O9.5 star and an electron density ~10^6 cm^{-3} (Cesaroni et al., 2019). It appears from the projected distribution of RRL emission that the HCHII region is both expanding and drifting towards the SW. In addition, a significant temperature gradient is evident, the electron temperature dropping to a few thousand K in a thin layer at the outer border of the HII shell. Considering the hot core as molecular material at the interface between the HII shell and the ambient molecular gas, firstly observation of the rotational transitions of CH_3CN and associated isotopologues are sufficiently unblended and intense from which to derive column density, rotation temperature, velocity, line width, and source size (Moscadelli et al., 2018, 2021). The molecular gas moves in line with the ionised gas, sharing the same direction and a similar global momentum. Shock and high-density tracers, such as SO_2 and CH_3CN, show well-defined south-west to north-east gradients, parallel to the ionised flow. Over the larger neutral molecular core of which the hot core is a part, line emission shows a velocity gradient directed approximately south-north, and this is interpreted as a relatively slow expansion of the A1 core rather than as a gravitationally supported toroidal rotation.

11.4 Multiple Sub-cores

Observations with the Submillimetre Array (SMA) on Mauna Kea, Hawaii, actually identify multiple dust sub-cores labelled A1, A1b, A1c, A1d, A2, and A2b, as shown in Figure 11.6 (Beltran et al., 2011). All of these align south-east to north-west coincident with the A molecular outflows of Figure 11.3, suggesting a preferential formation direction, the five masses ranging from 7 $M_\odot$ to 22 $M_\odot$. A black triangle close to A1 in the left-hand panel of Figure 11.6 locates the G24 HCHII. Two velocity components are revealed towards A1, one of which peaks close to the millimetre continuum peak and the HCHII location, having a gradient also closely coincident with that of CH_3CN emission.

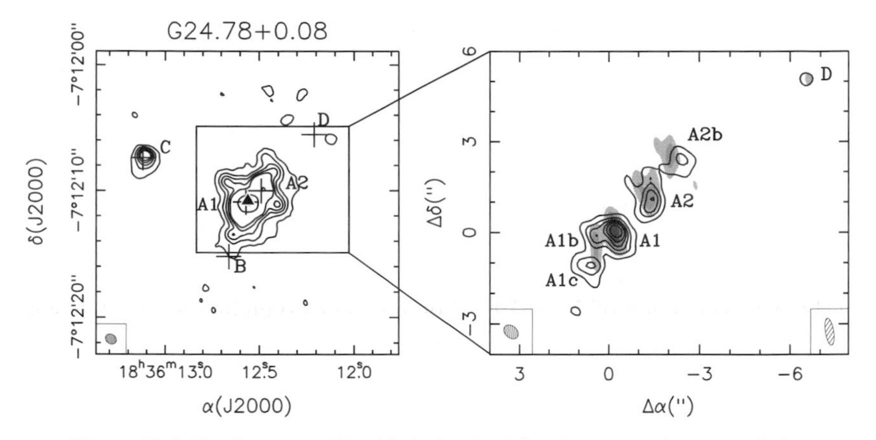

Figure 11.6 The five cores identified. On the left, 1.3 mm continuum emission obtained with the SMA. On the right, the same 1.3 mm continuum overlaid on 1.4 mm emission obtained with the PdBI (greyscale). (Beltran et al., 2011)

From position-velocity plots along outflow A and the ^{13}CO (2-1) averaged blue-shifted and red-shifted emission, this outflow appears driven by core A2, with A1 driving no outflow. While not evident in the A outflows, the highly collimated, 'knotty' outflow at C (particularly evident in a ^{12}CO position-velocity plot) suggests an episodic flow, where the knots are made of swept-up ambient gas.

What we now know with greater confidence is that the high-mass star-forming region G24.78+0.08 has an overall bolometric luminosity ~2 × 10^5 $L_\odot$, lying at a distance of 6.7±0.7 pc from the Solar system (Moscadelli et al., 2021). Its four principal molecular cores are distributed over a 0.1 pc scale, parts of which appear as massive rotating toroids of molecular gas and dust (Beltran et al., 2006).

11.5 The A1 HCHII

Inside the most prominent core, A1, 21 cm to 7 mm observations with the VLA reveal a bright combined hot core and HCHII association at the ~1,000 AU scale (Beltran et al., 2007; Cesaroni et al., 2019). Red-shifted absorptions in NH_3 also show mass infall from the hot core to the HCHII on a scale ~5,000 AU (Beltran et al., 2006). This is molecular gas and dust disk material accreting onto the YSO. Simultaneously, although spatially opposed to this, 1.4 mm detections towards the HCHII with ALMA have achieved a 0.2″ resolution equivalent to ~1,300 AU (Moscadelli et al., 2018), and analysis of H30α lines shows a fast bipolar flow of ionised gas (V_{LSR} range ~ 60 km s^{-1}) which, with maser velocities, suggest the ionised gas is expanding at >200 km s^{-1} into the surrounding molecular gas. From the free–free continuum flux, the spectral type of the energising star has been taken to be at least O9.5 with a mass ~20 $M_{\odot}$ (Beltran et al., 2007). Additional ALMA observations have achieved a 0.05″ resolution which brings a focus ~300 AU, sufficient to map the velocity of the ionised gas and the kinematics of the surrounding molecular cloud.

Figure 11.7 shows the velocity distribution of H30α across A1 (panel (a)) and the maser distributions with their proper motions (panel (b)). The ionised gas appears to have three velocity components. At the centre of the HCHII region (<500 AU), there are two perpendicular V_{LSR} gradients at PA = 39° (22 km s^{-1}) and PA = 133° (3 km s^{-1}), one much steeper than the other. At larger radii (a few thousand AU), there are slow and wide-angle motions of ionised gas along the south-west to north-east direction. The combined observations are interpreted as a disk-jet system – the fast motion along the 39° axis being the ionised jet, the slower perpendicular motion the rotation of an ionised disk about the same axis. Assuming a rotation in gravito-centrifugal equilibrium, a V_{LSR} gradient ΔV ~12 km s^{-1} across a distance ΔS ~850 AU suggests a central mass equal to $\Delta V^2 \Delta S/8G$ ~17 $M_{\odot}$, where G is the gravitational constant (Moscadelli et al., 2021). Jet-disk systems are likely to regulate mass accretion and, if the jet is the fastest and most collimated flow portion of a disk wind, the slower and less collimated wind will also correlate with the slow, wide-angle expansion of ionised gas observed at the larger scales (a few thousand AU).

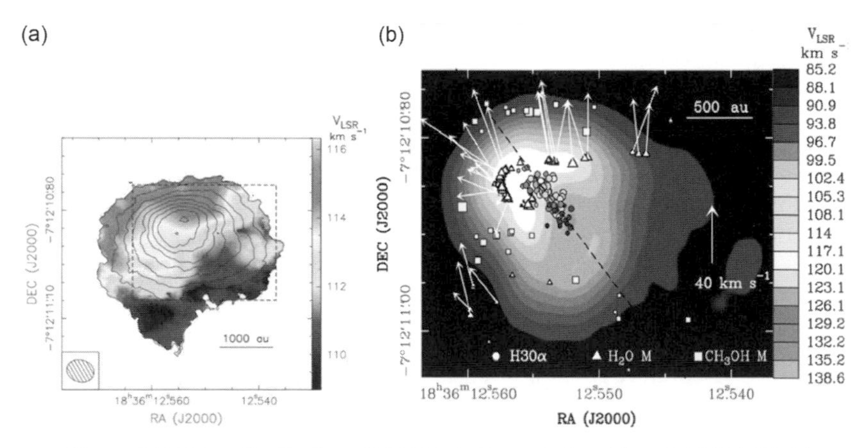

Figure 11.7 V_{LSR} distribution of ALMA H30α emission (a) and ALMA/VLA/VLBI (b) 7 mm continuum (background greyscale) with H_2O and CH_3OH masers (triangles and squares) (Moscadelli et al., 2021).

The apparent ionised disk and disk-wind relationship within an HCHII region is supported by numerical modelling of massive star formation that considers feedback from photoionisation and radiation forces from the star (e.g. Tanaka et al., 2016). The HII region is modelled as forming when the YSO has accreted between 10 and 20 $M_\odot$. While some part of the outer disk, being shielded by the inner disk and disk wind, remains neutral, almost the whole outflow is ionised in 10^3–10^4 yr. For low-to-moderate rates of accretion, the thickness of the residual flaring neutral disk around an ionising star of 20 $M_\odot$ is only a few hundred AU at a 1,000 AU radius. In the G24.78A1 case, the ALMA angular resolutions ~400 AU have not yet identified such a thin disk buried in the ionised emission.

Comparing the detections shown in both panels of Figure 11.7, the interaction between the HCHII region and the surrounding molecular environment is traced with collisionally pumped water and methanol masers. CH_3OH masers emerge at larger separation from the ionised gas than H_2O masers and show slower velocities, tracing the expansion of the ambient medium swept away by the protostellar outflow (Moscadelli et al., 2018). The arc of water masers sits just ahead of the north-east lobe of the fast ionised outflow observed in the H30α line. Since water masers are generated by shocks produced when ionised flow collides with denser molecular gas and have large proper motions (>40 km s^{-1}) expanding away from the HCHII centre, an estimate of ionised flow velocity ~200 km s^{-1} at ~500 AU from these observations (Moscadelli et al., 2018) agrees well with theory (Tanaka et al., 2016). Nonetheless, the protostellar outflow emerging from the HCHII region in G24 A1 has yet to expand on a large scale. The water maser arc indicates an outflow still trapped by surrounding dense material, with the YSO much closer to the edge of the molecular core in the north-east than in the south-west where ionised gas is escaping more freely (Moscadelli et al., 2018). Where such asymmetric morphology occurs, it seems likely that mass-accretion rates will vary with timescales of a few hundred years (Meyer et al., 2017), causing outflow cavities to be replenished periodically, a scenario that has been both modelled (Peters et al., 2010a,b) and observed in other HCHII and UCHII regions (Rodriguez et al., 2007).

11.6 The A1 Molecular Disk

The neutral gas surrounding the HCHII region in A1 has been observed in molecular line emission from our familiar dust-surface-engendered desorbed COMs, including CH_3OH, CH_3CN, CH_3CH_2CN, and CH_3OCHO, suggesting temperatures across the core that perhaps reflect shocks generated by the fast expansion of ionised gas (Moscadelli et al., 2018). High-energy transitions of the vibrationally excited tracer CH_3CN are potentially radiatively pumped such that the observed rotational temperature can differ significantly from the actual gas kinetic temperature. Therefore, since averaging $^{13}CH_3CN$ emission over a 0.1″ radius centred on the HCHII source suggests a rotational temperature ~330±80 K for a column density of 5×10^{15} cm^{-2}, how otherwise might we establish the validity of this as a reflection of kinetic temperature? If we look to $^{12}C/^{13}C$ isotopic ratios for the G24.78 Galactic region, we can derive an average ratio ~35 (Wilson & Rood, 1994). Recent ALMA studies of CH_3CN abundances in massive hot cores then indicate a typical mean value for $[CH_3CN]/[H_2]$ ~10^{-9} (Pols et al., 2018). For a 10^{15} cm^2 column density derived from observation, we obtain an H_2 column density of 10^{26} cm^{-2} (Moscadelli et al., 2021). From position-velocity plots, $^{13}CH_3CN$ emission is mainly found at radii between 500 and 1,000 AU. Taking 1,500 AU as a crude estimate of the disk diameter, the H_2 number density in the disk would be ~4×10^9 cm^{-3}. Within such a high-density gas, molecular rotation transitions should be principally excited by collisions, in which case the rotational temperature should be a good approximation to the gas kinetic temperature. Again, these values of particle density and temperature are consistent with simulations for a 20 $M_\odot$ star at radius <1,000 AU and temperature 100–400 K (Tanaka et al., 2016).

The disk is also indirectly identifiable from the kinematic signatures of position-velocity plots obtained from ALMA observations of molecular tracers such as CH_3CN (Moscadelli et al., 2021). All the tracers show an increase in velocities in moving from larger radii (~4,000 AU) to smaller (~500 AU), consistent with Keplerian rotation around a 20 $M_\odot$ star. The molecular emission lines fade close to the YSO, where ionised gas reaches velocities much higher than those of the neutral gas. As the gas temperature of the disk decreases with increasing radius, we hope to see high-excitation emission lines emerging from warmer gas within the inner portion of the disk, just as we also expect emission from less abundant (optically thinner) species emerging. ALMA observations have shown exactly that excitation and optical depth tracing of the smaller disk radii (Moscadelli et al., 2021).

To estimate the disk mass and the potential final mass of the YSO, we must assume the disk is close to being edge-on so that its projection on the sky falls inside a rectangle centred on the HCHII region, oriented at the velocity gradient angle PA = 133°. With a major side length of 8,000 AU (twice the radial extension of the molecule velocity profile) and a minor side length of 2,000 AU, this would be large enough to encompass the whole HCHII region. Integrating the 1.4 mm continuum emission observed by ALMA over this rectangle, Moscadelli and colleagues (2018) derive a flux of 120 mJy. After correcting for a dominant free–free contribution from the HCHII region of ~86 mJy, the residual flux from dust emission is ~34 mJy. Assuming a dust opacity of 1 cm^2 g^{-1} at 1.4 mm (Ossenkopf & Henning, 1994), a gas-to-dust ratio of 100, and an average dust temperature of 200 K over the disk (Tanaka et al., 2016), an upper limit for the total mass of molecular gas in the disk is derived as no more than 2.5 $M_\odot$ (Moscadelli et al., 2021). The mass of the ionised disk should be a negligible part of this. The radius of the ionised disk is $R_b = GM_*/(2c_s^2)$, for a gravitationally trapped HII region, where M_* is the stellar mass and c_s is the sound speed in the ionised gas (Keto, 2007). Unfortunately, at an anticipated sound speed value of 13 km s^{-1} and stellar mass of 20 $M_\odot$, the R_b value reduces to 54 AU, which for so small a radius would be an ionised gas density >10^{12} cm^{-3} (far too high) and yield a non-negligible ionised disk mass ~1 $M_\odot$. Without better information, we can only approximate that the mass of molecular and ionised gas in the disk is ~10 per cent of the mass of the central star, consistent with earlier modelling, while acknowledging the possibility that the more massive star in any cluster is just as likely to accrete additional material from other nearby molecular cores as well as its own.

11.7 Summary

G24.78 offers a multiple-core and sub-core complex in which both ultra- and hyper-compact HII locations are identified, along with outflows, accretion disks, and hot cores. Molecular emission lines as well as radio recombination lines and free–free emission offers evidence for thermal, pressure, and dynamical (including infall and rotation) kinematics. Molecular line signatures trace HII/hot core interactions, and also enable estimates of the physical parameters of HMSF accretion disks (such as density, temperature, mass, and radius). Chapter 12 takes the detail of HII/hot core interactions one step further.

12
G34.26+0.15 in Aquila

12.1 Introduction

In contrast to the low-mass case, high-mass protostars begin hydrogen nuclear burning before accretion is complete. From within the dense natal core, therefore, irradiated photons with energies in excess of the Lyman limit (13.6 eV; $\lambda < 91.2$ nm) strip electrons from protons in the close atomic hydrogen cloud. The correlation between the internal nuclear activity of a protostar and its ionising flux is rather uncertain, but the high surface temperature alone, which is largely a function of accretion scale and rate, can be sufficient to engender ionising photons. The limit of ionising luminosity determines the extent of the resulting HII region; in simple spherical terms, it will extend out to what is called the Stromgren radius. The ionised gas continually recombines with free electrons, and the earliest detections of ionised hydrogen, and therefore the OB stars arising from HMSFRs, were through detections 50 years ago of both free–free continuum at radio wavelengths and the recombination lines (RRLs) to which we have already referred in Chapter 10. The free–free continuum arises from the deceleration of an electron in the electric field of a proton. The recombination transitions arise from the de-excitation cascades of post-combination electron and proton interactions, emitting across frequencies from radio to UV.

12.2 G34.26+0.15

For a terrestrial night-sky directional location for G34.26, we may look to the bright star Altair in the constellation of Aquila, a star visible through the summer in the northern hemisphere and due south at its zenith, with its two companions β and γ Aquilae and the prominent variable ε Aquilae. Altair is yellowish white in the optical with a true luminosity over 10 times that of our Sun. In addition to giving a directional prompt towards G34.26, Altair is interesting for a distance comparison: the G34.26 UCHII region lies about 3.7 kpc away (~1.2×10^5 light yr) from Earth, which is a thousand times further away than Altair.

Figure 12.1 shows a GLIMPSE (Galactic Legacy Infrared Mid-Plane Survey Extraordinaire) image at 8 μm of the 30 to 35 degrees Galactic longitude sector of the Galactic plane in which G34.26+0.15 is located. Although difficult to see in this black-and-white image, a small white circle with a central black dot does mark the UCHII site. The GLIMPSE observations were made with NASA's infrared Spitzer Space Telescope (SST) launched in 2003 (retired in 2020) which drifted in an Earth-trailing orbit about 2×10^5 km in our wake. The survey spanned the entire 360 degrees of Galactic longitude, oscillating between two and four degrees of latitude above and below the Galactic plane.

A much closer image of the G34.26 location, this one viewed in the visible with the Hubble telescope, is shown in Figure 12.2. As an HMSFR, the complex shows a molecular envelope cloud across ~80,000 AU in the plane of the sky. The compact HII source was the third-brightest 100 μm object in the IRAS Point-Source Catalogue of 1985 (after Sgr B2 and NGC 2024) and was first observed in one of the earliest ever Galactic radio surveys over 50 years ago (Altenhoff et al., 1970). Radio continuum observations actually distinguish two very condensed UCHII regions (named 'A' and 'B'), a more evolved compact HII region 'C' with classic 'cometary' morphology (Wood & Churchwell,

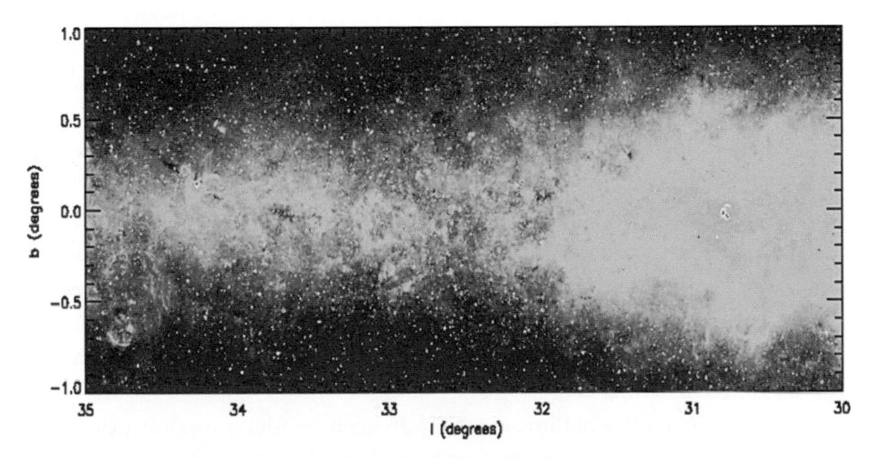

Figure 12.1 Galactic longitude 30–35 degrees from the 8μm GLIMPSE survey. A small white circle with a central black dot marks the G34.26+0.15 UCHII location (Bob Benjamin/University of Wisconsin-Madison).

Figure 12.2 A Hubble Space Telescope image of the G34.26+0.15 reflection nebula. The supernova remnant W44 (G34.7-0.4) lies 3 kpc off-image to the south-east (NASA/JPL-Caltech).

1989), and an extended ring-like HII region, 'D' (Reid & Ho, 1985; Mookerjea et al., 2007; Li et al., 2017). These features are marked on the 2 cm continuum emission map of Figure 12.3.

The scale of Figure 12.3 equates to a 1 cm square centred on the bright nebula in the previous Figure 12.2. It also equates to the box in the centre of Figure 12.4 but rotated clockwise through 90°. The central 'cometary'-shaped feature in Figure 12.3 is further delineated in Figure 12.5. In characterising the HMSF region schematically shown in the 2 cm continuum emission map of Figure 12.3, infalling gas (at a few 10^{-3} $M_{\odot}$ yr^{-1}) flows from the wider surrounding molecular cloud towards the embedded sources A, B, and C. This infalling gas was first observed in ammonia, NH_3, through both emission and absorption (Heaton et al., 1989), and subsequently a hot core with a rich and complex gas-phase chemistry was revealed through multiple molecular line emission (Macdonald et al., 1996). Located at the interface with UCHII-C, this was one of the earliest ultra-compact hot core (UCHC) sources identified (Churchwell et al., 1992; Olmi et al., 1993) and the first to be modelled in detail (Macdonald et al., 1995, 1996; Millar et al., 1997).

There are substantial outflows in G34.26 seen in SiO emission extending north-west, south-east, and north-east from the HII regions (Hatchell et al., 2001), marked as blue- or red-shifted in Figure 12.3. One major outflow is also detected in 4.5 μm emission with the VLA, flowing south from the central

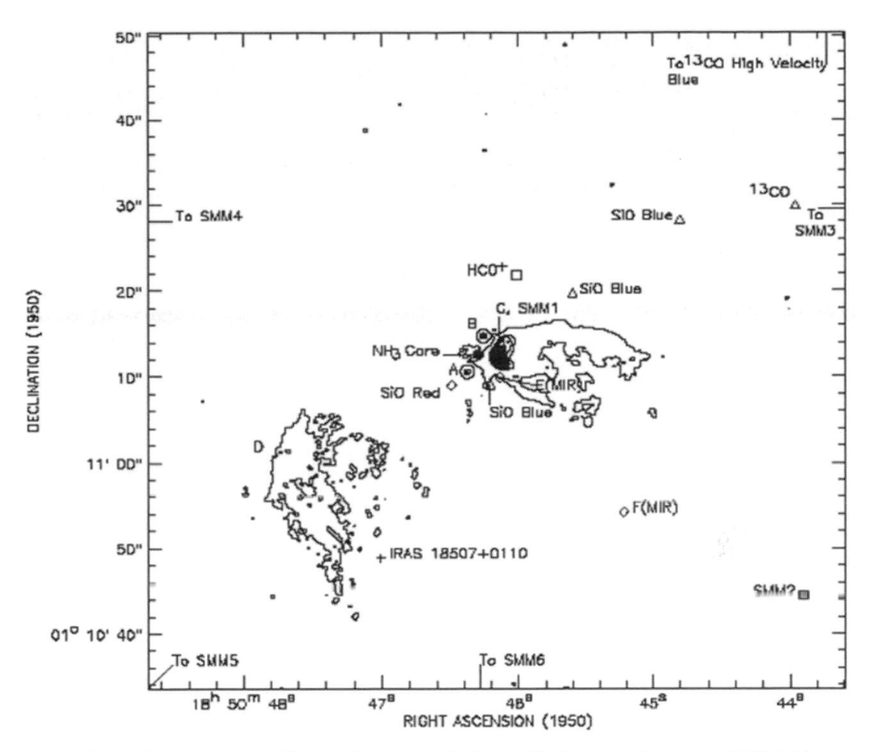

Figure 12.3 A 2 cm radio continuum emission with the main features of G34.26 marked (Campbell et al., 2004).

triangle of compact HII features (A, B, and C). The extended HII region, D, is detected as a bright feature in 8 μm emission to the south-east in Figure 12.4.

12.3 UCHIIs A and B

The UCHII source C with its cometary morphology is shown in the continuum image of Figure 12.5, along with UCHII sources A (*lower*) and B (*upper*) to the east. High-resolution observations of the A and B sources at 7 mm using the VLA are imaged in Figure 12.6.

The resolved intensity profiles of both sources shown in Figure 12.7 reveal a limb brightening that seems indicative of the presence of inner cavities. The profiles show intensity of emission against angular dimensions of each object A and B in arcseconds (Θ″) out from the source centres at zero Θ. The profiles show that the intensity of emission drops to zero in each source by a radius ~0.4″ defining the source sizes. By 'limb' we mean towards the edge of the

Figure 12.4 A VLA composite image of G34.26 in 8 μm, 4.5 μm, and 3.6 μm, with a 7 mm radio continuum overlaid (Sewilo et al., 2011).

source, so in each case the intensity peak is not at the centre but at a radius ~0.2″. Since both sources are at the low-density end of highly compact (HC) HII regions and their emission is deemed optically thin, whether or not there is an inner 'cavity' (a lower-density region) between the centre and the edge or, perhaps more likely, simply an inner clumpy distribution of gas and dust, the profile certainly does not appear to be consistent with a uniform distribution of ionised gas increasing in density towards the centre.

As a first approximation, using a spherical ionised shell model bounded by inner and outer radii, shown schematically in Figure 12.8, and with an electron density gradient defined as $n_e = Ar^{-\alpha}$, the parameters also tabulated in Figure 12.8 generate the best-fit lines drawn in Figure 12.7. Along with inner and outer radii, R_1 and R_2 respectively, the electron density parameter α defines the profile (Avalos et al., 2006). Radius R_1 may correspond to the outer limit

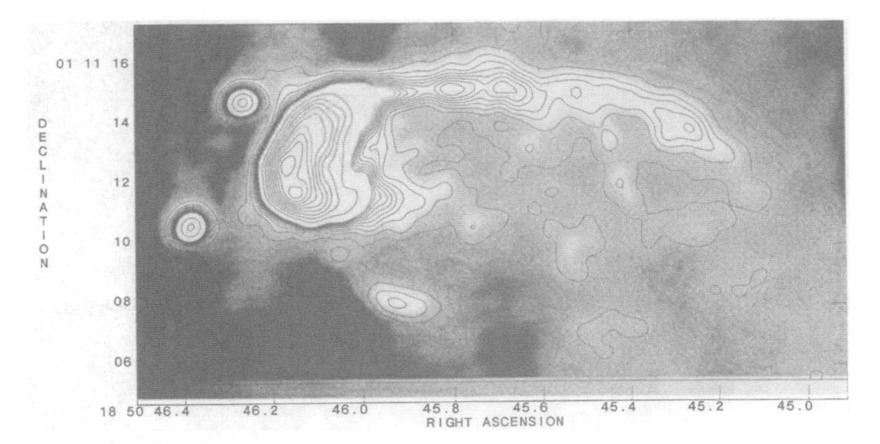

Figure 12.5 An early 8.4 GHz VLA continuum image of UCHII features A, B, and C (Van Buren et al., 1990).

of a young stellar or photoevaporated disk wind, and R_2 to the interface between the ionised region and the neutral envelope. The best-fit indication is of an inner radius ~400 AU and an outer radius ~1,000 AU, with a shallow (α~0.3–1.0) rather than steep (α = 2) electron density gradient. With a shell width $(R_2 - R_1)/R_2$ ~0.1–0.6, this is a thicker shell than expected if swept up simply by stellar winds. This still-unexplained characteristic of these two highly compact HII sources has also been shown in the profiles of hypercompact HII regions G24.78+0.08 A1 (Beltran et al., 2007) and G28.20-0.04 N (Sewilo et al., 2008).

12.4 UCHII-C

The ionised source UCHII-C has a prototypical cometary morphology; in other words, it has a parabolic shape with a well-defined ‘leading edge’ and a surface brightness decreasing from the compact ‘head’ towards the east and the diffuse ‘tail’ to the west. A 2 cm continuum map of VLA detections is shown in Figure 12.9. Such morphology is usually taken to arise from the bow shock interaction between an ambient molecular cloud and the wind from an energetic young star(s) moving supersonically through the cloud (Wood & Churchwell, 1989; van Buren et al., 1990). In G34.0.26+0.15 such a stellar wind would be blowing from a western star(s) location moving eastwards into higher-density gas. However, it seems equally likely that HCHII A and B harbour ionised stellar winds that interact with the western dense cloud to form the cometary or bright-limbed morphology seen

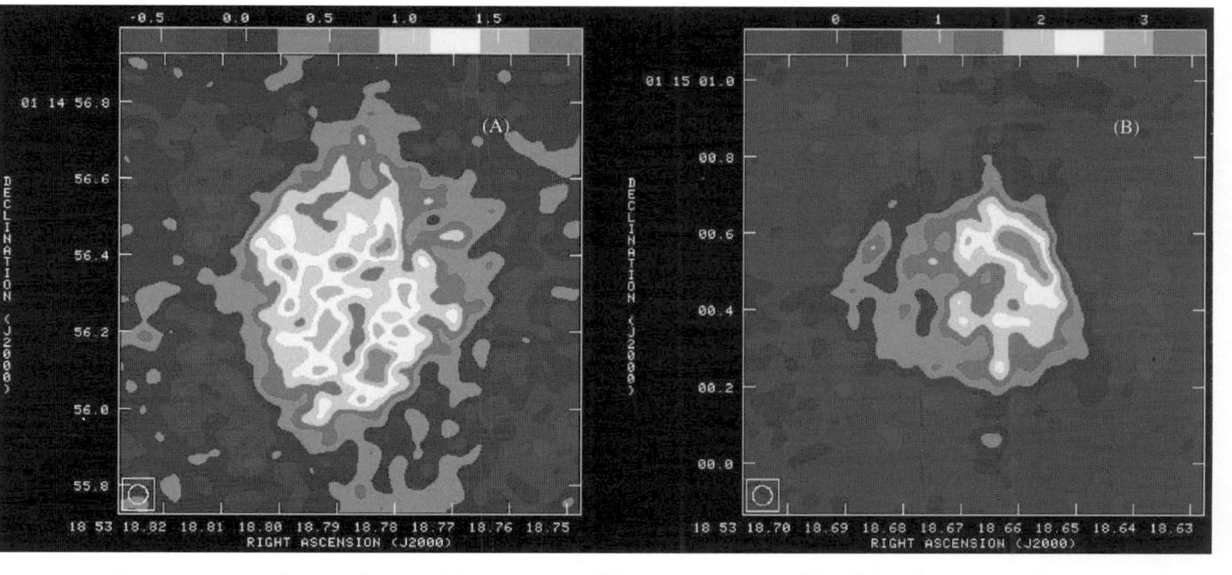

Figure 12.6 VLA observations (at 7 mm) of highly compact HII sources A and B in G34.26 (Avalos et al., 2009).

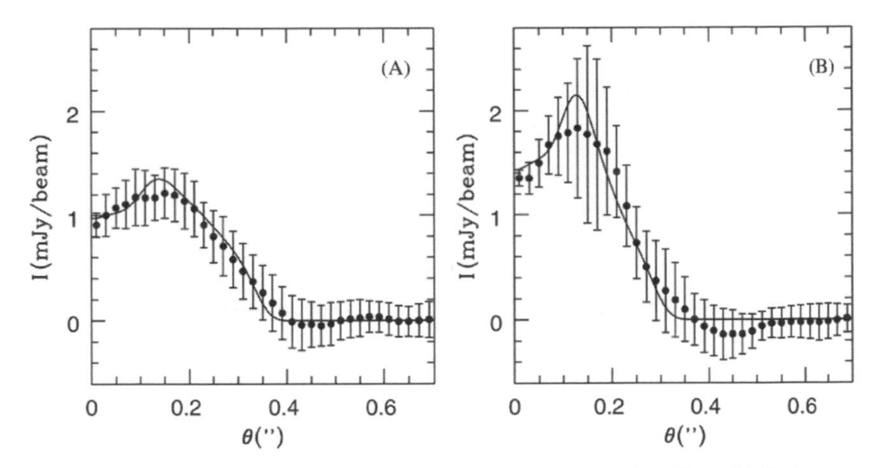

Figure 12.7 Intensity profiles at 7 mm of G34.26+0.15 A and B. The solid circles correspond to the observations, the solid lines to best-fit models (Avalos et al., 2009).

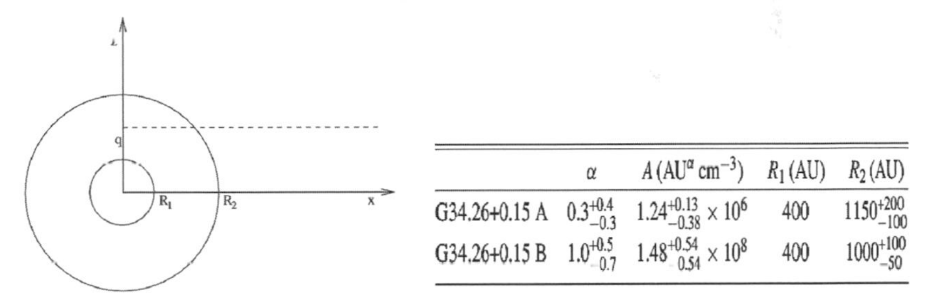

	α	$A\,(\mathrm{AU}^{\alpha}\,\mathrm{cm}^{-3})$	R_1 (AU)	R_2 (AU)
G34.26+0.15 A	$0.3^{+0.4}_{-0.3}$	$1.24^{+0.13}_{-0.38} \times 10^6$	400	1150^{+200}_{-100}
G34.26+0.15 B	$1.0^{+0.5}_{-0.7}$	$1.48^{+0.54}_{-0.54} \times 10^8$	400	1000^{+100}_{-50}

Figure 12.8 Schematic for an ionised shell model with parameters derived for HCHII A and B in G34.26 (Avalos et al., 2006, 2009).

in the centimetre continuum emission of UCHII-C. The existence of shell-like structures seen from the centimetre continuum emission in each of the highly compact HII regions A and B to which we have referred, plus the outflow activity including that which extends from component B (Hatchell et al., 2001), lend support to the existence of a stellar wind–driven origin for UCHII C in an east-to-west flow.

The HMSFR encompassing all three HII sources, A, B and C, is therefore subject to a complexity of velocity gradients associated with observed masers, the ultracompact molecular core, the UCHII-C feature, and either one or both of stellar candidates A and B (Heaton et al., 1985; Garay & Rodriguez., 1990; Gaume et al., 1994; Watt & Munday, 1999; Mookerjea et al., 2007). Of the H_2O, CH_3OH, and OH masers observed, the H_2O maser spots are found distributed ahead of and projected onto the cometary arc of component C, while the OH masers appear more tightly confined to the

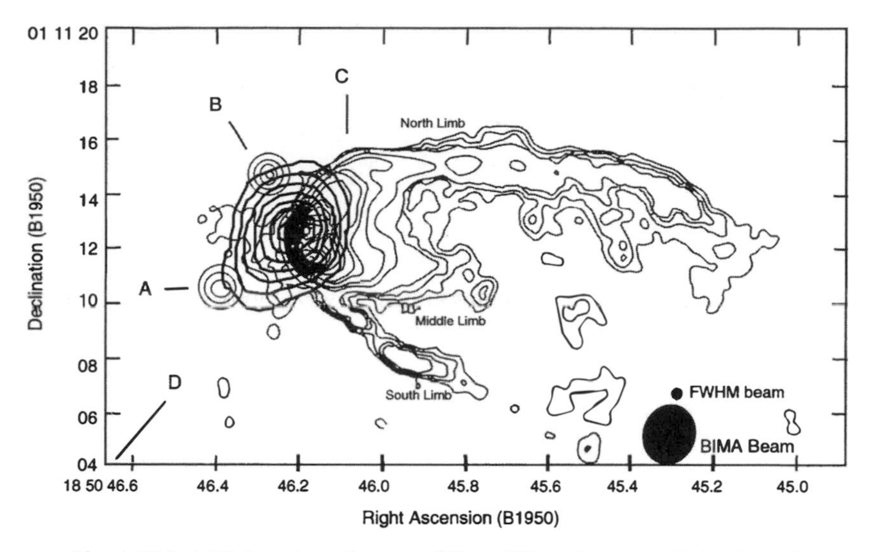

Figure 12.9 A VLA contour diagram of 2 cm HII continuum emission towards G34 UCHII C (Gaume et al., 1994) with CH_3CN (6-5) distribution in bold contours overlaid (Watt & Mundy, 1999).

parabolic arc along its eastern edge (e.g. Gasiprong et al., 2002; Sewilo et al., 2011). In fact, observations of maser activity strongly suggest the existence of four separate YSO outflows, each with associated H_2O maser feature clusters within 5″ of each other (~0.09 pc at 3.7 kpc) and located in the triangle between components A, B, and C (Imai et al., 2011). These H_2O maser clusters have persisted observationally for over 10 years, in contrast to masers typically associated with low-mass YSOs that appear and disappear on time scales usually much less than one year (Claussen et al., 1996; Imai et al., 2011). Given the likely lifetime of active H_2O maser regions associated with high-mass star formation as ~10^5 yr (Genzel & Downes, 1977), the observations suggest that these stars were born close together in time as well as space. On the larger scale, the existence of a global outflow has been confirmed, traced by ^{13}CO and SiO emission which, having a total detectable length of 0.6 pc suggests a kinematical age of 10^5 yr (Matthews et al., 1987; Hatchell et al., 2001). Using NH_3 absorption towards G34.26, derived infall rates in the inner region are found to be several times 10^{-2} $M_\odot$ yr^{-1}, high enough to overcome likely radiation pressure (Wyrowski et al., 2012; Hajigholi et al., 2016). While the NH_3 absorption transitions only probe the dynamics of the location where the outer hot core meets the envelope, the mass-accretion profile suggests accretion in the inner part of the hot core may have ceased, although the absence of detections could reflect the ionisation or dissociation of NH_3 close to the central source.

12.5 The Hot Core

As for hot core material and its relationship to the ionised hydrogen in G34.26, molecular line observations indicate that a UCHC lying between components A, B, and C, on a scale of a few times 0.01 pc and closest to the 'nose' of UCHII C in the plane of the sky, sits within an encompassing compact core (r ~ 0.1 pc) and a surrounding 'halo' cloud (r ~3–4 pc) (Matthews et al., 1987; Heaton et al., 1989; Garay & Rodriguez, 1990; Keto et al., 1992; Heaton et al., 1993; Hofner & Churchwell, 1996; Thompson et al., 1999; Mookerjea et al., 2007). Figure 12.10 shows the beam-averaged profiles from an early VLA detection of NH_3 inversion lines (Heaton et al., 1993). The NH_3 emission lines appear distributed around the leading edge of UCHII C shown by the dashed contours.

Based on the location of NH_3 peak emission and a lack of NH_3 absorption against the bright surface of UCHII-C, the hot core gas is envisaged in the plane of the sky as both beside and probably extended behind the ionised gas of component C. Indeed, within the larger velocity gradient flow shown in Figure 12.11, the NH_3 and HCO^+ maps of Figure 12.12 show these hot core

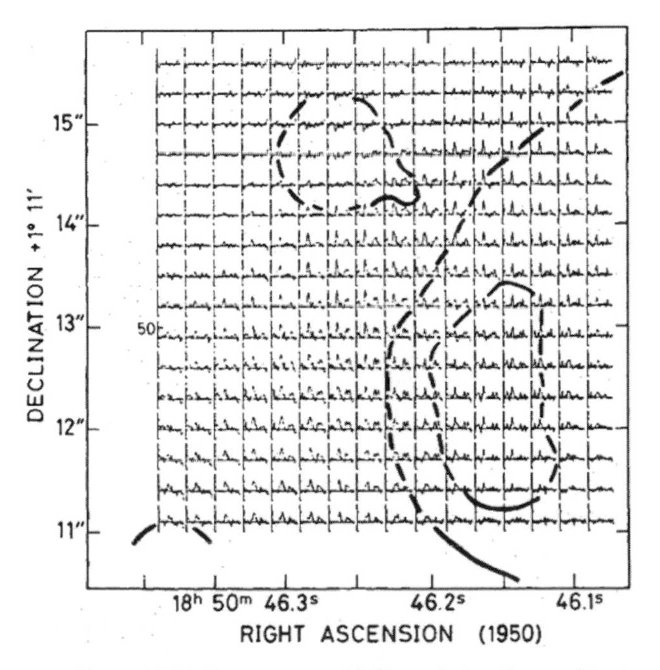

Figure 12.10 Beam-averaged NH_3 emission line profiles towards the G34.26 UCHC detected with the VLA. The 2 cm continuum contours delineate the leading edge of UCHII C and define the extent of neighbouring HCHII-B, plus UCHII-A appearing in the bottom left corner (Heaton et al., 1989).

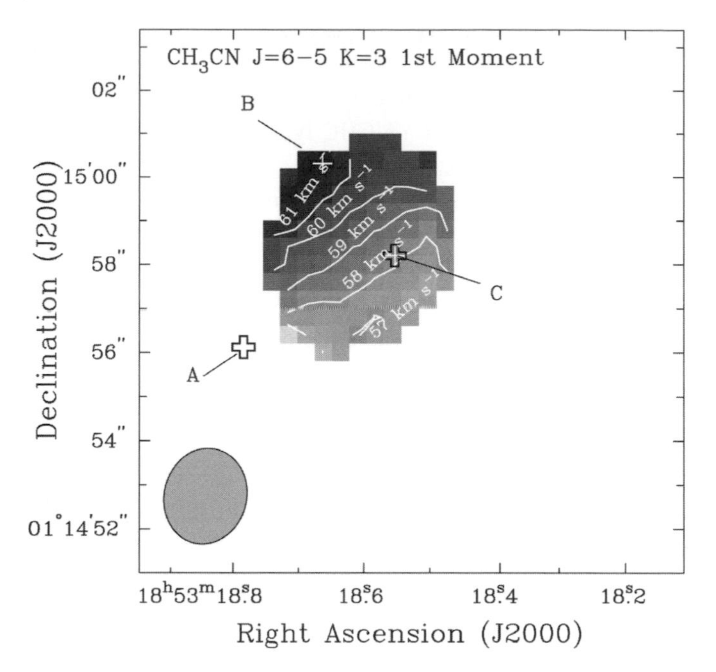

Figure 12.11 A mean velocity moment map of CH_3CN emission associated with the hot core and its surroundings close to UCHII C (Watt & Mundy, 1999).

species apparently 'wrapping around' the leading edge of UCHII C, with the ionisation front 'advancing' into the cloud.

A 330–360 GHz survey of the G34.26 UCHC detected 338 lines from 35 separate molecular species, and this was followed up with the first depth-dependent model of a hot core attempted, in a gas-phase chemical network of 225 species and 2,184 reactions (Macdonald et al., 1995; Millar et al., 1997). In contrast to the 'one-point' models of the day (single sets of species with initial abundances and fixed physical parameters of density, temperature, CR ionisation rate, etc. evolving over time), the observationally constrained three-component structure of this source (halo, compact core, ultracompact core) enabled calculations of abundance over time per unit volume to be made at multiple radial depths. Results were then integrated over depth to calculate column densities through the whole cloud along the line of sight for comparison with observations. Figure 12.13 shows how the column densities accumulate with depth, species differing in their depth-dependent contribution to total column density.

The correspondence between model and observations is remarkably close for most species. This suggests that the physical parameters listed in Table 12.1 represent a reasonable first approximation to the reality.

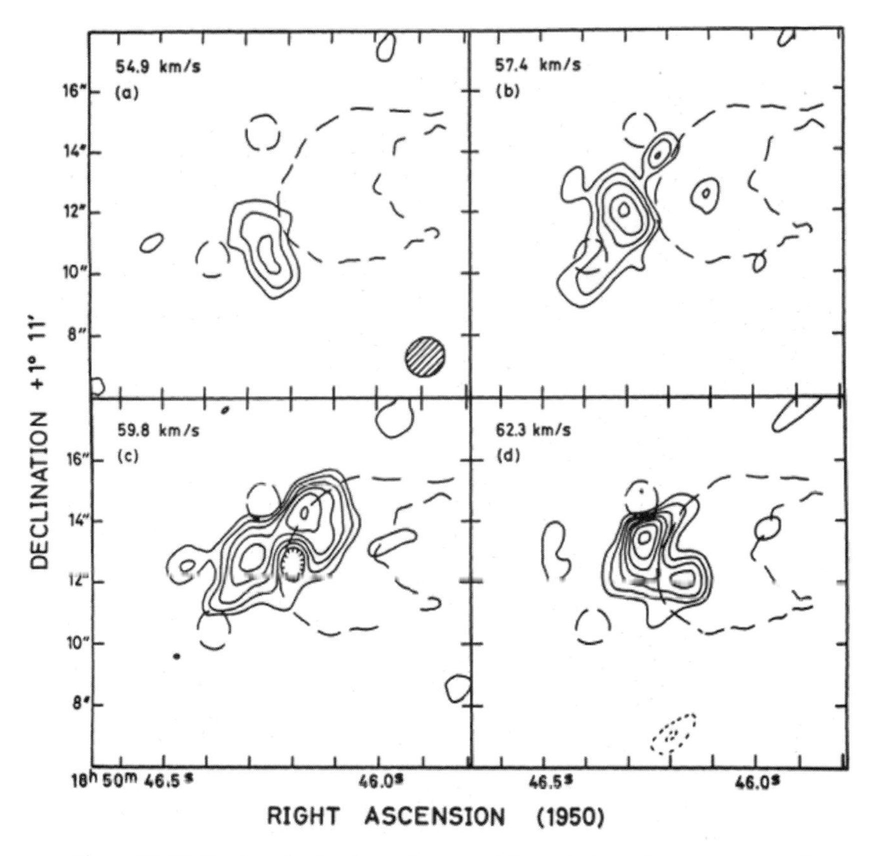

Figure 12.12 Single-velocity channel maps of the main hyperfine component of NH_3 (3,3) emission (Heaton et al., 1989).

12.6 Displaced CH_3CN

Observationally, peak NH_3 emission is offset slightly to the north of an axis of symmetry through the apex of the adjacent cometary UCHII region, 2″ east of the UCHII continuum peak and coincident with the 450 μm dust continuum. This is therefore taken as the centre of hot core activity. While the UCHC appears to be a single coherent clump, the emission peak in CH_3CN is observed in its turn to be offset from the NH_3 and dust continuum peak (Macdonald & Habing, 1994; Macdonald, Habing, & Millar, 1995). Using VLA and BIMA observations, the small but potentially significant displacement between peak NH_3 and peak CH_3CN emission ~1″ (0.015pc) correlates with the systemic velocity gradient in CH_3CN emission of 56 km s^{-1} pc^{-1} over an angular distance of 0.089 pc (Watt & Mundy, 1999; Wyrowski et al., 2012). The velocity

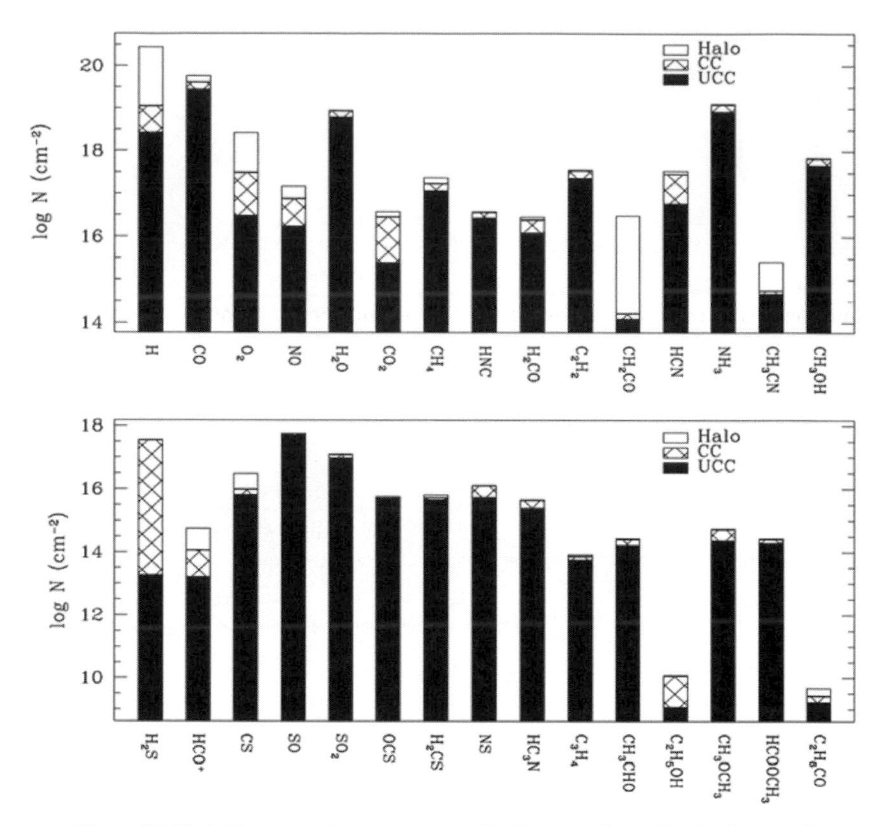

Figure 12.13 A histogram showing the contributions to column density from each of the three physical environments for a sample of species in the G34.26 UCHC model, at a core age of 3×10^3 yr and a halo age of 10^5 yr (Millar et al., 1997).

Table 12.1 *Physical parameters in the G34.26 model (Millar et al., 1997).*

Component	$n(H_2)$ (cm^{-3})	T (K)	Outer radius (pc)	$N(H_2)$ (cm^{-2})	Av (mag)
Ultracompact core	2×10^7	300	0.01	6×10^{23}	640
Compact core	10^6	$30r^{-0.4}$	0.1	2.7×10^{23}	288
Halo	10^4r^{-2}	$30r^{-0.4}$	3.5	2.9×10^{23}	310

difference for the two species is consistent with a flow of gas, increasing from 56 km s^{-1} in the south-west (nearest to the UCHII region) to 61 km s^{-1} in the north-east. The kinematics associated with G34.26+0.15 are obviously complex, but the velocity gradient fits with the directions and magnitudes of earlier measurements of molecular gas in G34.26 over a range of scales between 0.1 pc and 0.75 pc (Carral & Welch, 1992; Heaton et al., 1989, 1993; Akeson &

Carlstrom, 1996). Overall, the hot gas is showing a south-west to north-east red-shifted velocity gradient centred near the LSR velocity of 58 km s^{-1}.

The question arises as to whether the spatial separation between the NH_3 and CH_3CN peaks is indicative of a coherent bound core whose 'outer' and 'inner' materials are subject to different conditions (Watt & Mundy, 1999); or whether outer material is being swept away (Macdonald et al., 1995); or is it perhaps indicative of spatially separate clumps displaying distinctly different temperature and density profiles (Mookerjea et al., 2007). Although the exact contributions to the relative motion of the YSOs and the ambient cloud, including the UCHC, remain uncertain, there seems no doubt that the UCHC material is primarily moving both eastward and away from us along the line of sight. We might imagine a laminar flow of material parallel with the leading edge or 'nose' of the cometary ionisation front. Of the two other YSO sources (associated with HCHII-A and -B), were component B to be in front of a single coherent, or multicomponent, UCHC as viewed on the plane of the sky (i.e. closer to us on the line of sight), its location to the north-east of the NH_3 molecular core might provide a stellar wind likely to impact the far side of the UCHC and contribute to the red-shifted CH_3CN velocity gradient observed (Macdonald & Habing, 1994; Watt & Mundy, 1999). The CH_3CN peak downstream of the NH_3 peak and moving to the north-west would be moving away from us obliquely to the line of sight and around the 'bow shock' front of UCHII-C.

An early notion was that NH_3 as an established ice evaporate and potential precursor to CH_3CN through gas-phase reactions was being processed within the flow, a displacement between peaks of 0.04 pc equating to a dynamical time $\sim 4 \times 10^4$ yr for a relative velocity ~1 km s^{-1} (Macdonald et al., 1995). However, among the generally successful match of gas-phase modelling to the 330–360 GHz survey of the UCHC, CH_3CN stands out as almost uniquely poorly reproduced. This led to the suggestion that CH_3CN might most likely be a grain-surface product (Millar et al., 1997). However, in the G34.26 case, the observed displacement, if real, would be difficult to interpret if both NH_3 and CH_3CN are grain mantle species evaporated contemporaneously into the hot core gas. Sublimation temperatures are too close and mantle ice composition too complex, in spite of CH_3CN having the larger dipole moment, for significant distinctions on the observed scale to arise. We certainly could not anticipate a CH_3CN peak downstream from an NH_3 peak if both species are evaporates. Given the close proximity of UCHCs generally to newly formed OB stars, an impinging FUV field inevitably contributes to the chemical processing of gas-phase species at a core margin. Therefore, as an alternative to a predominantly grain-surface origin for CH_3CN in this case, perhaps photochemical processing of evaporated NH_3 and other precursors contribute to CH_3CN production within the gas phase at the margins of the hot core, a model which neither requires nor precludes the physical displacement apparent in G34.26. Simple calculations do indeed seem to satisfy the observational column densities (MacKay, 1999), but we will return to the issue in detail when we look at anomalous CH_3CN abundance in relation to a PDR closer to home in the Horsehead Nebula in Chapter 15.

12.7 Summary

The three UCHII regions identified at the G34.26 location and first detected over 30 years ago present the evidence for envelope infall, protostellar outflows, expanding ionised gas, and associated molecular hot core chemistry, now familiar in many HMSF cloud complexes. In the case of the prototypical 'cometary' UCHII region 'C' in G34.26, the interface between ionised hydrogen and hot molecular core material is well observed and a rich hot core chemistry has been both detected and modelled. The interesting question of CH_3CN formation is raised and will be reconsidered in the low-flux PDR conditions of the Horsehead Nebula in Orion described in Chapter 15.

PART V

Photodissociation

13
ATLASGAL PDRs

13.1 Introduction

As we have highlighted in the preceding HMSF chapters, where observed molecular emission indicates interaction between HII and hot core regions, we acknowledge that photodissociation (or photon-dominated) regions (PDRs) are those that typically lie between a fully ionised gas close to a high-mass star and the surrounding molecular cloud. The incident FUV stellar flux controls the degrees of partial ionisation, the various energy balances, plus the atomic and molecular composition of the evolving PDR gas and gas–grain interactions. From a theoretical point of view, any region sharing bright atomic fine-structure cooling lines in CII, CI, OI, or SiII, whatever the strength of impinging UV flux, will show PDR character, so this would include the diffuse and translucent atomic phases of the ISM. However, the more generally considered PDRs are precisely those associated with dense gas close to HII regions and subject to significant incident FUV radiation from young high-mass O- and B-stars (Tielens & Hollenbach, 1985; Tielens, 2006). While some of the photochemistry between the two density and flux extremes may be shared, the conditions associated with star formation are much more complex to unravel. While astronomers wait with much anticipation for the opportunities that the James Webb Space Telescope (JSWT) launched in late 2021 will bring (which will include the study of PDRs in the infrared spectrum both within and well beyond our own Galaxy), ground-based telescopes continue to reveal a wealth of new information.

The most detailed studies of PDRs have naturally concentrated on a few nearby regions, such as the Orion Bar and the Horsehead nebula, which we will consider in the two following chapters, each having a molecular cloud structure engendered by significantly differing FUV fluxes (Rodriguez-Franco et al., 1998; Teyssier et al., 2004; Pety et al., 2005; Gerin et al., 2009;

Cuadrado et al., 2015). Investigations have also begun on numerically large, robust surveys of sample PDRs that are not biased by initial conditions or evolutionary stage, such as the following.

13.2 ATLASGAL

The ATLASGAL survey already considered in Chapter 6 offers a catalogue of sources from which a 3 mm line survey of 409 selected dust clumps has provided valuable data on a range of PDR targets through selected molecular line emission, using the IRAM (Institut de Radioastronomie Millimétrique) 30 m telescope (Kim et al., 2020). Based on previous observations (e.g. Rizzo et al., 2005; Boger & Sternberg, 2005; Gerin et al., 2009), in one study eight molecules were adopted as typical PDR tracers: HCO^+, HOC^+, C_2H, c-C_3H_2, $H^{13}CN$, $H^{15}CN$, $HN^{13}C$, and CN, along with $C^{18}O$ and $H^{13}CO^+$ as two general probes of density in the dust clumps. The sample is divided first into two groups: the HII and non-HII, based on detections of mm-RRL (hydrogen radio recombination lines) and radio continuum emission, with the latter further split into IR-bright and IR-dark. Table 13.1 illustrates the classification in five of the 409 sources in the sample.

Of those clumps previously detected through RRLs at (sub)millimetre wavelengths and radio continuum, the former (IR bright) are associated with more energetic UV fields from predominantly O-stars, whereas the latter (IR dark) may have the weaker UV fields generated by early B-stars. Clumps with neither emission indicator are likely embedded massive young stellar objects (MYSOs) with

Table 13.1 *Dust temperature, distances, and classifications (Urquhart et al., 2018; Kim et al., 2020).*

ATLASGAL Identification	RA α(J2000)	Dec δ(J2000)	Distance (kpc)	T_{dust} (K)	Type	Classification
AGAL006.216-00.609	18:02:02.9	−23:53:13	3.0	14.6	non-HII	24 dark
AGAL008.049-00.242	18:04:35.2	−22:06:40	10.9	18.2	non-HII	IR bright or HII
AGAL008.671-00.356	18:06:19.0	−21:37:28	4.4	25.9	HII	IR bright or HII
AGAL008.684-00.367	18:06:23.0	−21:37:11	4.4	24.5	non-HII	24 dark
AGAL008.706-00.414	18:06:36.6	−21:37:16	4.4	11.9	non-HII	IR bright or HII

even weaker radiation fields. Those IR-dark clumps (without even an embedded MYSO) may be affected exclusively by cosmic ray (CR) impacts within their interior or additionally by the external radiation from nearby star formation. The ATLASGAL compact source catalogue therefore offers a wide range of UV field strengths and physical conditions. Of the trace molecular species chosen, they are not only linked to known UV photochemistry pathways but also suffer little from high extinction while offering velocity field data with which we can investigate kinematics and turbulence. One group of emissions identifies PDR surface layers (A_V ~2–5 mag and high UV exposure), including the reactive ions CO^+ and HOC^+ and the small hydrocarbons C_2H and c-C_3H_2. Another group identifies greater depths of PDR (A_V ~5–10 mag and low UV exposure) bordering cold, dense, neutral cloud, such as radicals HCO, C_2H, and CN.

To compare the relative abundances of the selected molecules between the different dust clump environments, column densities are estimated. Given the constraints in this survey of an optically thin single line per species, the necessary assumptions include that of LTE, a reasonable approximation because critical densities for the observed transitions are lower than those typical of HMSFR H_2 densities (>10^5 cm^{-3}). Also, the assumption is of gas and dust temperature equilibrium, which is likely given the high particle densities. The estimated column densities are beam-averaged values, and it is assumed that the medium is spatially homogenous and larger than the beam size. Table 13.2 gives some representative examples of sources and the derived column densities of the observed molecules, and Table 13.3 takes the average N (cm^{-2}) values for each species across the whole sample and converts those to abundance values relative to H_2 (X) and $C^{18}O$ (f).

Table 13.2 *Column densities (cm^{-2}) towards five sources in the ATLASGAL PDR survey (Kim et al., 2020).*

ATLASGAL Identification	$C^{18}O$ ($\times10^{16}$)	HCO ($\times10^{13}$)	$H^{13}CO$ ($\times10^{13}$)	C_2H ($\times10^{15}$)	c-C_3H_2 ($\times10^{13}$)	CN ($\times10^{15}$)	$HC^{15}N$ ($\times10^{13}$)	$H^{13}CN$ ($\times10^{13}$)	$HN^{13}C$ ($\times10^{14}$)
AGAL006.216-00.609	1.08	-	0.41	0.79	2.56	0.86	0.05	0.67	0.07
AGAL008.049-00.242	0.37	-	0.08	0.12	0.57	0.23	-	0.16	0.02
AGAL008.671-00.356	-	2.12	0.96	1.73	3.67	-	-	-	0.17
AGAL008.684-00.367	2.90	-	1.00	-	2.42	1.31	0.21	3.37	0.13
AGAL008.706-00.414	1.26	-	0.31	0.42	1.81	0.61	-	0.98	0.06

Table 13.3 *Average column densities (N), and fractional abundances relative to H_2 and $C^{18}O$ (X and f, respectively) for HII, IR-bright, and dark non-HII regions (Kim et al., 2020).*

Molecule	HII			IR bright non-HII			IR dark non-HII		
	N(cm^{-2})	X	f	N(cm^{-2})	X	f	N(cm^{-2})	X	f
$C^{18}O$	9.5(15)	1.2(−07)	-	6.5(15)	1.5(−07)	-	5.9(15)	1.4(−07)	-
HCO	1.4(13)	1.9(−10)	1.3(−03)	8.9(12)	1.6(−10)	1.1(−03)	9.0(12)	2.2(−10)	1.0(−03)
$H^{13}CO^+$	2.7(12)	3.6(−11)	2.6(−04)	1.5(12)	3.8(−11)	2.6(−04)	1.2(12)	3.0(−11)	2.1(−04)
C_2H	5.8(14)	6.4(−09)	5.3(−02)	3.5(14)	6.4(−09)	5.3(−02)	3.14(14)	6.0(−09)	5.4(−02)
c-C_3H_2	1.2(13)	1.7(−10)	1.1(−03)	7.0(12)	1.6(−10)	1.1(−03)	6.4(12)	1.6(−10)	1.1(−03)
CN	5.2(14)	6.7(−09)	5.4(−02)	3.2(14)	7.2(−09)	5.0(−02)	2.7(14)	6.4(−09)	4.8(−02)
$HC^{15}N$	9.8(11)	1.3(−11)	1.1(−04)	8.7(11)	1.1(−11)	1.0(−04)	5.6(11)	1.2(−11)	8.2(−05)
$H^{13}CN$	1.5(13)	1.1(−10)	1.1(−03)	1.6(13)	9.6(−11)	1.7(−03)	4.3(12)	6.0(−11)	9.3(−04)
$HN^{13}C$	3.0(12)	3.9(−11)	2.8(−04)	2.0(12)	4.6(−11)	3.3(−04)	1.8(12)	4.7(−11)	3.5(−04)

While the relative abundance values correlate with clump density and can be matched to computational modelling of the chemistry, column density ratios between related species, as we have frequently seen in earlier chapters, are often a more direct way to diagnose the chemical activity. The specific benefit is that these ratios are insensitive to beam filling factors or exact H_2 column density or variables in physical conditions, none of which are necessarily known. Let us consider two particular ratios in the ATLASGAL PDR survey.

13.3 Formation and Destruction of HCO/HCO$^+$

While HCO and $H^{13}CO^+$ abundances in cold gas show little variation, in PDR conditions their abundances are much influenced by photoinduced processes (Gerin et al., 2009; Goicoechea et al., 2009). While the primary destruction route for $H^{13}CO^+$ in PDRs is a fast dissociative recombination with electrons to give ^{13}CO and H, various photon-dependent pathways are open for its HCO precursor formation, particularly two gas-phase and one photodissociation-initiated route. The first gas-phase sequence is favoured in FUV-shielded conditions and is simply an electron exchange with a heavier element, a metal such as Mg or Fe (Schenewerk et al., 1988; Gerin et al., 2009). More likely perhaps is the reaction of atomic oxygen with carbon radicals, such as CH_2 (Watt, 1983; Leung et al., 1984; Schenewerk et al., 1988). One photon-dependent route is the gas–grain reaction through FUV photodissociation of formaldehyde: H_2CO (Schilke et al., 2001). Equally, the FUV-induced ice-mantle photodesorption of HCO formed on grains is likely (Willacy & Williams, 1993; Bergin et al., 1995; Gerin et al., 2009). Thermal desorption of grain-surface HCO is possible above ~30 K, so warm material close to HII regions would be susceptible. However, the high observed abundance seen towards PDR regions strongly suggests the greater likelihood of direct photodesorption of HCO, or photodissociation of H_2CO, as the principal formation routes.

Figure 13.1 shows clearly how the HCO/HCO$^+$ column density ratio decreases as H_2 density increases. In other words, the production of HCO depends on the photon penetration of PDR conditions, and the inclusion of the upper star symbol in the diagram representing the same ratio in the precisely constrained PDR peak of the nearby Horsehead Nebula (Chapter 15) confirms that expectation. The lower star (Horsehead DCO$^+$ peak) identifies the cold, dense gas hidden from the nebula's PDR margin. Figure 13.1 also shows clumps symbolised in two sizes; for example, the HII values in large and small circles. The larger indicate sources having HCO relative abundance $>10^{-10}$, which has previously been taken as indicative of FUV production

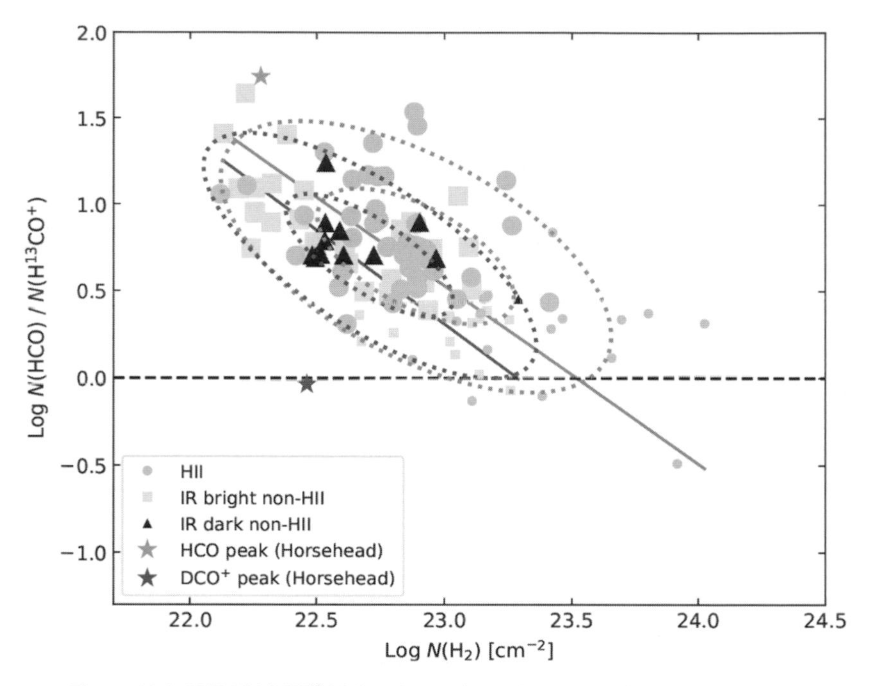

Figure 13.1 N(HCO)/N($H^{13}CO^+$) ratios against N(H_2) across the sampled sources. Of the two Horsehead stars, the upper star is the PDR peak, and the lower star is the cold, dense peak (see Chapter 18 for details). The dotted line ellipses (1σ and 2σ) represent HII and non-HII distributions (Kim et al., 2020).

(Gerin et al., 2009). The solid diagonal defines N(HCO)/N($H^{13}CO^+$) = 1 which, together with the >10^{-10} abundance, have been taken as defining parameters in modelling FUV photochemistry against observations towards both the Horsehead nebula and Mon R2 (Gerin et al., 2009; Ginard et al., 2012).

The same column density ratios set against dust temperature in Figure 13.2 seems to indicate clearly that thermal desorption is not the route to HCO production in any of the sources, including the PDRs, since its production is not enhanced by higher dust temperature. However, there is high HCO abundance evident from Table 13.3 in both HII and non-HII sources, so we cannot rule out grain thermal desorption processes from consideration, particularly with current low-angular-resolution detections averaging emissions and temperatures over large clumps with single-dish beams. Greater resolutions will be necessary to separate HCO emission from PDRs and cold gas regions (as will be evident in the Horsehead chapter), and observations of the COMs molecules such as H_2CO, CH_3O, and CH_3OH that previous chapters have identified as readily formed on grain mantles through hydrogenation of CO-ices will be necessary.

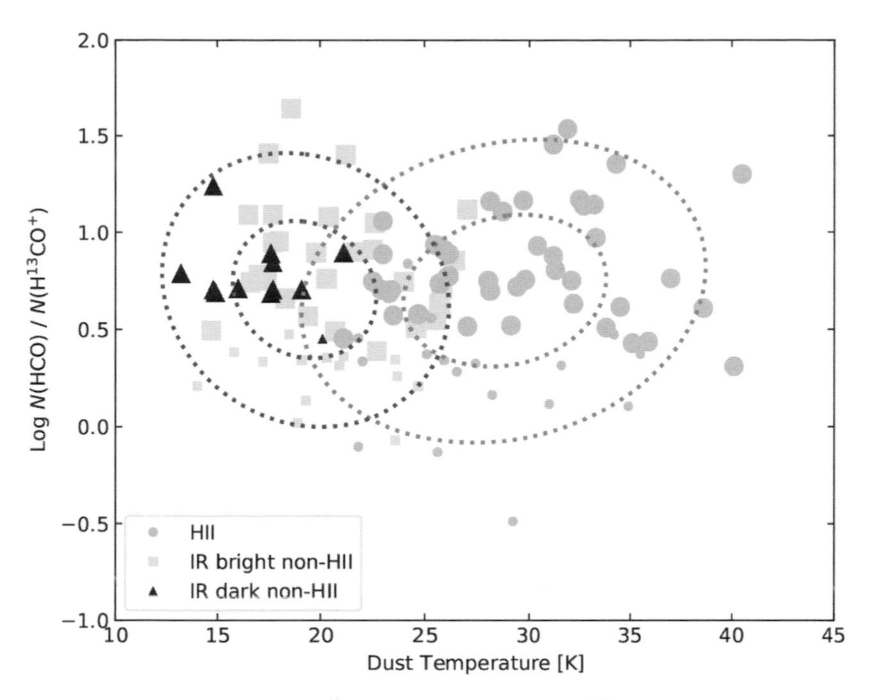

Figure 13.2 $N(\mathrm{HCO})/N(\mathrm{H^{13}CO^+})$ ratios against T_{dust} (K) across the sampled sources. The lack of column density ratio enhancement with dust temperature suggests thermal desorption is not a factor (Kim et al., 2020).

13.4 Small Hydrocarbons in PDRs

Small hydrocarbon formation in PDRs is not well understood, but both gas-phase and grain-surface reactions have been proposed. In strongly UV-illuminated PDRs with high gas temperatures, the formation of small hydrocarbons will dominate the chemistry within the outer atomic layer, where ionised carbon and electrons are abundant (Cuadrado et al., 2015). The initiating step will almost certainly be $C^+ + H_2 \rightarrow CH^+ + H$, with hot gas at several hundred degrees or FUV-pumped, vibrationally excited H_2 supplying the required activation energy (Black & van Dishoeck, 1987; Agundez & Wakelam, 2013). The rapid (zero barrier) reactions of CH^+ with H_2 generate CH_2^+ and CH_3^+ that are neutralised through electron recombination and then go on to build carbon-chain molecules through the ion-neutral reaction with carbon ions; for example, $CH + C^+ \rightarrow C_2^+ + H$, as illustrated in the schematic of Figure 13.3. We will look again at this scenario in more detail in Chapter 14 on the Orion Bar.

Although these small hydrocarbons are well-known PDR tracers, their abundances do vary from one source to another (e.g. Pety et al., 2005;

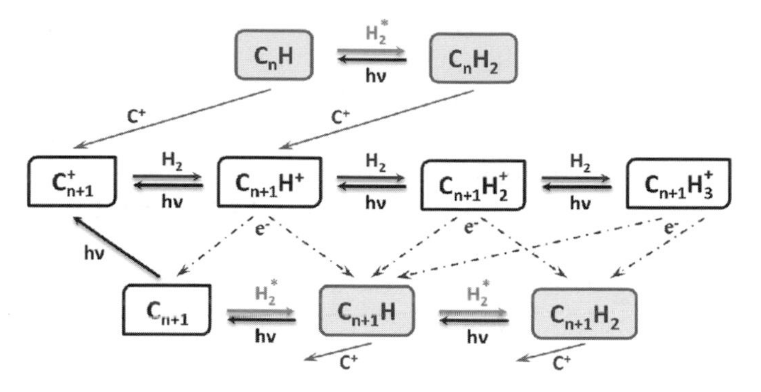

Figure 13.3 Schematic for the formation of small-carbon-chain molecules in high UV conditions (Cuadrado et al., 2015).

Mookerjea et al., 2012; Pilleri et al., 2013; Cuadrado et al., 2015; Tiwari et al., 2019). The ATLASGAL PDR survey offers an even more generalised insight into their distribution that equally finds small hydrocarbons to be ubiquitous and with varying abundances, but not immediately correlating with exclusively PDR processes. However, the IRAM 30 m beam size ~30″ (0.73 pc at a median distance of 5 kpc for the sources) is targeting highly illuminated or dense PDRs surrounding young HII regions that are spatially both radially small and thin shell–like in depth (~0.001 pc). The contribution of small hydrocarbons in the cold lower-density molecular envelope surrounding such PDRs and HII regions may well dominate the observed emission. In such circumstances, relative abundances become less useful in identifying specific PDR conditions than a selected column density ratio between closely correlated species.

For example, C_3H_2, whether as its linear or cyclic isomer, is undoubtedly photodissociated in a strong UV radiation field, as is expected to be the general case for all small hydrocarbons. However, C_3H_2 is observed in PDRs such as Mon R2 (Pilleri et al., 2013), even though gas-phase modelling fails to reproduce the measured abundances (e.g. Tiwari et al., 2019). Its presence cannot be explained, therefore, with gas-phase reactions only, and we must also look to grain-surface processes. For example, the photodissociation of small PAHs (polycyclic aromatic hydrocarbons) or VSGs (very small grains) is plausible in PDR conditions (Fuente et al., 2003; Pety et al., 2005), and laboratory experiments have indeed identified small-hydrocarbon production from small PAHs (e.g. Useli-Bacchitta & Joblin, 2010). The production of $C_3H_3^+$ (Figure 16.3) has also been shown to be enhanced by PAH-related photochemistry in which fragmentation or the ejection of electrons feeds the dissociative recombination to c-C_3H_2 (Mookerjea et al., 2012; Pilleri et al., 2013).

Taking the column density ratio between c-C_3H_2 and C_2H as a function simply of cloud density in the ATLASGAL PDR survey, shown in Figure 13.4, indicates no particular correlation. Therefore, we might reasonably assume that the presence of UV radiation and C^+ ions is probably a necessary additional condition for the formation of both C_2H and c-C_3H_2. Yet, the $N(HCO)/N(H^{13}CO^+)$ ratio being >1, together with a >10^{-10} HCO abundance criterion previously identified as indicative of FUV conditions (and symbolised with the larger circles in the figure), seems also to offer no clear correlation. What are we to make of this anomaly?

In lower-density envelopes (<10^5 cm^{-3}) with lower kinetic temperatures (<35 K) compared with the inner regions of prestellar cores, many species are locked in the icy mantles of dust grains, only to be sublimated during the core collapse. Figure 13.5 shows the c-C_3H_2/C_2H ratio against dust temperature, and the ratio does remain fairly constant over the dust temperature range of zero to 40 K (albeit with a significant scatter), which is what we should expect if these species are largely locked in ice mantles at these temperatures (Viti et al., 2004; Pilleri et al., 2013).

The dust temperature increase for those clumps identified with HII locations is, of course, evident in Figure 13.5 (the abundance of greyscale circles on the right-hand side), and if we focus on the distinctions between envelope gas and UV-illuminated gas, expressed this time as bolometric luminosity over mass, Figure 13.6 shows that relationship. The vertical scatter of column density

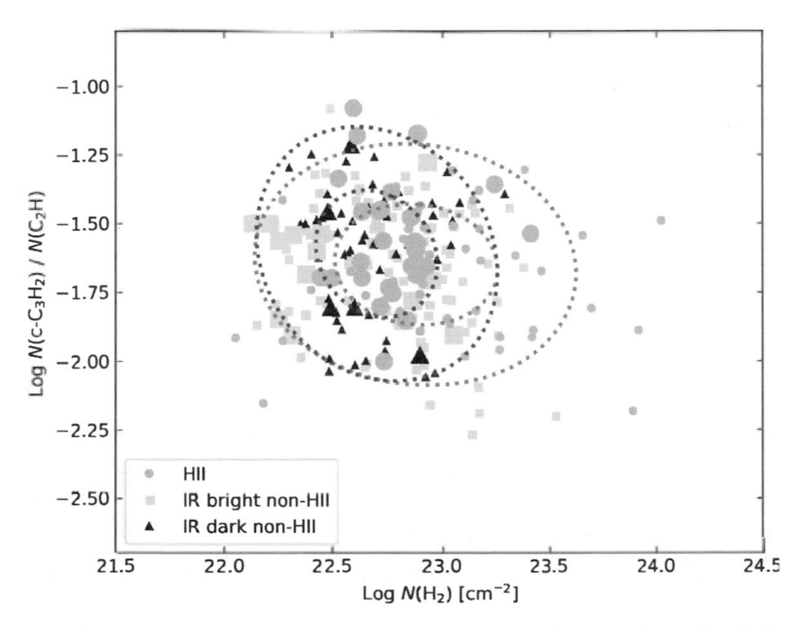

Figure 13.4 N(c-C_3H_2)/$N(C_2H)$ ratios as a function of $N(H_2)$ (Kim et al., 2020).

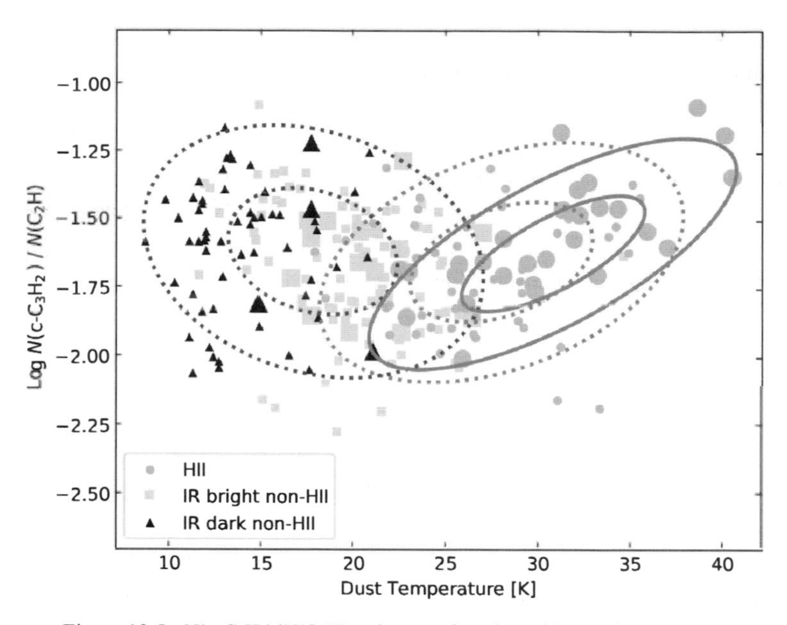

Figure 13.5 N(c-C$_3$H$_2$)/N(C$_2$H) ratios as a function of T_{dust} (Kim et al., 2020).

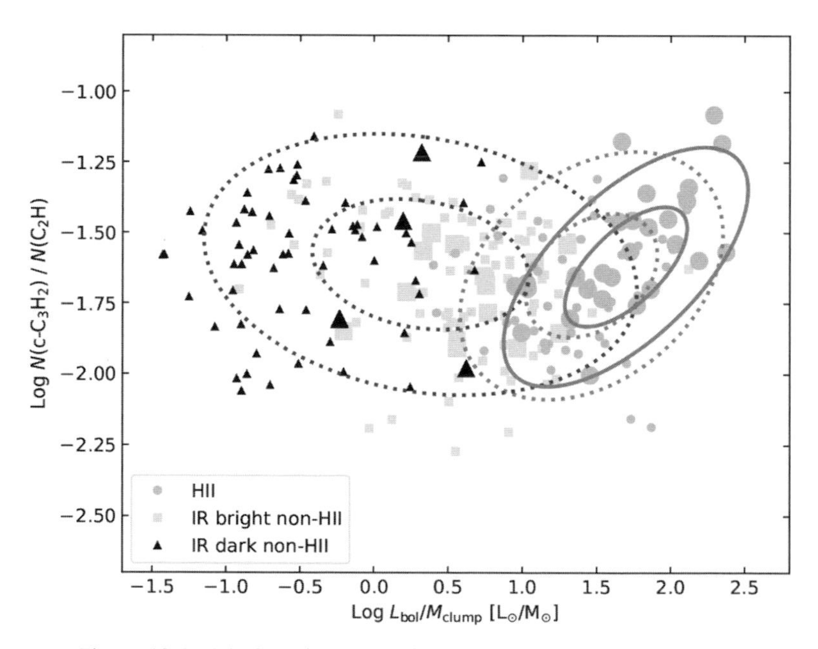

Figure 13.6 N(c-C$_3$H$_2$)/N(C$_2$H) ratios as a function of L_{bol}/M_{clump} (Kim et al., 2020).

ratios is again little changed, although slightly higher ratios are suggested towards the more evolved clumps (larger circle symbols).

We note that C^+ is known to exist in moderate amounts quite deep within molecular clouds so cannot be excluded from small-hydrocarbon formation in such regions along our line of sight (e.g. Perez-Beaupuits et al., 2012). With PAH dissociation in high-UV flux regions enhanced and contributing to small-hydrocarbon production, an A_V value >3.5 mag has been taken as the point at which dissociation outweighs gas-phase reactions in C_2H_2 production, one of the precursors to C_2H and c-C_3H_2 (Murga et al., 2020). In low-flux conditions, pure gas-phase pathways dominate, as the Horsehead case in Chapter 15 will make clear.

13.5 Summary

In summary, the ATLASGAL PDR survey confirms the high detection rates of the chosen PDR tracers towards HII sources. While previous chemical modelling of specific sources shows that, in a cold lower-density envelope, the abundances of C_2H and c-C_3H_2 vary little, subsequently, during cloud collapse (with density increase, temperature rise, and the emergence of HII regions), from 10^5 yr on in the models the column density ratio increases steeply. However, without knowing initial conditions, one cannot predict these time-dependent outcomes in reality. As for the observed abundances of some high column density tracers ($H^{13}CO^+$ and $HC^{15}N$) in the survey, they are almost constant over the range of H_2 column densities, while others (HCO, CN, C_2H, and c-C_3H_2) fall as H_2 increases. The HCO detections are confirmed as arising from clumps likely associated with PDRs, and higher HCO abundances are undoubtedly linked in the models to ongoing FUV chemistry. That said, the actual values derived from observation will definitely benefit from higher-resolution confirmations in the future. Let us look in some detail at two PDR sources in the following two chapters: the Orion Bar subject to a high FUV radiation flux, and the Horsehead nebula subject to a much lower field.

14

The Orion Bar in M42

14.1 Introduction

PDRs undoubtedly emit most of the IR radiation from the ISM of star-forming galaxies. They do so in mid-IR bands from PAHs, in IR from rovibrational lines of H_2, and in the fine-structure lines of CII (158 μm) and OI (63 μm). Having absorbed FUV photons, grains are also heated and re-emit in the mid- and far-IR (the warm dust continuum). At the same time photo-electrons are ejected from grains and PAHs that heat the gas. As a result, PDRs are seen to be stratified according to gas and dust column density that attenuates the FUV photon flux, identified in transition zones between atomic and molecular dominance, most notably $H^+/H/H_2$ and $C^+/C/CO$. At the deepest PDR cloud depths, thc flux of stellar FUV photons is almost completely attenuated, and gas-phase molecules and atoms freeze onto the grains. A much slower chemistry ensues, and weak ionisation is driven by cosmic-ray particles. While PDR observational diagnostics of density, temperature, and FUV field (through the CII 158 μm lines, PAH bands, H_2 lines, and far-IR continuum emission) can be used to estimate Galactic star-formation rates generally, more locally, PDR diagnostics can be used to determine the radiative feedback of massive stars on their natal clouds. Understanding how molecular clouds are evaporated by a strong FUV radiation field is important in determining lifetimes and how the star-formation process can be quenched in a given region. On the other hand, the dynamical effects induced by a strong stellar FUV field such as radiation pressure, along with the winds of massive stars and/or supernova explosions, will both trigger and regulate the formation of new generations of stars.

14.2 The Orion Bar

The ATLASGAL PDR survey supports the view that many compact HII regions are likely to be pressure-confined by surrounding high-density neutral gas. The classic example of just such a constrained dense PDR illuminated by a high UV flux that has been accessible for study over many years is the Orion Bar, oriented edge-on to our line of sight, 3° south of the Nebula in the Orion Complex. This is a bright rim margin in M42 at the interface between an expanding HII bubble around the Trapezium star cluster and the Orion A giant molecular cloud to the south-east (Figures 14.1 and 14.2). Multi-wavelength observations towards different positions in the Orion Bar have long been used in the development of PDR modelling and today are still used as a template to

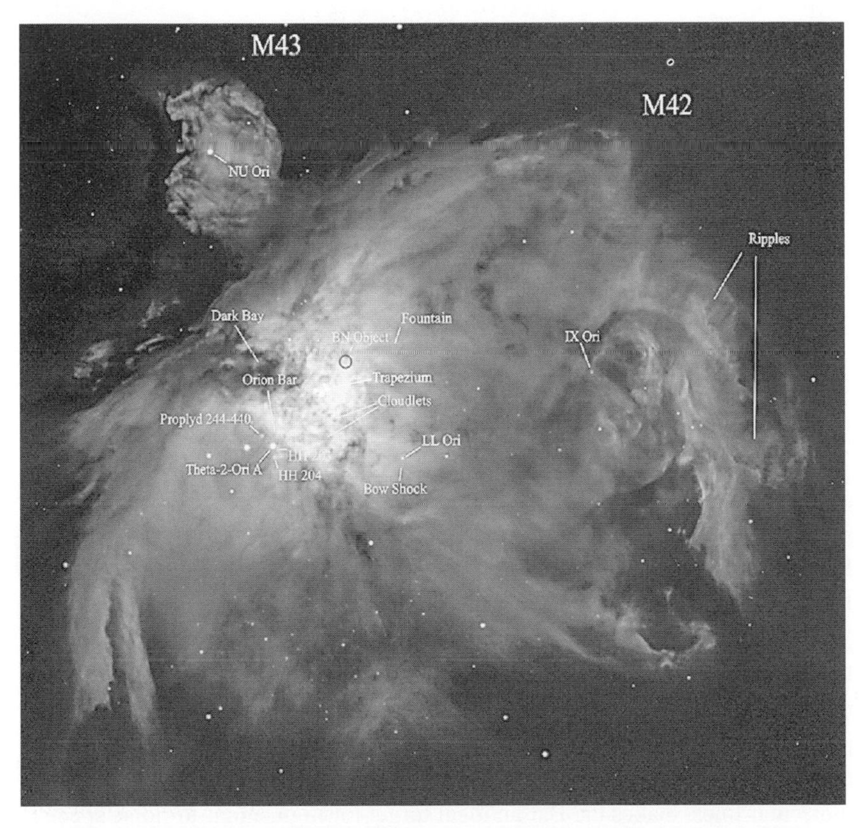

Figure 14.1 The Orion nebula (M42) in a narrow-band composite image (Hα, OIII, and SII) using the Australian National Observatory Siding-Spring remote telescope (T32). The bright Bar lies diagonally north-east to south-west close to the centre, just beneath the Trapezium cluster (Herbert Storzer, http://orion2nebula.net).

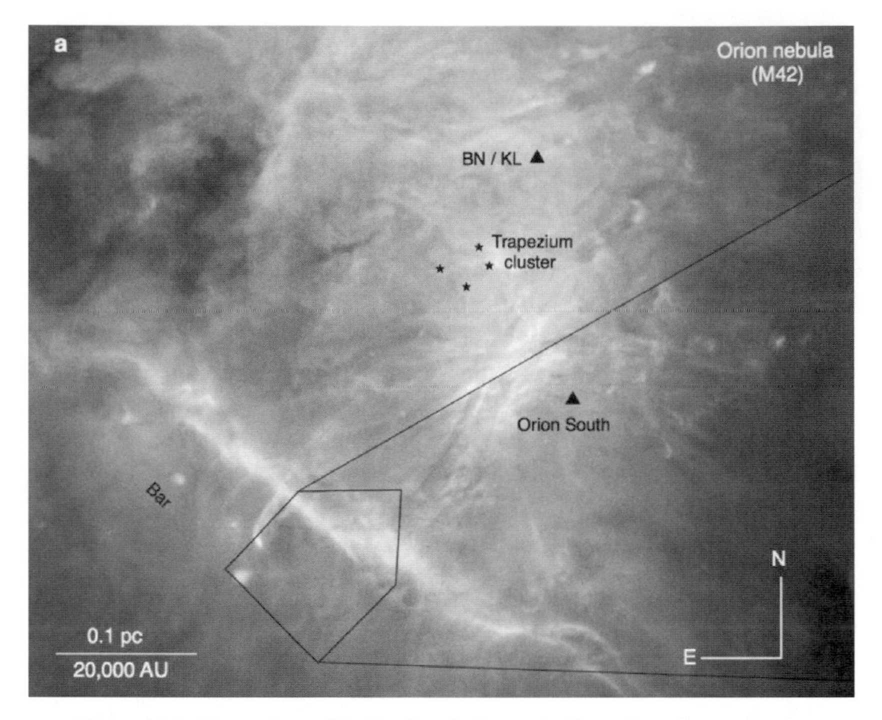

Figure 14.2 Closer view of the Bar in relation to the Trapezium cluster showing the distance scale in the plane of the sky (Goiceochea et al., 2016; VLT/MUSE).

understand unresolved emission from a wide variety of Galactic and extragalactic sources.

Figure 14.3a is a 1995 reworking of an early ionised hydrogen map, simple and clear in its delineation of a slice through the central M42 HII region in the plane of the sky. The reason that the Orion Bar emits so brightly is that we are looking here at an edge-on part of the HII cavity. Figure 14.3b then shows a Hubble telescope composite of a section of the same region in the optical.

With its convenient side-on orientation along the line of sight from Earth, the Bar face is strongly irradiated with a UV flux ~2-4 × 10^4 G_0 (G_0 being the mean interstellar FUV field in Habing units, equivalent to a few 10^8 photons cm^{-2} between 6 eV and 13.6 eV). Being close to the Solar system (just over 400 pc) and having high gas temperatures (T_k ~150–300 K), hence bright molecular lines, makes the Bar an ideal target for high signal-to-noise spectral imaging at different wavelengths. Detailed observations have been made with several different telescopes in recent years: for example, a line survey with the IRAM 30 m telescope of the Bar edge in 3 to 0.8 mm bands, complemented by 2′ × 2′ maps with 7″ resolution (Cuadrado et al., 2015, 2016, 2017). The

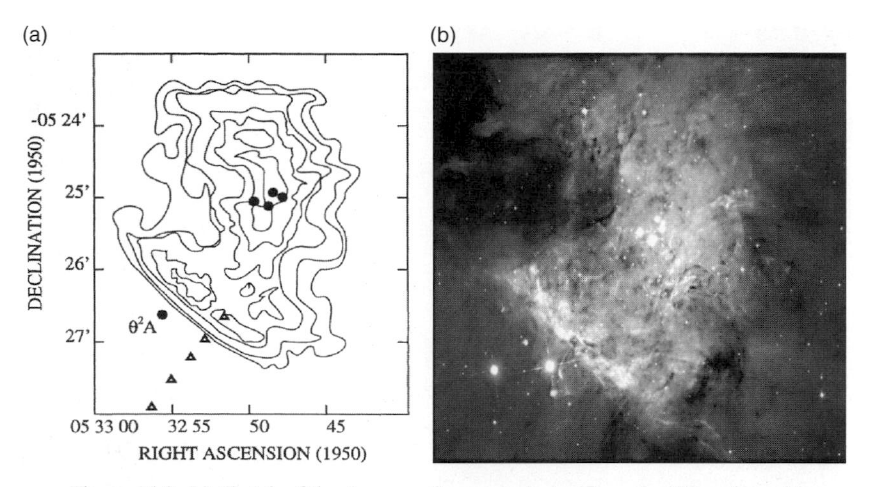

Figure 14.3 (a) Sketch of the 6 cm continuum emission from the HII region with the central Trapezium stars and close linear contours of the Bar to the south-east (sketch from Hogerheijde et al., 1995, based on Johnston et al., 1983); (b) Hubble optical image of the Bar reflecting light from the Trapezium cluster (NASA/ESA/Hubble ST/M.Robberto).

harshly irradiated leading-edge environment was found to have a richer chemistry than previously thought, with many reactive ions, such as SH^+, HOC^+, and C_3H^+, as well as unstable organic isomers such as cis-HCOOH not seen much elsewhere. Herschel/HIFI velocity structure mapping of the larger-scale gas environment (7.5′ × 11.5′) has used tracers such as CII line emission. Early ALMA mosaics of a small field scale (50″ × 50″) in several emission lines at 1″ resolution are shown in Figure 14.4.

While the strong ultraviolet field drives a hot ionised HII blister into the ambient cloud, the ionised gas is also observed to flow away from the Bar cloud rim surface at about 10 km s^{-1}, so there is both shock compression and ablation at this interface. Figure 14.5 sketches the main elements of a classic strongly FUV irradiated cloud edge which we can compare with the Orion Bar observations. The key physical elements are the ionisation and dissociation fronts that delineate the depths of photon impact on gas and dust in the cloud. The three panels of Figures 14.6 and 14.7 then show a map derived from HCO^+ emission and a gas temperature map derived from CO emission, together with a line intensity and line peak temperature plot of each. Note that the horizontal FUV flux impacts from the left in Figure 14.5 and 14.6, but is reversed, impinging from the right, in Figure 14.7. The separation of the two fronts in the Orion Bar averages about 15″ (~6,000 AU), sufficient to enable observers to identify compositional features of adjacent locations in considerable detail.

Figure 14.4 A combined VLT/ALMA image of ionised and molecular gas emission in the Orion Bar (Goiceochea et al., 2016).

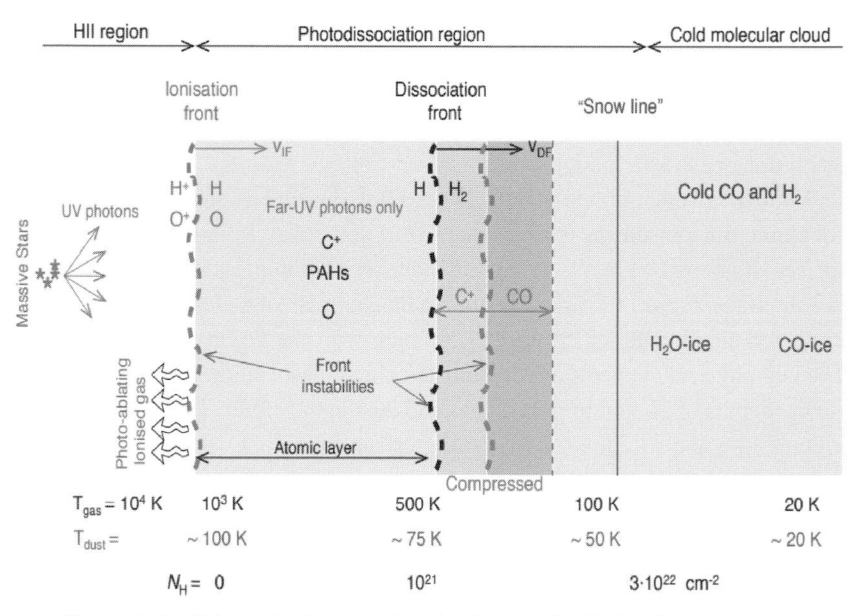

Figure 14.5 Schematic diagram of a classic strongly UV-irradiated molecular cloud margin (Goiceochea et al., 2016).

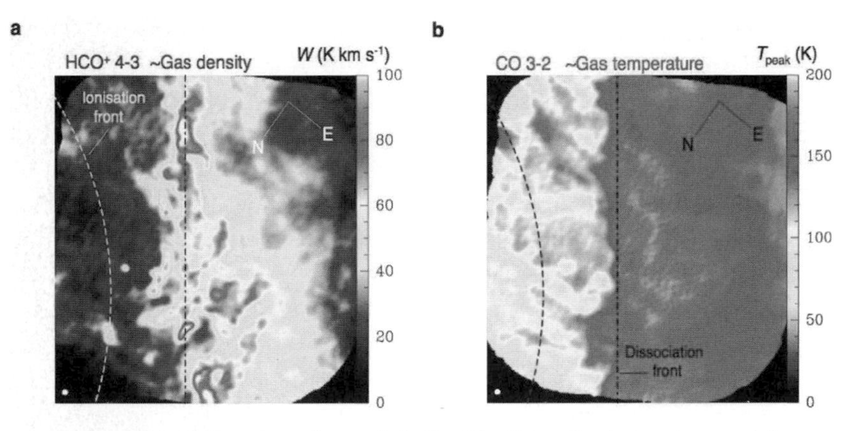

Figure 14.6 ALMA view of a small field of the Orion Bar in two tracers of the molecular gas: (a) HCO+ J=4-3 and (b) CO J=3-2. The flux of FUV photons decreases from left to right; the dotted arc represents the ionisation front, and the dotted vertical represents the dissociation front (Goiceochea et al., 2017).

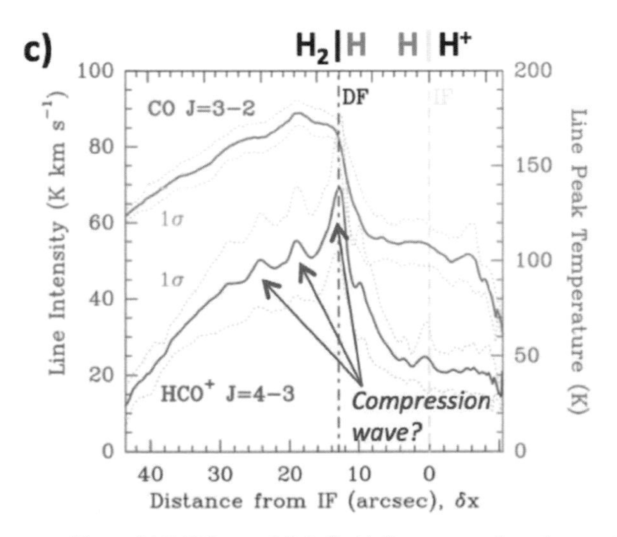

Figure 14.7 This panel (labelled 'c') corresponds to the previous figure's (a) HCO^+ and (b) CO maps, here showing vertically averaged intensity cuts perpendicular to the Orion Bar, distances from the IF right-to-left (Goiceochea et al., 2017).

14.3 PDR Ionisation

At the Bar surface where the HII region meets the initially neutral cloud, only FUV radiation penetrates, capable of dissociating molecular hydrogen but unable to ionise the resulting hydrogen atoms. The flux is, however, capable of ionising some other elements such as carbon (~11 eV) and silicon (~8 eV) atoms. The gas temperature drops from about 10^4 K to 10^3 K in crossing from the HII bubble into this atomised and partially ionised region and this is designated the outer PDR region where the impact of FUV flux on gas phase species gradually reduces with depth due to dust extinction and H_2 line absorption. Gas temperatures continue to drop with depth from that initial 10^3 K at the surface to about 100 K with corresponding dust surface temperatures dropping from 100 K to ~75 K (Hollenbach & Tielens, 1999).

The three gradients of FUV flux, temperature, and density engender a complex layered structure with different chemical mixes from cloud edge to the interior. Figure 14.8 shows firstly where the ionisation front can be identified through a neutral atomic emission line (oxygen's first ionisation energy is virtually the same as that of hydrogen). Although not shown, this is clearly distinguishable from emission in the visible from ions such as sulphur (ionisation energy ~10 eV) that characterise the HII region. Both the OI and SII emission arise from collisional excitation with electrons so at the ionisation front the transition boundary between H^+ and neutral H, where the electron abundance drops 10,000-fold, those emissions abruptly cease. Within the outer PDR, emission is also bright in mid-IR polycyclic aromatic hydrocarbons (PAHs) from which, along with those from small grains, energetic electrons are photo-ejected. These constitute the main source of gas heating competing with the cooling by far-IR, oxygen and C^+ fine structure lines. The majority of low-energy electrons are the product of carbon atom ionisation. The observed total particle density in the very outer PDR of the Orion Bar is ~5×10^4 cm^{-3}.

Before looking at the dissociation front and the remaining PDR compressed behind it, we can summarise the consensus view of PDR evolution in general. When an HII bubble initially collides with a dense neutral cloud, the ionisation

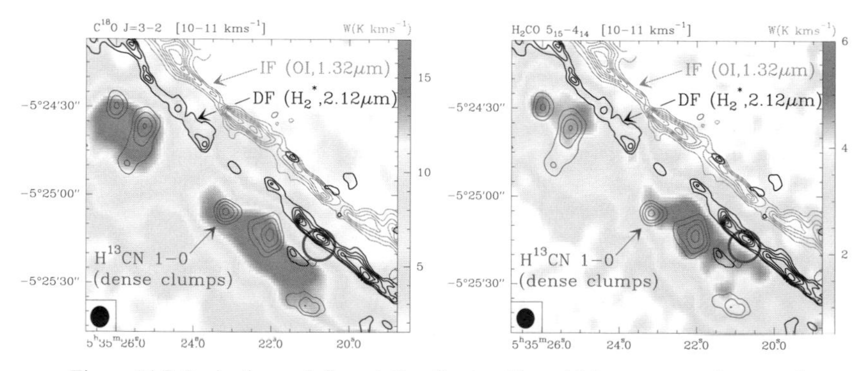

Figure 14.8 Ionisation and dissociation fronts with multiple contour and greyscale emission as discussed in the text (Cuadrado et al., 2017).

and dissociation fronts are deemed to be cospatial (Spitzer, 1978; Weilbacher et al., 2015). In less than a thousand years for the typical case, the expansion slows and the dissociation front gets ahead of the ionisation front within the interior of the cloud. A compressive wave emerges just ahead of the ionisation front and the density of the neutral gas ahead of it increases relative to the ionised gas behind it (Draine, 2011). For an ionisation front travelling at ~1 km s^{-1}, the Orion Bar separation of 15″ suggests a timescale of 25,000 to 50,000 years. In time, the compressing wave slowly enters deeper into the interior of the molecular cloud beyond the dissociation front.

One signature of FUV-irradiated gas is the ionisation fraction, defined as the abundance of electrons with respect to hydrogen nuclei ($x_e = n_e/n_H$) being higher than ~10^{-6} (Cuadrado et al., 2019). Cold molecular cores shielded from external FUV radiation show much lower ionisation fractions ($< 10^{-8}$), since the electron abundance is driven by a gentler flux of cosmic-ray particles rather than penetrating FUV photons (Guelin et al., 1982; Caselli et al., 1998; Maret & Bergin, 2007; Goiceochea et al., 2009). Not only does the electron density (n_e) control the rate of ion-neutral reactions (the dominant formation processes for most ISM gas-phase species), but ionisation fraction determines the coupling of matter with magnetic fields and, if high enough, influences the extent of inelastic collisions with high-dipole molecules providing additional rotational excitation. In these high x_e conditions, the observed molecular line emission is no longer determined by the most abundant collisional partner, H_2, so knowing n_e becomes essential in estimating gas density.

Carbon recombination lines (CRLs), in which a free electron recombines with carbon ions and cascades down from Rydberg electronic states to the ground state while emitting photons, are expected to arise from neutral gas close to the C^+/C/CO transition layer (e.g. Natta et al. 1994) and not from the hot ionised gas in the adjacent HII region (where electron temperature T_e ~10^4 K). CRLs are optically thin, with an intensity proportional to $n_e^2 T_e^{-2.5}$. Towards the Orion Bar, millimetre CRLs have been observed in the 85–115 GHz range, enabling derivation of n_e, T_e, and gas thermal pressure values (Cuadrado et al. 2019). Figure 14.9 shows the locations of those millimetre CRL detections towards the Bar. The molecular cloud in CO emission is shown in greyscale, the H_2 dissociation front in dotted contours, and ^{13}CII line emission in continuous contours. C41α and C39α CRLs observed with the IRAM 30 m telescope towards three positions within the PDR are circled and plotted.

From the observations, the inferred results put n_e at 60–100 cm^{-3} at the H/H_2 dissociation front. In most time-dependent models, the strong stellar UV field heats, compresses, and gradually evaporates the molecular cloud edge if the pressure of the surrounding HII medium is not significantly high. From

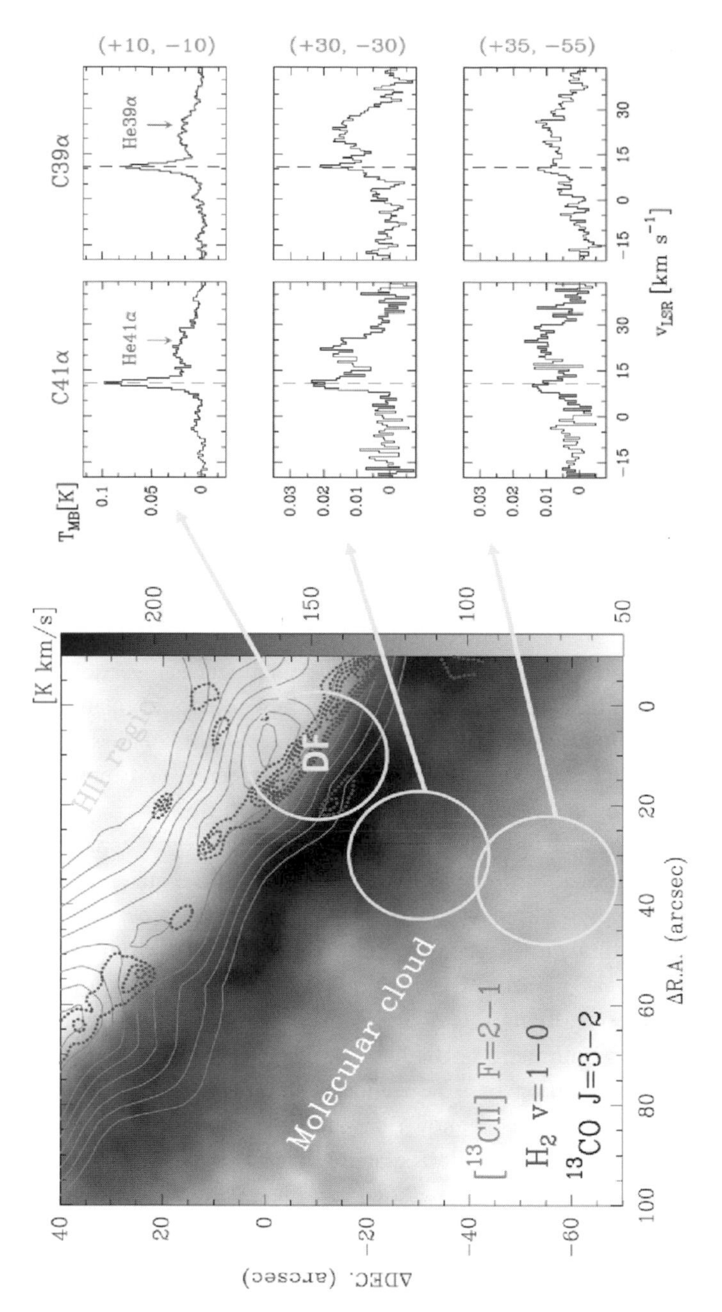

Figure 14.9 Locations of mm CRL detections towards the Bar: CO emission in greyscale, H_2 dissociation front (DF) in dotted contours, and ^{13}CII line emission in continuous contours. C41a and C39a detections circled and plotted (Cuadrado et al., 2019).

the derived n_e in the Orion Bar, the thermal pressure towards the dissociation front does indeed appear higher than that of the ionised gas at the ionisation front. Unfortunately, the IRAM single-dish observations (~25″ resolution) do not offer spatial distinctions between the ^{13}CII and mm CRLs emitting layers. An A_v ~1, roughly delineating the width of the H/H_2 transition layer, implies a 1.6″ and 3.2″ scale for densities 10^5 cm^{-3} and 10^6 cm^{-3}, respectively. A number of earlier observations have suggested a spatial coincidence between CI and CII with CO (Tauber et al., 1995; Wyrowski et al., 1997), observations at odds with the theoretical expectation (e.g. Tielens et al., 1993) and assumed to be a limitation of single-dish constraints. However, more recent ALMA images ~1″ resolution also show no appreciable offset between H_2^* (vibrational) emission and the edge of the HCO^+ and CO emission (Goiceochea et al., 2016). These observations do suggest that the classic C^+/C/CO sandwich structure of a PDR may not be discernible, perhaps not even exist in that form, in which case there would be no layer in the Bar where neutral atomic carbon is the most abundant carbon reservoir.

14.4 Molecular Dissociation

The main focus of PDR studies for the past 30 years, and the Orion Bar has been the classic example for study, has been understanding the relationship between carbon and hydrogen chemistries, specifically those transitions between C^+/C/CO in relation to the atom–molecule H/H_2 transition (e.g. van Dishoeck & Black, 1988; Hollenbach & Tielens, 1999; Papdopoulos et al., 2004; Rollig et al., 2007; Glover et al., 2010; Glover & Mac Low, 2011; Offner et al., 2013; Papdopoulos et al., 2018). In the standard PDR description, it is the FUV optical and IR photons that regulate the thermal state of the gas and its associated chemistry at low column densities, while CR penetration determines higher column density conditions (e.g. Strong et al., 2007; Grenier et al., 2015).

Two typical measures of PDR depth beyond which molecular photodissociation will not occur give closely comparable results. Both identify the point at which CR ionisation replaces FUV irradiation as the dominant process. The first is equated with the size of the region where oxygen chemistry is dominated by FUV photons. This is estimated from molecular O_2 formation and photodestruction rates ratios, a location that standard texts (e.g. Tielens, 2021) derive as having radius and visual extinction values estimated as

$$R_o \sim 0.36 \left(10^4\,\text{cm}^{-3} / n_o\right) \text{pc, and}$$
$$Av \sim 5.9 + 1/1.8 \ln\left[\left(G_o / 10^4\right) \left(10^4\,\text{cm}^{-3} / n_o\right)\right],$$

where n_o is the particle density (H nuclei) and G_o is the intensity of the radiation field in units of the average interstellar radiation field (the Habing field).

The second measure uses the limit at which photoelectric effects give way to CR ionisation, estimated from their two rates to give

$$R_I \sim 0.47\left(10^4\,\mathrm{cm}^{-3}/n_o\right)\mathrm{pc}, \text{ and}$$
$$Av \sim 7.7 + 1/1.8\,\ln[\left(G_o/10^4\right)\,(3\times10^{-17}\,\mathrm{s}^{-1}/\varsigma_{CR})],$$

where ς_{CR} is the CR ionisation rate.

It is evident that G_o/n is the controlling parameter. If we define G_o for the 6–13.6 eV FUV field in Habing units, where one Habing is 1.6×10^{-3} erg cm^{-2} s^{-1}, for a stellar radiation source with luminosity L_* at distance d from the PDR,

$$G_o = 625(L_*\chi/4\pi d^2),$$

where χ is the fraction of stellar luminosity between 6 and 13.6 eV. Adopting a Stromgren solution for the size of the HII region and assuming pressure equilibrium between the ionised gas (HII at 8,000 K) and the PDR gives

$$G_o/n \sim 0.22\left(n/10^4\,\mathrm{cm}^{-3}\right)^{1/3}\mathrm{cm}^{-3}.$$

In the Orion Bar, observation shows a dissociation front clearly delineated by intense HCO^+ and CO line emission shown in the two panels of Figure 14.8. These emissions coincide with the brightest peaks of H_2^* vibrational emission (the H_2 to H transition) shown as contours tracing the dissociation front (DF) in both panels. The C^+/C/CO transition zones should exist deeper within the molecular cloud than the H/H_2 zone (both because of the lower ionisation potential of C^+ and CO failing to self-shield from photodissociation as effectively as H_2). As we have said, observations in the Bar show both zones much closer than formally expected (Tielens, 1993; Hollenbach & Tielens, 1999; Rollig et al., 2007). It seems that the carbon cycle transitions are much more sensitive to CR ionisation rate than are those of the H/H_2 transitions. Cosmic rays can destroy CO molecules indirectly, via He^+, even while H_2 column densities remain high (Bisbas et al., 2015). For intermediate CR ionisation rates, this resulting 'CO poor' molecular gas is C rich, while for high CR ionisation rates the gas becomes C^+ rich.

14.5 Inhomogeneities and Proplyds

ALMA mosaics identify a fragmented ridge of high-density substructures, as well as the photo-ablative gas flows we have mentioned, with instabilities at the irradiated molecular cloud surface. The dynamical effects of a high-pressure wave moving into the cloud perhaps generate these small-scale substructures that may be the seeds of future star-forming clumps in isolation or by merging. Gravitational collapse is not yet apparent from either their density distributions or their low masses. However, HCO^+ emission is resolved from Proplyd 203-506, just one of nearly 200 externally illuminated photo-evaporating disks around YSOs in the Orion Nebula. Its location within the atomic region (A_v <1) of the PDR between ionisation and dissociation fronts suggests a low-mass protostar and protoplanetary disk, perhaps emerging within the molecular cloud after surviving the passage of the dissociation front.

Within the FUV-irradiated gas, what we can say is that it is the formation of a compressed layer at the border of the PDR, together with dust drift driven by radiation pressure, that are responsible for the appearance of the Bar in the ^{13}CO and HCO^+ ALMA maps of Figure 14.6. Interaction between the heated PDR and the colder molecular clouds leads to density enhancement that slowly moves into the cloud. The mutual location of the transition regions depends on A_V, and its value grows over a short distance within the compressed layer, the regions growing ever closer together. The combined effects of chemical and thermal structure bring H_2, ^{13}CO, and HCO^+ peaks into coincidence, with high column densities bringing bright HCO^+ line emission (corresponding also to the total gas density peak in the compressed region) appearing on the illuminated side of the CO dissociation front. Dynamical models reproduce these observations in a way that stationary models do not (e.g. Kirsanova & Wiebe, 2019).

For carbon molecule formation, as we discussed in the previous chapter, reactions between C^+ and H_2 are the initiators, in evidence here since HCO^+ and H_2^* emission peaks spatially coincide. The multiple HCO^+ peaks in the Orion Bar region resolved by ALMA suggest a widespread structure on the small scale (~2″, corresponding to about 0.004 pc). As markers of the dissociation zone, the CO, H_2CO (both greyscale), and HCN (contours) sites in Figure 14.8 also show the inhomogeneity of gas deeper within the molecular cloud, although on a larger scale (5″–10″). As for the density and temperature associated with the HCO^+ and CO peaks, a density of 10^6 cm^{-3} and a temperature range of 200–300 K are deduced. These compare with the outer PDR previously noted, having a density around 5×10^4 cm^{-3} and the ambient molecular cloud ~10^5 cm^{-3}. The small-scale inhomogeneities translate into masses no more than a thousandth of a Solar mass, so well below any critical mass needed to make them gravitationally unstable. Compression and fragmentation at these irradiated edges of clouds are expected to be common phenomena, as are small-scale corrigations (ripples) along the cloud edge due to competing thermal and ram pressures.

Observations lead us to a dust attentuation value at the dissociation front of Av ~2 mag, with PDR conditions persisting into the cloud to a depth defined by Av ~4 mag. However, the clumpiness means that the average gas density of 10^4–10^5 cm^{-3} undoubtedly includes local concentrations $\geq 10^6$ cm^{-3} at the centres of which significant FUV irradiation will not penetrate. As for temperature in these inner PDR conditions, CH_3OH emission peaks suggest no more than 50 K; nonetheless many complex organic species (COMs) and their precursors have been observed (Cuadrado et al., 2017). The depth dependency for all photodissociation and photoionisation reaction rates results in different COMs having their maximum abundances at different locations, which can be expressed through expressions of the form $k = C\chi\exp(-\alpha Av - \beta Av^2)$ s^{-1}, where α and β are extinction parameters and χ the enhancement factor over the standard interstellar radiation field. Such expressions are the basis for the derivation of rate constants in PDR modelling of observations. Some observations cast doubt on the origin of COMs in the Orion Bar PDR as arising through ice-mantle desorption or grain sputtering. With exclusively gas-phase PDR models also failing to reproduce the observed abundances of even H_2CO and CH_3OH in the given conditions, it is suggested there still remain unknown hot-phase reactions to be discovered, or the possibility of COMs formation on warm bare grains (rather than within ices), or that PDR dynamics help to engender them or their precursors (having formed on grains in the cold, dense interior cloud) being desorbed and advected into the PDR (Cuadrado et al., 2017).

14.6 Sulphur

Earlier chapters have made clear that the astrochemistry of sulphur remains a longstanding problem, essentially because the molecular reservoir for this abundant element remains unidentified. The observed abundances of S-bearing molecules in diffuse and translucent molecular clouds represent less than 1 per cent of cosmic elemental sulphur nuclei (e.g. Neufeld et al., 2015). Naturally, the question remains: where is this sulphur stored? In colder dark clouds and dense cores, shielded from stellar UV radiation, most sulphur is expected in molecular form. With their low ionisation potential (10.4 eV), S atoms are a dominant source of electrons in molecular gas at intermediate visual extinctions (A_v ~2–4 mag) and readily reactive with other abundant atomic elements and small molecules. The accumulated sulphur molecular abundance, however, in dense molecular conditions is typically a factor of ~10^2–10^3 lower than its solar abundance (e.g. Fuente et al., 2019). Where, then, is it, and might observations in the Orion Bar PDR contribute to an answer?

Historically it has been assumed that sulphuretted molecules, most likely in hydride form (H_2S), must be depleted on to grain mantles at cold temperatures and high densities (e.g. Graedel et al., 1982; Millar & Herbst, 1990; Agundez & Wakelam, 2013). More recent modelling predicts reservoirs of neutral atomic sulphur in the dense gas phase (Vidal et al., 2017; Navarro-Almaida et al., 2020), or perhaps high levels of depleted organo-sulphur species (Laas & Caselli, 2019). Observationally, no ice-trapped sulphur species other than OCS has ever been reliably detected (Palumbo et al., 1997). A dense PDR such as the Orion Bar offers a potentially good environment in which to study the early stages in molecular sulphur chemistry because of its high-density but relatively quiescent gas; temperatures and densities are quite well constrained; no fast shocks or turbulence contribute to gas heating; and we have a quantified impinging FUV flux from which specific reaction rates with pumped H_2^* may be anticipated (e.g. Faure et al., 2017; Kaplan et al., 2017).

A range of sulphur-bearing molecules have been both sought and observed towards the Orion Bar using single-dish telescopes, including CS, $C^{34}S$, SO, SO_2 and H_2S (Hogerheijde et al., 1995; Jansen et al., 1995), SO^+ (Fuente et al., 2003), $C^{33}S$, HCS^+, H_2CS and NS (Leurini et al., 2006), and SH^+ (Nagy et al., 2013). Figure 14.10 shows a sample of these and related species in greyscale and spatial relation to the dissociation front close to the PDR surface.

Higher-angular-resolution interferometric detections have also been made with ALMA of SH^+, SO, and SO^+ (Goiceochea et al., 2017), together with NS^+ (Riviere-Marichalar et al., 2019), $H_2{}^{32}S$, $H_2{}^{34}S$, and $H_2{}^{33}S$ (Goiceochea et al., 2021) with the IRAM 30 m telescope. Figure 14.11 shows the Orion Bar environment around which selected observations of sulphur species have been targeted. The bright greyscale shows fine-structure line emission from the $^{13}C^+$ isotope (158 μm), lines that are optically thin and trace the distribution of dense PDR material at the FUV-illuminated rim, the 'layer' in which C^+ dominates the carbon distribution.

The distance between the ionisation front (the specific PDR-HII interface where H transitions to H^+) and a depth of a few A_v (at which C^+ becomes C and then CO) is ~15″ (~0.03 pc). This depth coincides with the SH^+ emission peak (white contours) in Figure 14.11, essentially coincident with observations of the H to H_2 transition identified as the PDR dissociation front (DF) proper. Although sulphur has a lower ionisation potential than carbon, the extent of the S^+ layer depends on the attenuation of the FUV field (Goiceochea & Le Bourlot, 2007), as well as radiative recombination and resonant dielectric recombination details. But there seems no doubt that this trace sulphur molecular ion is forming via gas-phase reactions of S^+ with vibrationally excited H_2^* molecules (e.g. Goiceochea et al., 2021). Detections of the product hydrides obtained with

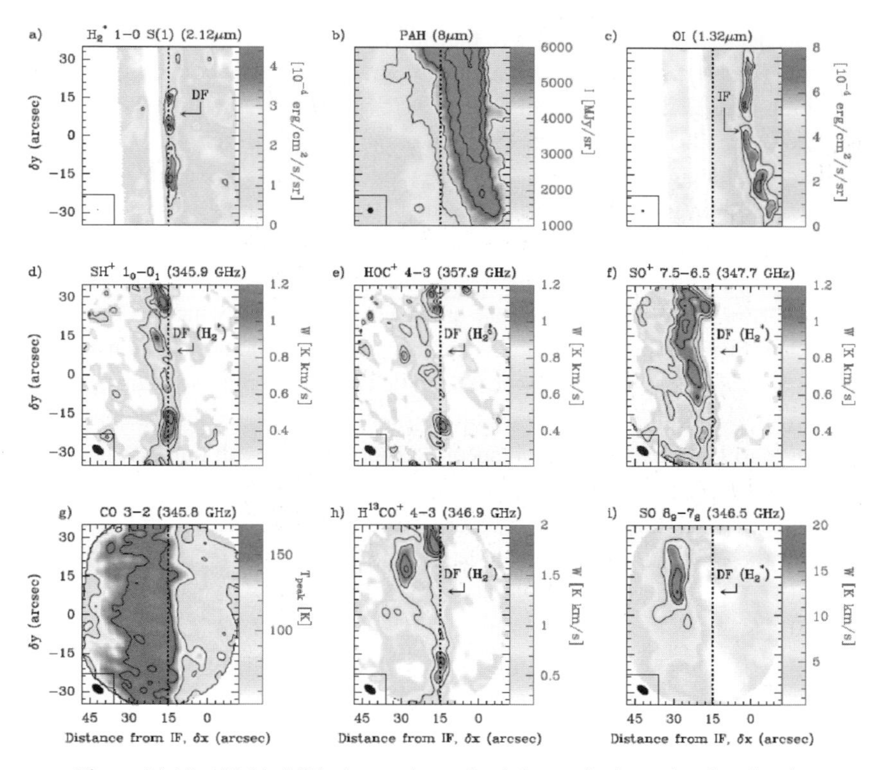

Figure 14.10 ALMA-ACA observations of sulphur and other related molecules towards the Orion Bar. The images show the FUV flux illuminating horizontally from the right, with the dotted line marked DF identifying the dissociation front 'inside' the immediate PDR outer margin (Goiceochea et al., 2017).

the IRAM 30 m telescope are then evident in Figure 14.12 (Goiceochea et al., 2021). The white dotted contours mark the infrared H_2^* DF tracer, the region where H_2S (and isotopologue $H_2{}^{34}S$) emission is brightest. Other molecular sulphur species, such as SO, emit from deeper inside the PDR, where the flux of photons has considerably decreased. Since the rare isotopologue $H_2{}^{33}S$ is also detected around the DF (where previously it has only been identified in the Sgr B2 and Orion KL hot cores), a high H_2S column density is implied.

Having therefore established the plausibility of S^+ reaction with H_2* under PDR DF conditions, and the high column density of simple sulphur hydrides, modelling of the many related pathways tries to account for the chemical mix, and one such model is summarised in Figure 14.13, with gas and grain reactions included, both endo- and exothermic.

This particular modelling of H_2S column densities echoes a comparable modelling of H_2O vapour in FUV-illuminated clouds (Hollenbach et al., 2009,

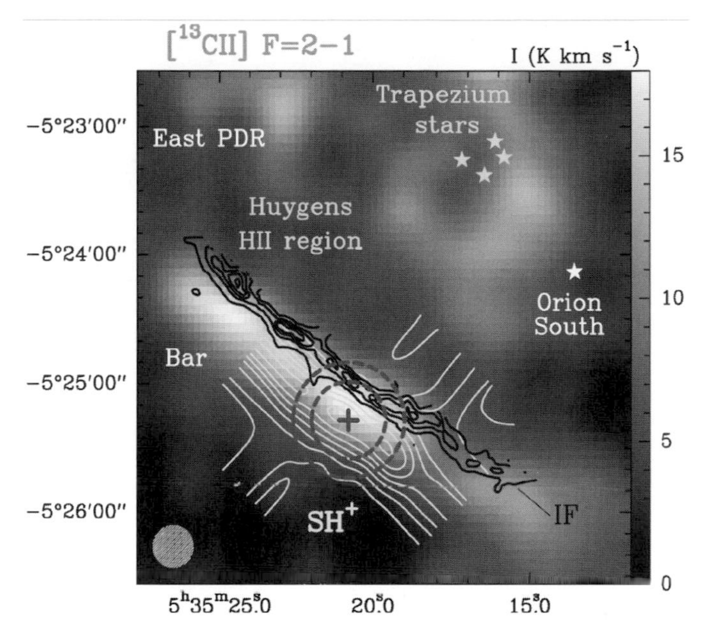

Figure 14.11 OI (black contours) and SH$^+$ (white contours) against 13C+ (bright background) (Goiceochea et al., 2021).

2012) in finding a column density (which for H_2S is ~10^{14} cm^{-2} up to A_v ~10 mag) and its abundance peak being close to the irradiated cloud edge (a few 10^{-8}) is little affected by variations in G_0 and n_H. Both the photodesorption rate of grain-surface s-H_2S and its gas-phase photodissociation rate are G_0 dependent. Increasing the FUV flux simply shifts the position of the H_2S peak abundance to a larger A_v until the rate of S atoms sticking on grains balances the H_2S photodissociation rate. On the other hand, the model shows that the formation rate of s-H_2S mantles depends on the product of atomic sulphur abundance and grain abundance, which product is proportional to n_H^2, whereas the H_2S desorption rate depends on grain abundance alone, which is proportional simply to n_H. Therefore, the higher the density, the more dominant is formation over destruction. As for the rates themselves, ab initio quantum calculations have been carried out to determine the vibrationally state-dependent rates. In contrast to the H_2O case, in O^+ reaction networks in which hydrogen abstraction reactions are exothermic (Gerin et al., 2010; Neufeld et al., 2010; Hollenbach et al., 2012), for sulphur hydrides hydrogen abstraction reactions have endothermicities that are significantly higher than kinetic temperatures even in PDRs, as noted in the schematic of Figure 14.12. Some of the resulting rate coefficients are shown in Table 14.1.

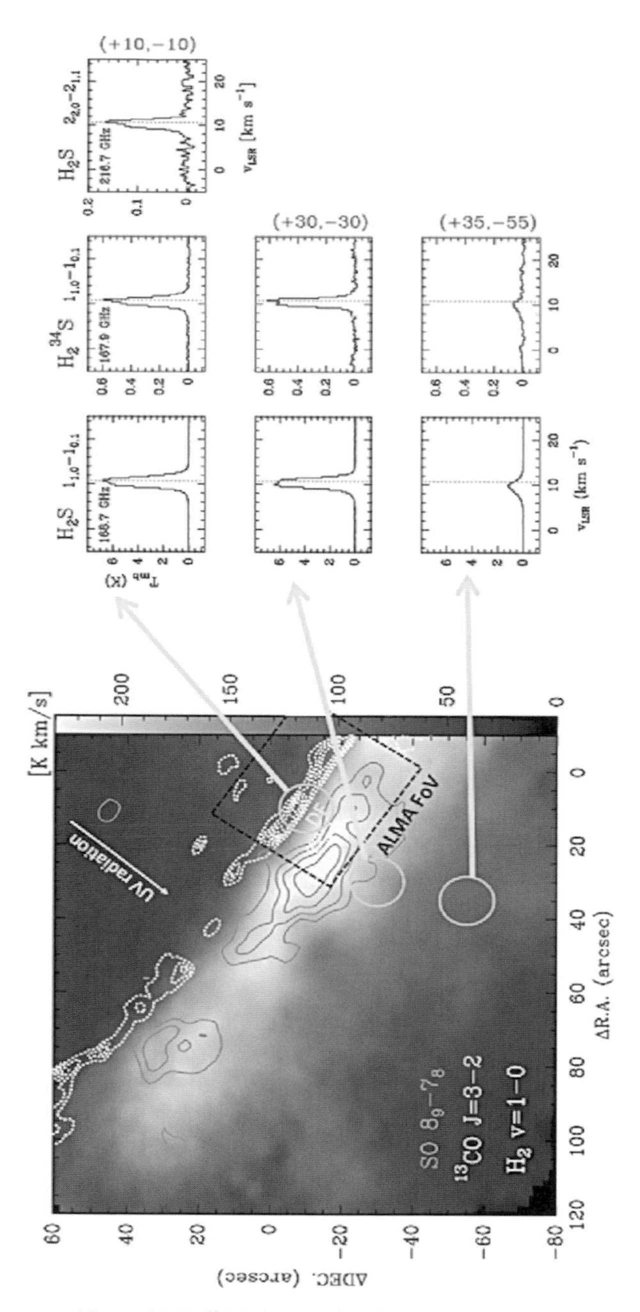

Figure 14.12 ^{13}CO (greyscale), SO (continuous contours), and H_2^* (dotted contours). Circles identify locations of H_2S detections shown in the righthand panel (Goiceochea et al., 2021).

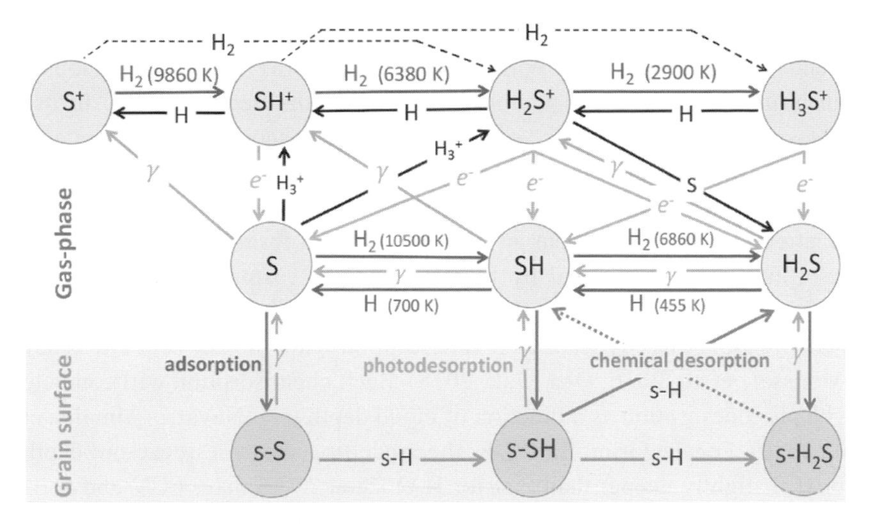

Figure 14.13 The main gas and grain reactions in sulphur hydride formation in the Meudon gas–grain PDR code. Endothermicities are given in K units, 'γ' is a FUV photon, and 's-' means solid (Goiceochea et al., 2021).

Table 14.1 *Selected rate coefficients for the Meudon PDR model, in which $k(T) = \alpha(T/300\ K)^{\beta} exp\ (-\gamma/T)$ (Goiceochea et al., 2021).*

Reaction	α (cm^3 s^{-1})	β	γ (K)
$SH^+ + H_2\ (v = 1) \rightarrow H_2S^+ + H$	4.97e^{-11}	0	1,973.4
$SH^+ + H_2\ (v = 2) \rightarrow H_2S^+ + H$	5.31e^{-10}	−0.17	0
$SH^+ + H_2\ (v = 3) \rightarrow H_2S^+ + H$	9.40e^{-10}	−0.16	0
$SH^+ + H \rightarrow S^+ + H_2$	1.86e^{-10}	−0.41	27.3
$SH^+ + e^- \rightarrow S + H$	2.00e^{-07}	−0.50	–
$H_2S^+ + H \rightarrow SH^+ + H_2$	6.15e^{-10}	−0.34	0
$S + H_2\ (v = 2) \rightarrow SH + H$	~8.6e^{-13}	~2.3	~2,500
$S + H_2\ (v = 3) \rightarrow SH + H$	~1.7e^{-12}	~2.0	~1,500
$SH + H \rightarrow S + H_2$	5.7e^{-13}	2.48	1,600
	7.7e^{-14}	0.39	−1.3
$S^+ + H_2\ (v = 2) \rightarrow SH^+ + H$	2.88e^{-10}	−0.15	42.9
$S^+ + H_2\ (v = 3) \rightarrow SH^+ + H$	9.03e^{-10}	−0.11	26.2
$S^+ + H_2\ (v = 4) \rightarrow SH^+ + H$	1.30e^{-09}	−0.04	40.8
$S^+ + H_2\ (v = 5) \rightarrow SH^+ + H$	1.21e^{-09}	0.09	34.5

In quantifying a grain-surface production route for H_2S, atomic sulphur is allowed to deplete onto grains as the temperature drops. One can adopt a timescale for this such as $x(S)^{-1}\ n_H^{-1}\ T_k^{-1/2}$, where $x(S)$ is the abundance of neutral sulphur atoms with respect to hydrogen nuclei. At the edge of the Orion Bar ($A_v < 2$ mag) both gas and dust temperatures are too high to allow much grain surface formation, whether reactions of S or S^+ (the latter more abundant in the bright rim). Within the cloud, H atoms (which are obviously much more abundant than S atoms) stick on grains frequently and diffuse across a grain surface rapidly, on a timescale much shorter than S atom desorption. The reaction of s-H and s-S to form an s-SH radical is without an energy barrier (e.g. Tielens & Hagen, 1982; Tielens, 2021). Further hydrogenation is similarly without an energy barrier to form s-H_2S. Subsequent desorption may be thermal by FUV photons or cosmic rays. Laboratory experiments have also shown that excess exothermicity from the reaction of s-H with s-SH directly leads to H_2S desorption with ~60 per cent efficiency (Minissale et al., 2016; Oba et al., 2018). Such chemisorption will compete with photodesorption as a function of cloud depth (e.g. Navarro-Almaida et al., 2020). For the Orion Bar PDR, the modelling shows a freeze-out depth for H_2S slightly deeper than that for H_2O ice at A_v >6 mag. FUV and thermal desorption of H_2S ices in the laboratory show pure s-H_2S ices thermally desorbing ~82 K and at higher temperatures for H_2S-H_2O ice mixtures (e.g. Cruz-Diaz et al., 2014).

In conclusion, comparison of observations with chemical models suggests firstly that SH^+ emitting layers at the edge of the Orion Bar (A_v < 2 mag) are characterised by no, or very little, gas-phase sulphur depletion. At intermediate PDR depths (A_v < 8 mag) the s-H_2S formation and depletion proposal matches the observed gas-phase column density (a few 10^{14} cm^{-2}) in the Bar and, as independent of G_0 and n_H, at the edges of other observed mildly illuminated clouds. Deep inside the modelled cloud (A_v >8 mag), H_2S appears likely to form through direct chemical desorption and/or photodesorption by secondary FUV photons, with a variety of currently unpredictable abundances for other sulphur-bearing species (Goiceochea et al., 2021). Have we solved the larger sulphur question of ‘missing’ sulphur in the galactic ISM? No, but we have taken another small step in understanding the reaction network possibilities for sulphur chemistry. The low flux observations in the Horsehead nebula in the following chapter will add yet more intriguing sulphur chemistry.

14.7 Summary

The Orion Bar has long been the canonical high-flux PDR exemplar. In addition to a detailed description of the source, the estimation of physical parameters such as ionisation fraction and observational indicators such as carbon recombination lines have been discussed. High-resolution observations point to the sensitivity of carbon chemistry to CR ionisation and the apparent merging of C/C^+/CO transition and H/H_2 transition zones not readily predicted by theory. A wide range of molecular sulphur observations has also offered the opportunity to rethink gas–grain reaction networks and model their consequences, and the following chapter looks at the low-flux PDR case of the Horsehead nebula, through which the sulphur question will be further explored.

15
The Horsehead Nebula in Orion

15.1 Introduction

The small number of PDRs having both proximity and favourable edge-on orientation that have been studied in detail include the Orion Bar of the previous chapter, plus NGC 7023 (Fuente et al., 1993, 1996, 2000), Mon R2 (Ginard et al., 2012, Trevino-Morales et al., 2014, 2016; Pilleri et al., 2012, 2013, 2014), and the Horsehead nebula. The Horsehead PDR offers a distinctively different set of observations to the others, associated as they are with warmer regions and higher FUV fluxes. The fact that dust temperatures in the Horsehead are low (~20–30 K), being below or close to the sublimation temperatures of many species, has implications for some of the molecular reaction network issues raised in the Orion Bar chapter in relation, for example, to COMs and sulphuretted species. The relatively rich observational chemistry of the Horsehead PDR suggests rich surface processing on irradiated grains is likely, while simultaneously the low dust temperatures suggest thermal desorption must be inefficient.

15.2 Low-Flux Photodissociation

The line of sight to the Horsehead nebula in the night sky is just south of ζ Orionis (Alnitak), the easternmost star of Orion's Belt. The nebula is observed nearly edge-on at a distance of just over 400 pc, at the western side of the L1630 molecular cloud, which is a component of the Orion B cloud within the large Orion Molecular Cloud complex (Figure 15.1). The nebula is illuminated by an O9.5V star to the south-west in the plane of the sky, the principal young high-mass star in the σ-Orionis cluster. Given its proximity, the nebula offers observers one of the brightest Galactic photodissociation locations in the mid-infrared range.

Figure 15.1 The Horsehead nebula with reflection nebula NGC2023 to the north and σ Orionis to the west (NASA/ESA/Hubble Heritage/STScI/AURA).

Much of the detailed study of this source over the past two decades has been conducted using the now familiar IRAM 30 metre radio telescope and the Plateaus de Bure Interferometer (PdBI), one in the high Spanish Sierra and the other in the French Alps, part of a wide-band high-resolution survey at 1, 2, and 3 mm (WHISPER). The chemical complexity in the Horsehead gas is high, with COMs of up to seven atoms observed, including CH_3CHO and CH_3CCH (e.g. Guzman et al., 2014). Given its proximity, the Horsehead is a location in which the astrochemistry of a photodissociation region (PDR) subject to a relatively low-level impinging FUV flux is well constrained. Estimates of the impinging FUV radiation field put it at just 60 times the average interstellar field, which is a modest degree of illumination within a star formation region (and a thousand times less than that impinging on the Orion Bar of the previous chapter). At 400 pc distance, a nebula resolution of 10″ equates to ~0.02 pc in the plane of the sky, rather more than 4,000 AU (Gratier et al., 2013). On this scale, a steep gas-phase particle density gradient changes from a value $\sim 10^2$ cm^{-3} in the bright rim of the Horsehead 'mane', evident in Figure 15.2, rising to $\sim 10^5$ cm^{-3} towards the centre of the 'head' (Habart et al., 2005). Models of photodissociated gas-phase chemistry in these environments can be closely tested and compared with observations, just as the contrasting outcomes can be compared with those from high-FUV flux PDRs.

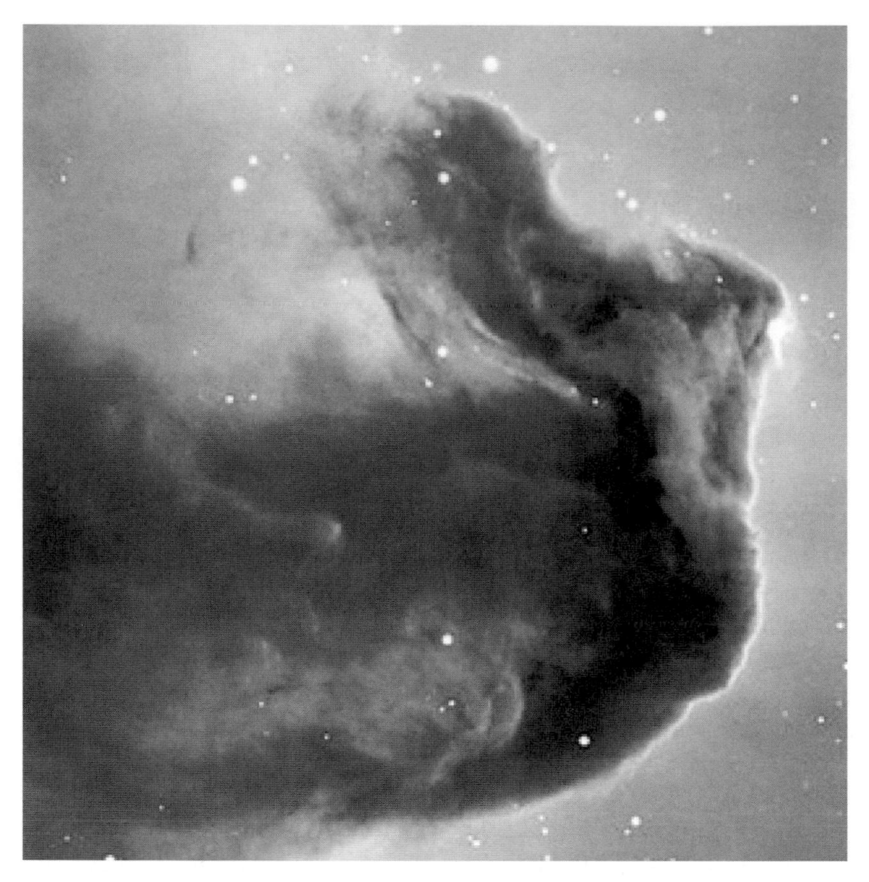

Figure 15.2 The bright rim and dark interior of the Horsehead PDR is evident in this Hubble image (NASA/ESA/Hubble Heritage/STScI/AURA).

Two useful molecules among the 30 or so detected towards the Horsehead Nebula that help to distinguish two important locations from one another are the dense cloud radical HCO and its ionized deuterated analogue DCO^+. A distance of 40″ (~0.08 pc which is ~16 k AU) separates the HCO peak emission location within the PDR margin from the DCO^+ in the dense centre of the head. With a vertical line defining the western edge of the Horsehead PDR margin, Figure 15.3 shows the two distributions and their peak locations, the right-hand cross identifying the HCO peak, and the left-hand cross the DCO^+ peak. The greyscale intensity shows none of the deuterated species within 20″ of the outer-edge PDR, whereas pockets of scattered HCO abundance across most depths suggest a clumpy structure with possible FUV penetration to some degree well beyond the margin. Nonetheless the distinction between the two emission peaks, PDR and dark cloud, remains and can be taken as a focus for

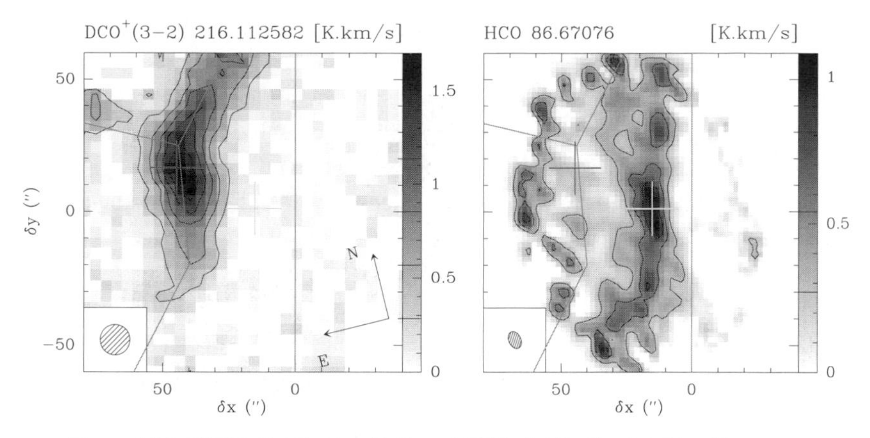

Figure 15.3 DCO+ distribution (*left*) and HCO distribution (*right*) with crosses marking peak emission locations for each. The vertical line represents the bright rim FUV-irradiated from the right (Gratier et al., 2013).

comparison with other molecules of interest under distinctively different conditions of temperature, density, and external irradiation.

Gas temperatures at the outermost edge (~100 K) where dust extinction is least ($A_v \leq 0.1$ mag) and irradiation strongest drop to ~60 K by the HCO emission peak. Gas density at this point is ~5×10^4 cm^{-3}, and dust temperatures even at the outer edges never exceed 20 K. Further inwards, the DCO^+ environment being shielded from FUV irradiation shows a gas temperature ≤ 20 K in a particle density that is doubled (~10^5 cm^{-3}), and with dust temperatures down to ~12 K (Goicoechea et al., 2009; Le Gal et al., 2017). The illumination of the Horsehead PDR is, in fact, inadequate to warm dust sufficiently to enable thermal desorption of anything but the lowest mass and least bound surface ice molecules. Considering only the dust temperature, it is difficult to see how most of the more complex COMs molecules locked into grain-surface ices could break into the gas phase. The Horsehead PDR therefore offers us the opportunity to study alternative mechanisms to simple thermal release.

15.3 Nitrile COMs

Among the COMs detected in the Horsehead nebula, formaldehyde (H_2CO) and methanol (CH_3OH) are key species in any subsequent synthesis of more complex organics in warm gas conditions, even perhaps the prebiotic (Garrod et al., 2008). Both of these precursor species are widely observed in the range of interstellar and circumstellar sources that we have already studied in earlier chapters, from hot cores to dark clouds to shocked regions, all of them dense cloud conditions with gas phase relative abundances for both species typically ~10^{-10} (Cuadrado et al., 2017). There are theoretically secure routes to H_2CO production in both the gas phase and by successive hydrogenation surface reactions (CO to HCO to H_2CO). For CH_3OH all pure gas-phase reactions studied to date have proved insufficient to account for observed abundances, and the simple successive hydrogenation on grain surfaces seems the likely dominant pathway (H_2CO to H_3CO to CH_3OH).

Gas-phase detections in the Horsehead have also included the four nitriles CH_3CN, CH_3NC, HC_3N, and C_3N (Gratier et al., 2013), species observed in a wide variety of other star formation sources (Purcell et al., 2006), including protoplanetary disks (Oberg et al., 2015). In the case of CH_3CN in the Horsehead nebula, relative abundances show a marked difference at the PDR peak (several times 10^{-10}) from that at the dark core peak (between 5×10^{-12} and 10^{-11}). In effect, CH_3CN is 30 times more abundant in the PDR than the dark core, while still three orders of magnitude more abundant than purely gas-phase computational modelling is able to reproduce (Gratier et al., 2013) and two orders of magnitude more abundant than combined gas–grain modelling (Le Gal et al., 2017). Relative abundances for HC_3N are similar at both locations (~10^{-11}), while its isomer HC_2NC is not detected at all. Finally, C_3N abundance is ~10^{-11} in the PDR while also not detected in the dark core. The most recent gas-grain modelling, while failing to match the observed CH_3CN abundance in the PDR (dust extinction Av < 2 mag), can nonetheless reproduce those HC_3N and C_3N observations. This support for grain-surface routes to molecular complexity does, of course, return us to the question of subsequent desorption mechanisms that we will consider shortly.

CH_3CN

Firstly, in considering the CH_3CN enhancement, we are reminded of the G34.26 UCHC situation described in Chapter 12, except that where that hot core conjectured PDR margin is a more speculative scenario, the Horsehead offers a self-evident condition in which to consider formation and destruction processes. We can first recognise that the gas densities in the Horsehead nebula are comparable to any average dense-cloud conditions. The average particle density, excluding the very outermost edges, is observed between $10^4 cm^{-3}$ and $10^5 cm^{-3}$. In such conditions, at a typical dense-cloud temperature ~10 K, the dominant gas phase formation route to CH_3CN has long been taken as the radiative association:

$$CH_3^+ + HCN/HNC \rightarrow H_4C_2N^+ + hv,$$

followed by dissociative recombination of the intermediate molecular ion:

$$H_4C_2N^+ + e^- \rightarrow CH_3CN/CH_3NC + H.$$

Pure gas-phase modelling suggests that the G34.26 hot core conditions (T_k ~100–300 K, n_H ~10^6–10^8 cm^{-3}, Av>10 mag) that we are comparing seem not to reduce the dominance of this cold dense cloud reaction pathway, but they do alter the availability of the major precursor species, CH_3^+ and HCN. One consequence of this is that the HCN/HNC branching ratio is higher in the cold dense case than the hot, which in turn determines the CH_3CN/CH_3NC product ratio. An increased ratio arises because the formation pathway in hot cores is thought to follow from evaporation of NH_3 from grain mantle ices, followed by its protonation by H_3^+ to NH_4^+, then by dissociative recombination to NH_2, which in turn reacts with atomic carbon to form HNC. A further exchange reaction can follow with atomic hydrogen to form HCN. The key to the abundance ratio between the two product isomers is the fact that this latter hydrogen exchange has a modest but significant activation energy barrier (100 K), ensuring the proportion of HCN increases at and above this temperature. While HCN is the simplest precursor to CH_3CN in hot core conditions, in passing we can note that its protonation is followed by dissociation to the CN radical which, reacting with hydrocarbons, may help to drive formation of other larger nitriles in the gas phase (such as HC_3N and CH_2CHCN) adding to any grain surface formation pathways for these complex nitriles (Charnley et al., 1992).

In the Horsehead PDR, of the two isomers CH_3CN and CH_3NC, 85 per cent is observed in the former state, suggesting the dominance of HCN as a precursor over its isomer through efficient proton exchange. Unfortunately, as we have noted, the kinetic temperature in the 'mane' (where CH_3CN is about 30 times more abundant), although three times warmer than the dark core 'head', is insufficient to enhance that proton exchange reaction. Additionally, the moderate FUV illumination translates into dust temperatures that we have also noted are low enough for the thermal evaporation of COMs to be relatively inefficient, even if the upper ice matrix loosens up enough for release of these species at temperatures well below pure sublimation values (Gratier et al., 2013). While CH_3CN is observed elsewhere in shock conditions (in which sputtering of ices and grain mantles by collision with energetic gas particles is common), the Horsehead PDR is not subject to shocks from nearby star formation, so we cannot take the abundance as tracing shock conditions either. What options remain? Photodesorption from grain-surface ices seems the most likely, although we have as yet no secure theoretical surface formation reaction pathway, nor indeed any observational evidence for CH_3CN as existing in ices anywhere. Perhaps there is a photochemically enhanced gas reaction pathway from desorbed CH_3OH and NH_3 precursors effective across a range of FUV field strengths (MacKay, 1999). However, while the latter might appear more appealing, it does have its flaws and we have to question every assumption.

For example, there is the possibility that CH_3CN forms by the UV photolysis of ethylamine ($CH_3CH_2NH_2$) ices, a complex precursor species inescapably formed on grain surfaces. Unfortunately, modelling of that idea to date seriously underestimates the amount of product that could form on dust having just a 20 K surface temperature (Danger et al., 2011). Of course, the model is dependent on our assumptions about ice composition and our assumed rate of the photolysis process. Perhaps then the rate is an underestimate of the reality. It is also suggested that gas-phase formation of H_2CCN followed by adsorption onto grains followed by hydrogenation could be a route (Andron et al., 2018). There are also studies of chemical desorption, as we have seen in earlier chapters in which exothermic surface reactions engender sufficient localised hot spots to evaporate products. While promising, to date these appear to offer opportunities for only a selected number of species. For all these possibilities, neither the surface reaction rates as a function of surface composition, nor binding energies on essentially unknown surface structures that then determine desorption rates, are well defined. Recent modelling with new binding energy computations continue to grossly underproduce the observed COMs observations (Le Gal et al., 2017). There is modelling evidence of enhanced grain-surface formation of CH_3CN as cloud density increases in the low-illumination case, but, of course, such modelling is a function of the surface reaction set one includes which, if rather speculative in some cases and lacking solid laboratory data to support it, offers a rather circular argument.

One particularly challenging laboratory result even for the assumption that ubiquitous CH_3OH is formed through the straightforward successive hydrogenation sequence offered earlier, followed by desorption into a gas-phase abundance level $\sim 10^{-10}$, is the complete failure of FUV irradiation experiments at less than 11 eV to photodesorb methanol intact at all (Cruz-Diaz et al., 2016). The counter evidence, nonetheless, remains overwhelming for CH_3OH as having a predominantly grain-surface origin followed by desorption. That cannot be said as yet for CH_3CN, even though they share a similar binding energy. While their detected gas-phase abundances are closely equivalent in the Horsehead PDR, there is well over an order of magnitude difference in CH_3OH's favour in the dark core (Guzman et al., 2013; Gratier et al., 2013).

15.4 Dust Extinction

In a dense gas with typically 10,000 particles per cubic centimetre or more, the dust-to-gas particle ratio is about 1:100. In a PDR, the impinging FUV flux will create an outer layer of warm dust radiating across infrared wavelengths. Within that layer, at depths to which stellar photons still penetrate, dust temperatures will be much less than gas temperatures because the dust has many accessible low-lying modes for efficient cooling, in contrast to the gas that has only relatively high-lying fine structure levels. As we have noted, in the Horsehead a gas temperature at the HCO PDR peak location ~60 K coexists with a dust temperature ~20 K, and at the DCO^+ dark core peak a gas at ~20 K with dust ~12 K.

Returning to the particular anomaly of CH_3CN, the effect of dust extinction in computational modelling has highlighted the sensitivity of CH_3CN abundance to gas density in PDR conditions at low dust extinction. A two-orders-of-magnitude abundance range can be shown to result from the correlated factors of impinging FUV flux versus dust extinction (Esplugues et al., 2019). To invoke only two variables in any formation reaction network is an obvious approximation which may bear little relation to likely reality, yet the rates of photodissociation and photoionization reactions in the gas do depend critically on dust extinction parameters, such as absorption cross sections and grain-scattering properties. The resulting depth dependency of any photon-driven interaction will be expressed through a cumulative extinction defined by A_V. The simple rate coefficient expression that we mentioned in the Orion Bar chapter ($k = C\chi\exp(-aA_V - bA_V^2)$ s^{-1}) includes *a* and *b* as extinction parameters and C the enhancement factor over the standard interstellar radiation field (Roberge et al., 1991). In the CH_3CN case, the formation sensitivity is the result of a more efficient pathway to the precursor HCN with density, while the precursor CH_3^+ (as a direct product of photonized CH_3OH photodissociation) is clearly dust extinction–dependent. PDR modelling typically focuses on steady-state conditions, where rates of formation and destruction are in dynamic equilibrium, conditions that can be rapidly attained in photon-dominated gas. With photodissociation occurring at rates around 10^{-10} s^{-1}, this can mean a steady state is reached in less than a thousand years even in low-radiation fields. The Horsehead nebula is likely to have a lifetime greater than this by several orders of magnitude. The critical factor in the chemical mix is the attenuation of impinging photons that results from dust extinction, and this increases the depth-dependent photoreaction timescale within the cloud to differing degrees for each individual reaction. For a given gas and dust density gradient from edge to centre of the Horsehead PDR, it is the relationship between photoreaction timescales and the impinging FUV flux that determines effective PDR depth, and that depth will be different for different species and their distinctive reaction pathways.

15.5 The Warm High-Density Case

Ammonia (NH_3) line emission was originally favoured by radio astronomers as a probe of warm high-density conditions and still is over linear molecules such as CO because, as a symmetric top molecule, NH_3 has a more complex set of available rotational transitions. CH_3CN subsequently proved even more useful. Its J and K quantum numbers describe total angular momentum and the projection of that along the molecule's axis of symmetry. As a prolate symmetric top, CH_3CN has a characteristic splitting of each of its rotational J levels into doubly degenerate K components. With each J-ladder observed simultaneously in a relatively narrow frequency band, these probe a range of energies typically from 10 K to several hundred degrees. Provided the emission lines are thermalised and CH_3CN level populations are closer to thermal equilibrium in higher-density circumstances, the molecule offers itself as a reliable diagnostic of kinetic temperatures.

In the Horsehead nebula many more CH_3CN lines are observed at the warmer PDR location than towards the cooler dark core. The integrated intensities of matched lines (the isotopologue CH_3NC having a very similar J and K energy structure) enable observers to deduce a weighted average line ratio of CH_3CN/CH_3NC ~0.15 at the PDR position (Gratier et al., 2013). In the dark core, where gas density is double, level populations for the J = 5-4 K-ladder are thermalised and both gas kinetic temperature and column density can be quite accurately derived. In lower-density conditions, level populations do not follow local thermodynamic equilibrium (LTE) distributions for this ladder, so other K-ladders closer to thermalisation are chosen. Unfortunately, column densities are then underestimated and, if the excitation is strongly subthermal (due to even lower densities), the gas temperature is also underestimated. Nonetheless, given the total particle density of the Horsehead dark core, the column densities for CH_3CN are estimated with some confidence.

In the conjectured higher-flux PDR conditions of the G34.26 UCHC discussed in Chapter 12, a gas-phase reaction network modelled for the higher-density case does show significant enhancements of precursor HCN and hence daughter CH_3CN over a hot core lifetime of $\leq 10^5$ yr (Esplugues et al., 2019). With pen-and-paper approximations one can conjure an expression for this CH_3CN abundance in PDR conditions as a function of precursor abundances and the key rate coefficients for photodissociation and photoionization, as well as contributing particle collision reactions (MacKay, 1999; Gratier et al., 2013). While such analytical expressions derived for the G34.26 case imagined as a PDR at the hot core margins appear to match the observations, a comparable expression for the Horsehead PDR fails to do so. The research question therefore remains regarding CH_3CN formation networks, particularly in PDR conditions.

15.6 Sulphur

The sulphur reservoir question that we considered in some detail in the previous chapter on the Orion Bar, and that we have noted as being somewhat problematic in several other chapters, also benefits from Horsehead nebula observations. A significant inventory of Horsehead sulphur species and their abundances exists, as exemplified by Table 15.1. As for their origin, estimates of adsorption energy predict sulphur atoms sticking to grains at temperatures below ~22 K, and high atomic hydrogen abundance with easy mobility on grain surfaces seems likely, leading to rapid sulphur hydride formation. The most favoured molecular reservoir for sulphur, H_2S, has been observed to be the most abundant sulphur-bearing molecule in certain cometary ice emission (Bockelee-Morvan & Biver, 2017). However, a firm detection in interstellar ices has yet to be reported. Were H_2S the principal reservoir, it is typically assumed that H_2S abundance in ices would be 10 times lower than that of H_2O, since either UV photons or cosmic rays will process H_2S to other sulphur species, such as OCS or SO_2, both of which have been detected in ices (Geballe et al., 1985; Palumbo et al., 1995; Boogert et al., 1997). Laboratory experiments using UV photons, X-rays, and ions have also shown that energetic processing of H_2S-bearing ices readily generates sulphur–sulphur bonds, with H_2S_2 and S_2H the main products detected through their IR bands (Jimenez-Escobar & Munoz Caro, 2011). These polysulphanes (S_xH_y) could therefore also be the missing stable ice reservoir for sulphur, efficiently converting to atomic sulphur once desorbed into the gas phase (Druard & Wakelam, 2012; Fuente et al., 2017).

Towards the two locations in the Horsehead nebula, the central panel of Figure 15.4 shows an integrated intensity map of H_2S distribution, with the FUVstellar radiation impinging horizontally from the right as in Figure 15.3. The PDR-illuminated edge runs approximately vertically down the right-hand side. Only one H_2S line frequency is observed, so its exact abundance at either location is uncertain, but at no more than a few $\times 10^{-9}$ it is undoubtedly

Table 15.1 *Rotational temperatures, column densities,and abundances derived from rotational diagrams for sulphur-bearing species in the two principal locations in the Horsehead nebula (Riviere-Marichalar et al., 2019).*

	PDR			CORE		
Species	T_{rot} (K)	N_X (cm^{-2})	N_X/N_{Htotal}	T_{rot}	N_X (cm^{-2})	N_X/N_{Htotal}
CS	6.3	$(3.0\pm0.6) \times 10^{13}$	$(8.0\pm2.0) \times 10^{-10}$	5.9	$(5.3\pm1.0) \times 10^{13}$	$(9.0\pm2.0) \times 10^{-10}$
SO	11.6	$(3.2\pm0.3) \times 10^{13}$	$(8.4\pm0.8) \times 10^{-10}$	8.2	$(5.0\pm0.5) \times 10^{13}$	$(8.6\pm0.9) \times 10^{-10}$
p-H_2CS	4.2	$(5.0\pm3.0) \times 10^{11}$	$(1.3\pm0.8) \times 10^{-11}$	8.6	$(9.0\pm1.0) \times 10^{11}$	$(1.6\pm0.2) \times 10^{-11}$
o-H_2CS	8.8	$(9.0\pm3.0) \times 10^{11}$	$(2.4\pm0.8) \times 10^{-11}$	8.2	$(2.5\pm0.8) \times 10^{12}$	$(4.3\pm1.4) \times 10^{-11}$
SO_2	6.5	$(3.0\pm0.6) \times 10^{12}$	$(8.0\pm1.6) \times 10^{-11}$	6.9	$(2.9\pm0.6) \times 10^{12}$	$(5.0\pm1.0) \times 10^{-11}$
CCS	10.3	$(3.0\pm1.0) \times 10^{11}$	$(8.0\pm2.6) \times 10^{-12}$	8.1	$(1.5\pm0.7) \times 10^{12}$	$(2.6\pm1.2) \times 10^{-11}$
HCS^+	11.6	$(5.0\pm0.6) \times 10^{11}$	$(1.3\pm0.2) \times 10^{-11}$	10*	6×10^{11}	1.0×10^{-11}
SO^+	11	$(6.3\pm1.8) \times 10^{11}$	$(1.7\pm0.5) \times 10^{-11}$	9	$(4.5\pm0.5) \times 10^{11}$	$(7.8\pm0.9) \times 10^{-12}$
HDCS	10*	$(6\pm1) \times 10^{11}$	$(2\pm0.4) \times 10^{-11}$	10*	$(1\pm0.2) \times 10^{12}$	$(2\pm0.4) \times 10^{-11}$
NS	10*	$(4\pm0.8) \times 10^{12}$	$(1\pm0.2) \times 10^{-10}$	10*	$(3\pm0.6) \times 10^{12}$	$(5\pm1) \times 10^{-11}$
NS^+	10*	$(1\pm0.2) \times 10^{11}$	$(3\pm0.6) \times 10^{-12}$	10*	$(6\pm1) \times 10^{10}$	$(1\pm0.2) \times 10^{-12}$
DCS^+	10*	$<1 \times 10^{11}$	$<3 \times 10^{-12}$	10*	$<1.0 \times 10^{11}$	$<2 \times 10^{-12}$
CCCS	10*	$<3.2 \times 10^{11}$	$<8 \times 10^{-12}$	10*	$<3.2 \times 10^{11}$	$<5 \times 10^{12}$

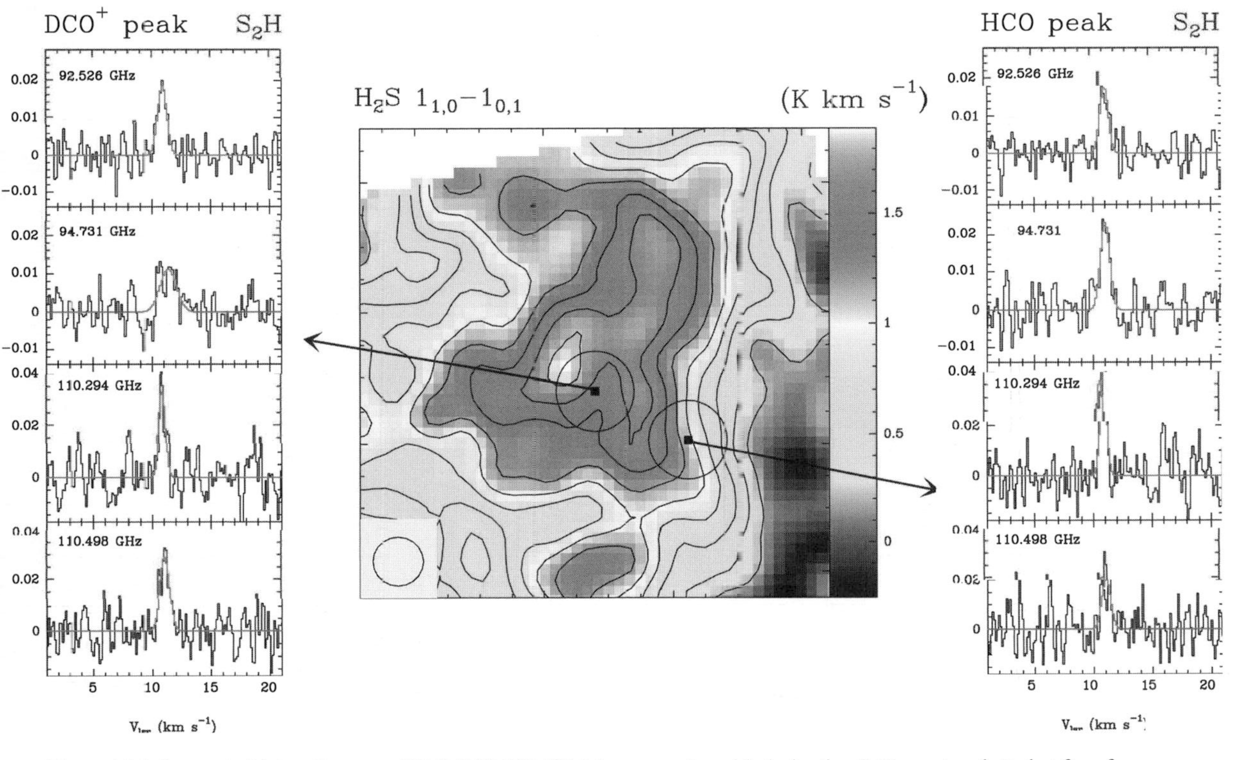

Figure 15.4 Integrated intensity map of H_2S (168.763 GHz) in greyscale, with derivative S_2H spectra plotted at four frequencies towards the two targeted locations (Fuente et al., 2017).

quite low across both targeted sites in the nebula. Column densities observed in other sources show comparable H_2S values. For example, the mildly illuminated rims of TMC-1 and Barnard 1b clouds (G_0 ~10) show column densities~10^{14} cm^{-2} and abundances ~10^{-9} are inferred for the shielded interior of these dark clouds (A_v >10 mag) (Navarro-Almaida et al., 2020). In warmer star-forming environments, observations towards hot cores detect H_2S abundance increasing up to ~10^{-8} (e.g. van der Tak et al., 2003; Herpin et al., 2009). Generally speaking, H_2S gas-phase abundances appear therefore to be broadly similar in a wide range of environments. In other words, this molecule seems not particularly sensitive to temperature and density differences. This may reflect several efficient chemical as well as FUV photo- and CR-induced secondary FUV desorption mechanisms. Whatever the mechanism(s), if the desorption of grain surface H_2S is widely efficient, the gas-phase observations are as we might expect. Even in extragalactic observations of H_2S emission in the starburst galaxy NGC 253 (see Chapter 18), where limited resolution suggests that extended molecular gas is illuminated by stellar FUV radiation, the inferred abundance is also comparable at ~10^{-9}.

Returning to Figure 15.4, the outside panels show spectra of the H_2S derivative S_2H detected through four emission frequencies at both locations. Given a significant desorption contribution from direct FUV irradiation, we might expect to see column densities of H_2S or its derivatives at higher values in the PDR peak location than in the core. However, photodissociation of desorbed H_2S or, indeed, polysulphanes,will result in SH, S, and S^+ products. As in the Orion Bar case, S^+ also appears to be a dominant form of sulphur in the outer gaseous layers of the Horsehead PDR cloud, as indirectly identified through SO^+ and NS^+ observations. NS^+ has long been thought to form through reaction of atomic nitrogen with S^+, SH^+, and SO^+ ions in FUV-illuminated gas (e.g. Sternberg & Dalgarno, 1995) and Figure 15.5 shows a modelled network involving all three radical ions.

The most abundant directly observed sulphur species in the Horsehead PDR apart from H_2S are CS and SO, of order 10^{-10} in relative abundance. As with the H_2S observations, the column densities of species in the lower part of Table 15.1 (asterisks in the rotation temperature column) are the most uncertain, arising from detections of single transitions only. Figure 15.6 shows integrated intensity mapping of CS, SO, C_2S, and NS^+ at the two locations, PDR and core, with the IRAM 30 m telescope. Since their total sulphur abundance is much less than the elemental abundance ratio with respect to hydrogen derived from solar photosphere estimates, they cannot alone represent the main sulphur reservoir. Comparing the PDR emission peak with the dark core peak, C_2S and o-H_2CS present fractional abundances a factor greater than two higher in the

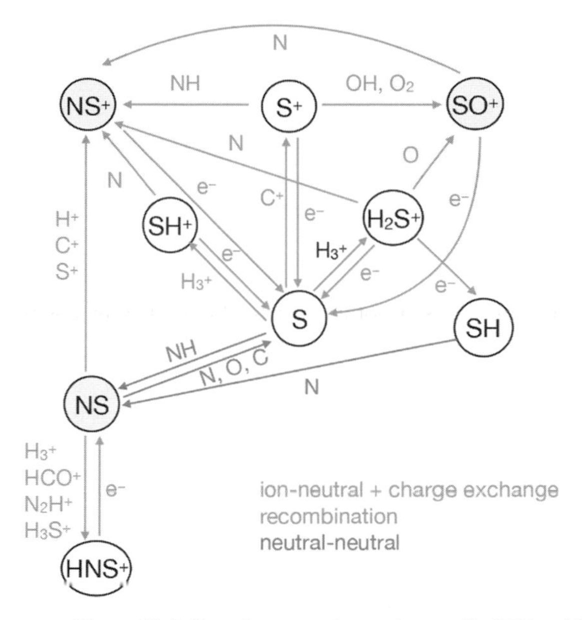

Figure 15.5 Gas-phase reaction pathways for NS^+ and SO^+ involving H_2S photo-dissociation products (Riviere-Marichalar et al., 2019).

core than in the PDR. In contrast, the oxygen-sulphur species SO, SO_2, and OCS present similar abundances towards both positions (as does HCS^+). If the inferred S, HS, H_2S, and implicit polysulphanes are indeed the main solid-state reservoirs, then they, along with S^+, would dominate the gas phase as progenitors. The sulphur radical ions SO^+ and NS^+, and neutralised radical NS, are all more abundant towards the PDR, as is S_2H, previously proposed as a tracer of low-UV PDR conditions where grain-surface chemistry and gas-phase photochemistry coexist (Fuente et al., 2017). The case for atomic sulphur and the sulphur hydrides being the principal solid-state reservoir for sulphur in dense molecular cloud conditions certainly begins to gain credence.

However, if we are to draw general conclusions about sulphur chemistry and the sulphur reservoir widely in the ISM, there are also wider comparisons to be made. Table 15.2 shows fractional abundances derived from observations of the Horsehead, the dark cloud TMC 1, the Orion KL Hot Core, and various Solar System comet sources.

Observations of sulphur-bearing volatiles in comet emission do show high abundances of H_2S and a relative elemental ratio against, for example, oxygen as expected against Solar values (Bockelee-Morvan & Biver, 2017). It is also the case that H_2S is the dominant gas-phase species observed towards the Orion KL hot core considered in Chapter 9. While in dark cores such as TMC

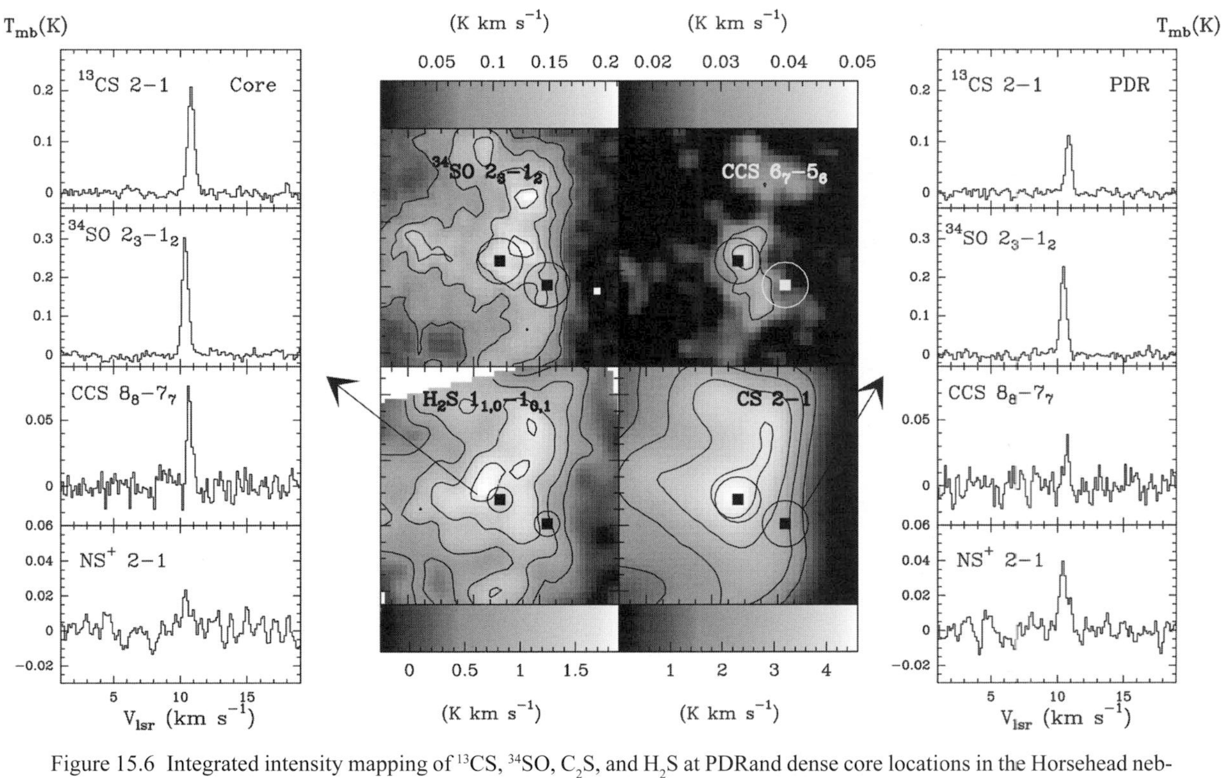

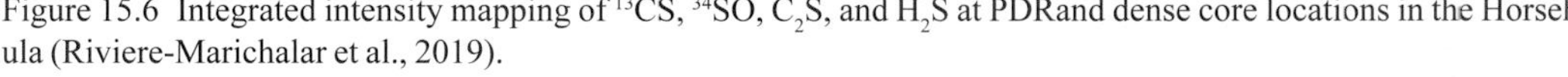

Figure 15.6 Integrated intensity mapping of ^{13}CS, ^{34}SO, C_2S, and H_2S at PDRand dense core locations in the Horsehead nebula (Riviere-Marichalar et al., 2019).

Table 15.2 *Comparison of fractional abundances (Riviere-Marichalar et al., 2019).*

	HH-PDR	TMC 1-CP	Orion KL	Comets (%wrt H_2O)
CS	$(4.5\pm1.1) \times 10^{-10}$	1.4×10^{-9}	7.0×10^{-8}	0.02–0.2
SO	$(3.9\pm0.5) \times 10^{-10}$	7.5×10^{-10}	8.0×10^{-8}	0.04–0.3
o-H_2CS	$(2.4\pm0.3) \times 10^{-11}$	3.0×10^{-10}	5.6×10^{-9}	0.009–0.09
OCS	$(5.0\pm0.8) \times 10^{-11}$	1.1×10^{-9}	5.0×10^{-8}	0.03–0.4
CCS	$(8.0\pm2.6) \times 10^{-12}$	3.5×10^{-9}	8.0×10^{-11}	-
SO_2	$(1.2\pm0.2) \times 10^{-10}$	1.5×10^{-10}	1.9×10^{-7}	0.2
SO^+	$(1.7\pm0.5) \times 10^{-11}$	-	-	-
HCS^+	$(1.1\pm0.2) \times 10^{-11}$	1.5×10^{-10}	-	
NS	$(1.0\pm0.2) \times 10^{-10}$	4.0×10^{-10}	3.4×10^{-9}	0.006–0.012
NS^+	$(3.0\pm0.6) \times 10^{-12}$	2.4×10^{-12}	-	
H_2S	$(3.1\pm0.5) \times 10^{-10}$	$<2.5 \times 10^{-10}$	1.5×10^{-6}	0.13–1.5
S_2H	$(8.7\pm3.1) \times 10^{-11}$	-	-	-
Sulphur budget	$\sim1.5 \times 10^{-9}$	$\sim8 \times 10^{-9}$	1.9×10^{-6}	4×10^{-6} ($n(H_2O)/n_H = 2.7 \times 10^{-4}$)

1, only about 0.05 per cent of the cosmic abundance of sulphur is observed in gas-phase molecules detected at millimetre wavelengths, with S, HS, and S_2 either unobservable or extremely difficult to quantify, the assumption that most sulphur is locked in ice remains, as yet, difficult to confirm.

15.7 Summary

The comparison of low- against high-flux PDR conditions is presented. Contrasting observations of selected species between the PDR margin and the inner dark cloud allow chemical modellers to test formation and destruction reaction networks against quite closely constrained physical conditions. The anomalous abundance of CH_3CN identified in Chapter 12 is considered here in the Horsehead context in the presence of other nitrile COMs observed. Comparison of sulphur chemistries in the Horsehead nebula also enables the latest ideas on the ISM sulphur reservoir to be tested.

PART VI

External Galaxies

16

Extragalactic Surveys: CANON and PHANGS-ALMA

16.1 Introduction

The extragalactic molecular astronomy of star formation has until recently been at a stage comparable to that of its Galactic progenitor 40 years ago. Back then, an inventory of Milky Way molecular clouds and those of our three immediate galactic neighbours (the two Magellanic Clouds and M31) using CO (J = 1-0) emission lines was coming to completion. Instrumentation limits restricted other observations exclusively to relatively simple molecules in cloud structures close to Earth (such as within the Taurus Molecular Cloud and the Orion Complex). As the instrumentation has developed, extragalactic studies have moved out to reach all of the Local Group galaxies (e.g. Rosolowsky et al., 2007; Bolatto et al., 2008; Wong et al., 2011; Druard et al., 2014). Unfortunately, the local systems are not representative of the galaxies where most stars form in the 'present-day' universe. Locally, they are typically low-mass galaxies that show faint, isolated CO emission (Fukui et al., 1999; Engargiola et al., 2003; Schruba et al., 2017) distinct from the bright, spatially contiguous emission distributions observed in the more massive galaxies at greater distances where star-formation rates are undoubtedly much higher (Hughes et al., 2013). Beyond single-dish observations, interferometers force trade-offs between surface brightness sensitivity and resolution, so accessing an entire cloud population of a normal star-forming disk galaxy (which is typically over 10 times more distant than Local Group targets) is difficult. Most studies using the higher resolution of interferometers as they became available have therefore focused on single GMCs in single-galaxy targets (Colombo et al., 2014; Egusa et al., 2018; Faesi et al., 2018), usually the most massive clouds within the inner regions, as we shall see in the examples of Chapters 17 and 18.

The link between molecular gas and star formation is, of course, proven from the Galactic observations described in preceding chapters, where rates of star formation and their locations closely track gas distribution. However, variations in star-formation rates per unit molecular gas among the different types of galaxies that we do know, and within different regions of those same galaxies, are evident (e.g. Leroy et al., 2021, and references therein). As the dominant observable molecular species in dense molecular clouds (given undetectable cold H_2), CO emission as mentioned has been used to map the extragalactic interstellar medium, as it was in the early days of Galactic dense cloud studies. Change in that CO brightness is taken to reflect both changes in molecular gas density and changes in CO emissivity per unit of molecular gas according to excitation conditions.

The first small survey we will review, part of the CANON CO(1-0) Survey, has the simple virtue in spite of its relatively local nature of confirming some essential facts about molecular gas distribution in galaxies other than our own (Donovan-Meyer et al., 2013). Between 2007 and 2012, five galaxies at distances out to 7.5 Mpc (10 times further than M31/Andromeda) were observed with measurable low rotational CO fluxes. The flux of the central region of each galaxy was measured with the Nobeyama Radio Observatory 45 m dish (NRO45) in Japan, and targeted interferometer resolutions of GMCs in the tens of parsecs range were observed with the Combined Array for Research in Millimetre Astronomy (CARMA) in California.

The second survey is ongoing, benefitting from much higher-resolution ALMA observations out to a wider range of distant galactic environments. This PHANGS-ALMA survey is looking at ninety typical star-forming galaxies selected out to a distance of 17 Mpc, ensuring minimum angular resolutions ~1.5″, almost equivalent to those of CANON (~ 100 pc) but at much greater distance. Over 100,000 GMCs have been identified to date, with the expectation that around 300,000 will be observed by the program's end. All of the PHANGS galaxies studied are relatively massive and generally face-on spirals visible from the Southern Hemisphere.

16.2 CANON

Figure 16.1a shows a Hubble image of one of the five CANON sample, NGC 6946, known as the Fireworks galaxy, ~7.7 ± 0.3 Mpc (about 25,000,000 light years) from Earth.

Figure 16.1b shows the CANON GMC targets scattered around the central region and its immediate disk (Donovan Meyer et al., 2013).

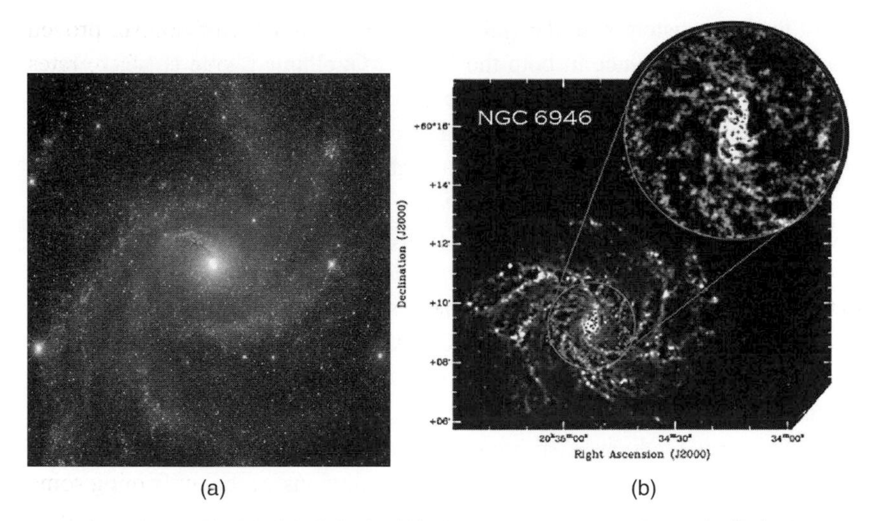

Figure 16.1 (a) NGC 6946 (ESA/Hubble & NASA, A. Leroy, K. S. Long); (b) the GMC targets in the CANON survey of NGC 6946 (Donovan-Meyer et al., 2013).

^{12}CO (J=1-0), with an energy of 5.5 K above ground, is collisionally excited at densities of only a few hundred particles cm^{-3} and it is observed in each of the GMCs of NGC 6946 targeted, as well as in the nearly 200 GMCs observed across the three brightest spiral galaxies. In pursuit of virial cloud masses for these GMCs, the conversion factor between CO flux and H_2 mass is a well-constrained parameter from Galactic observations and the CANON survey sought to confirm an equivalent extragalactic value while also validating the assumption of virialization. As we discussed in Chapter 6, at its simplest the balance between kinetic and gravitational potential energies determines the stability of a cloud and hence its potential as a star-formation nursery, with other contributions such as magnetic fields and external pressures contributing subsidiary support or constraint. We have seen in earlier chapters how complex the star-formation activity can be within a larger molecular cloud that, as a whole, must offer such stability. If we express the dynamic state of a GMC through a virial parameter (α_{vir}) as a function of kinetic and self-gravitational potential energies, we can relate it to cloud mass, radius, and velocity dispersion (Bertoldi & McKee, 1992).

CO luminosity measurements that lead to virial masses across the range of observed GMCs in our own Galaxy give results that agree to within a factor of two. Since alternative non-virial measures (e.g. via far-IR and HI mapping, or gamma ray tracing of mass) also produce closely comparable values, the CO luminosity conversion is taken to be reliable. An early extragalactic conversion factor within the GMCs of the M33 (Triangulum) galaxy was also found to be consistent with the Galactic value (Wilson & Scoville, 1990) and many observations since in both the Large Magellanic Cloud (LMC), Small Magellanic Cloud (SMC), and in nearby dwarf and irregular galaxies – in fact throughout the studies of Local Group galaxies – the average conversion factor has been found to be largely consistent with the Galactic value (e.g. Fukui & Kawamura, 2010). The CANON survey resolved 182 individual GMCs within the central kiloparsecs of three of their five galaxies. In addition, they detected GMCs within both the galactic centres and the inner disks of their sample having dimensions and population densities entirely comparable with corresponding cloud populations in our own Galaxy.

Figure 16.2 shows both CO luminosity against velocity dispersion and against radius for the three principal sources in the CANON survey.

With sizes and velocity dispersions proving closely consistent with those of the Milky Way disk GMCs,as do velocity dispersions against CO luminosity, the trend in CO luminosity against cloud size seems indistinguishable from that seen in the Milky Way across more than four orders of luminosity magnitude. A linear relationship between GMC virial masses and CO luminosities implies a constant CO–H_2 conversion factor and supports a virialization assumption. Figure 16.3 shows the consistency of the resulting mass–luminosity relationship, with the average conversion factor in each galaxy for comparable GMC mass consistent with that of the Milky Way to within a factor of two. The CANON observations therefore confirm some of the essential facts about molecular gas distribution in galaxies other than our own – assumed to be obvious perhaps, but now an assumption with evidential backup.

16.3 Comparative Resolution

From high-resolution CO imaging, one analytical approach enables us to quantify molecular gas surface density (Σ) and velocity dispersion (σ) along each sightline at a range of fixed spatial scales (Sun et al., 2018 and references therein), and from these we can deduce precisely those physical properties such as gas-surface density, velocity dispersion, dynamical state, and turbulent energy content that we wish to relate to potential star-formation rates. In our own Galaxy and its Local Group neighbours, most molecular gas exists in GMCs with masses ~10^4–10^7 $M_\odot$, tens of parsecs in size, and dominated by supersonic turbulence. Since these GMCs do not fill a galaxy's disk, low-resolution extragalactic observations present CO emission diluted by non–CO-emitting regions. The intrinsic distribution of molecular gas is clumped on scales much smaller than the previous generation of mapping surveys could accomplish.

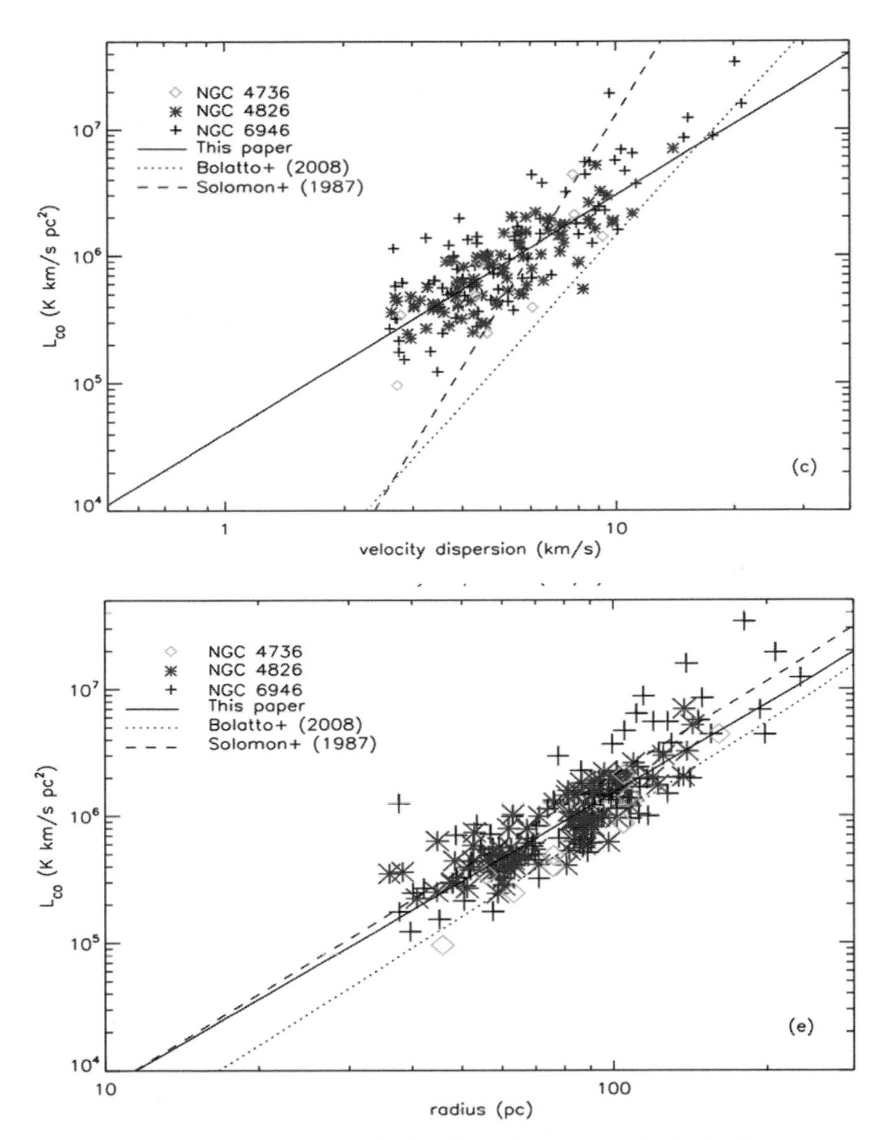

Figure 16.2 (*upper panel*) Velocity dispersion versus CO luminosity, and (*lower panel*) radius versus CO luminosity, for the GMCs in the CANON survey (Donovan-Meyer et al., 2013).

The cloud-scale imaging of the CANON survey offers CO (1-0) emission at 2″ resolution across the inner regions of spiral galaxies (Koda et al., 2009; Donovan-Meyer et al., 2012, 2013; Momose et al., 2013). Subsequently, the PdBI Arcsecond Whirlpool Survey (PAWS) mapped M51 at 1″ (~40 pc) (Schinnerer et al., 2013; Pety et al., 2013), improving on a CARMA 3″

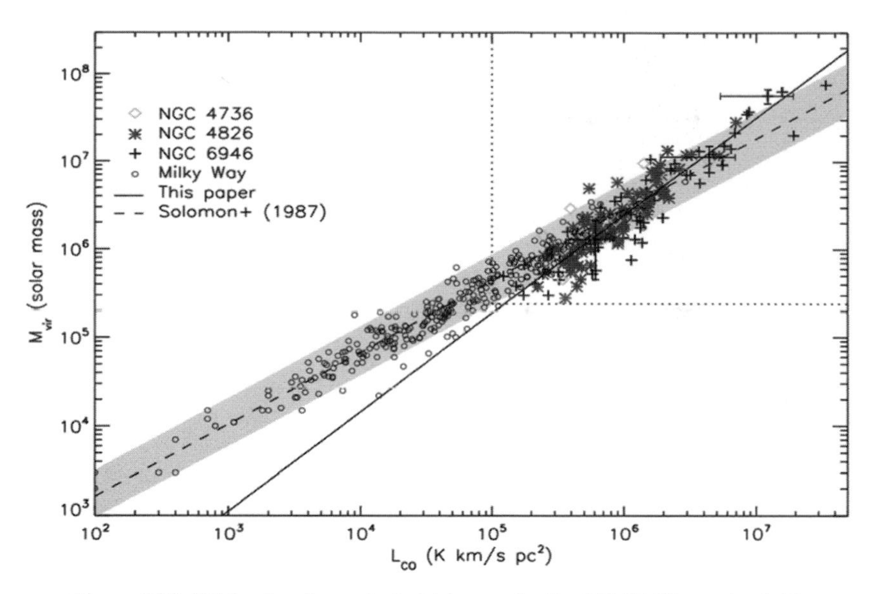

Figure 16.3 CO luminosity against virial mass for the 182 GMCs most reliably measured in three of the five CANON sample galaxies. The shaded region defines a factor of two range in the conversion encompassing most of the extragalactic and Milky Way GMCs (Donovan-Meyer et al., 2013).

observation of the same source (Koda et al., 2009). With improved resolution, a key discovery was that surface density and line width of molecular gas vary significantly and systematically as a function of location in a galaxy and between the galaxies observed, and Figure 16.4 illustrates this with mapping examples of NGC 1097.

ALMA observations transform molecular line studies of external galaxies, and make possible enhanced sample size, field of view, and sensitivity, towards more-typical star-forming galaxies, beyond the idiosyncrasies of the Local Group. Where PAWS observations required almost 130 hours on-source to map CO (1-0) emission from the inner 4.5′ × 3.0′ of M51 (Pety et al., 2013), ALMA can map CO (2-1) emission at 1″ resolution across a 2′ × 2′ field with twice the sensitivity in just two hours on-source main array time.

16.4 PHANGS-ALMA

The survey acronym PHANGS-ALMA stands for ‘Physics at High Angular Resolution in Nearby Galaxies with the Atacama Large Millimetre-submillimetre Array.’ In mapping CO (2-1) emission from galaxies up to 17

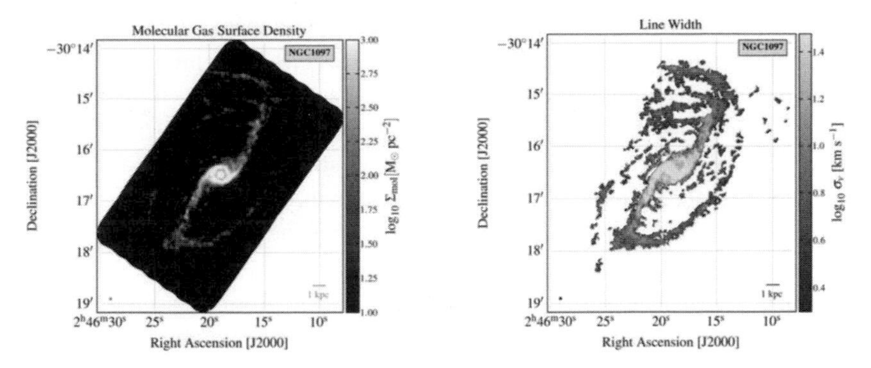

Figure 16.4 Representative galaxy NGC 1097 from the PHANGS-ALMA survey. CO (2–1)-based estimates of molecular gas surface density (*left panel*), and effective line width (*right panel*) (Leroy et al., 2021).

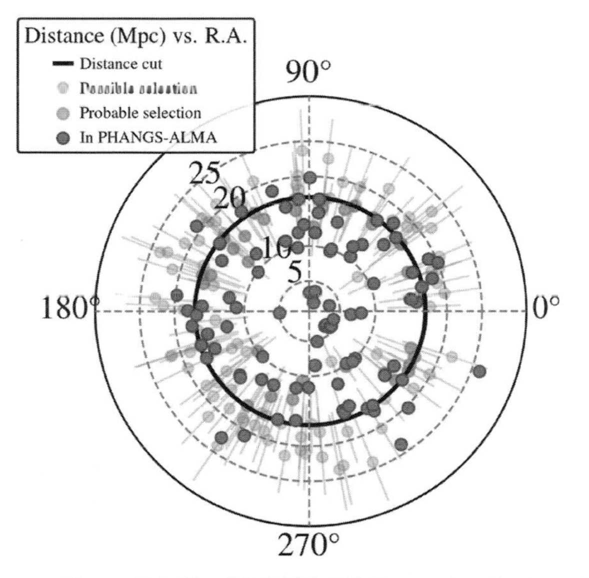

Figure 16.5 The PHANGS-ALMA sample selection, with the 'probables' and 'possibles' initially considered. The distances are in increments of 5 Mpc out to a sample cut-off at 17 Mpc (Leroy et al., 2021).

Mpc away, resolutions of 1″–1.5″ encompass active star-forming galaxies down to total stellar masses ~5 × 10^9 $M_\odot$. The 360° sample selection is shown schematically in Figure 16.5. The 12 m (fine-scale) and 7 m (large-scale) interferometric arrays combined with the total-power antennas offer sufficient sensitivity and resolution to detect individual GMCs across the star-forming disks of each target galaxy. One early analysis has looked at results from a sample of just eleven of the ongoing detections and compared them with results from four additional targets taken from earlier research, giving preliminary data spanning examples from dwarf spiral to starburst galaxies. Even this limited sample offers tens of thousands of independent measurements at spatial scales from 20 to 130 pc comparable to the characteristic dimensions of a Galactic GMC (Sun et al., 2018). The additional four sources considered here are M31, M33 and M51 from the Local Group, and the Antennae (the nearest major pair of merging galaxies).

Before looking at exactly what can be deduced from measurements of velocity dispersion (σ) and surface density (Σ) across this sample, let us note a few details about these sources. NGC 2835 and NGC 5068, for example, are low-mass disk galaxies with low surface CO brightness and low star-formation rates. NGC3351, on the other hand, is a strongly barred galaxy with a prominent central molecular disk, a gas-poor bulge, and a ring of molecular gas at greater galactocentric distance. Star formation occurs in both the central disk and the ring beyond the bulge. M51 (Whirlpool) is a classic spiral with abundant star formation in the central region as well as the spiral arms. M31 (Andromeda) has a high stellar mass but remains relatively quiescent in star-formation terms, although with double the total number of stars over the Milky Way for a spiral broadly comparable in size to our own. Its gas reservoir is dominated by a large HI disk, with most molecular gas distant from the galactic centre. M33 (Triangulum) is a Local Group dwarf spiral with an HI-dominant gas. Finally, the Antennae pair of interacting galaxies offer an overlap region dramatically different in molecular gas dynamical terms from the others in the sample. Table 16.1 summarizes a selection of parameters for the sampled galaxies.

Galaxies with the widest range of molecular gas surface density (derived from CO flux) and velocity distribution show multiple peaks in the distribution of both parameters. The highest-value peaks arise at the brightest locations in the innermost regions. If we define these innermost 'central' regions as within 1 kpc of the galaxy's nucleus, comparing these with the surrounding 'disk' distribution, strongly barred galaxies have noticeably higher peaks than those without bar-driven inner structure, while the latter still outdo their disk surroundings. Enhanced relative Σ and σ values are also evident in spiral arms (Sun et al., 2018). The correlation between Σ and σ is shown in Figure 16.6 across the sample of star-forming galaxies and for around 30,000 independent beams. It shows that the relationship appears to be a fundamental property of the ISM at cloud scales.

We are looking for comparable behaviour across the galaxy sample, a first step towards a universally applicable principle in which, for example, Σ and σ

Table 16.1 *The PHANGS-ALMA sample of galaxies under consideration (Sun et al., 2018).*

Galaxy	Morphology	Distance (Mpc)	Inclination (deg)	$\Sigma M_{stellar}$ (10^{10} $M_{\odot}$)	Telescope	Line	Resolution (arcsec/pc)	Channel Width (km s^{-1})	Sensitivity (K)
NGC 628	Sc-A	9.0	6.5	2.1	ALMA	CO(2-1)	1.0/44	2.5	0.13
NGC 1672	Sb-B	11.9	40.0	3.0	ALMA	CO(2-1)	1.7/98	2.5	0.09
NGC 2835	Sc-B	10.1	56.4	0.76	ALMA	CO(2-1)	0.7/34	2.5	0.27
NGC 3351	Sb-B	10.0	41.0	3.2	ALMA	CO(2-1)	1.3/63	2.5	0.12
NGC 3627	Sb-AB	8.28	62.0	3.6	ALMA	CO(2-1)	1.3/52	2.5	0.09
NGC 4254	Sc-A	16.8	27.0	6.5	ALMA	CO(2-1)	1.6/130	2.5	0.06
NGC 4303	Sbc-AB	17.6	25.0	7.4	ALMA	CO(2-1)	1.5/128	2.5	0.10
NGC 4321	Sbc-AB	15.2	27.0	7.9	ALMA	CO(2-1)	1.4/103	2.5	0.09
NGC 4535	Sc-AB	15.8	40.0	3.9	ALMA	CO(2-1)	1.5/115	2.5	0.08
NGC 5068	Scd-AB	9.0	26.9	1.1	ALMA	CO(2-1)	0.9/39	2.5	0.24
NGC 6744	Sbc-AB	11.6	40.0	8.1	ALMA	CO(2-1)	1.0/56	2.5	0.18
M51	Sbc-A	8.39	21.0	77	PdBI	CO(1-0)	1.2/49	5/0	0.31
M31	Sb-A	0.79	77.7	16	CARMA	CO(1-0)	5.5/21	2.5	0.19
M33	Scd-A	0.92	58.0	0.3–0.6	IRAM 30m	CO(2-1)	12/54	2.6	0.04
Antennae	Merger	22.0	-	-	ALMA	CO(3-2)	0.6/54	5.0	0.13

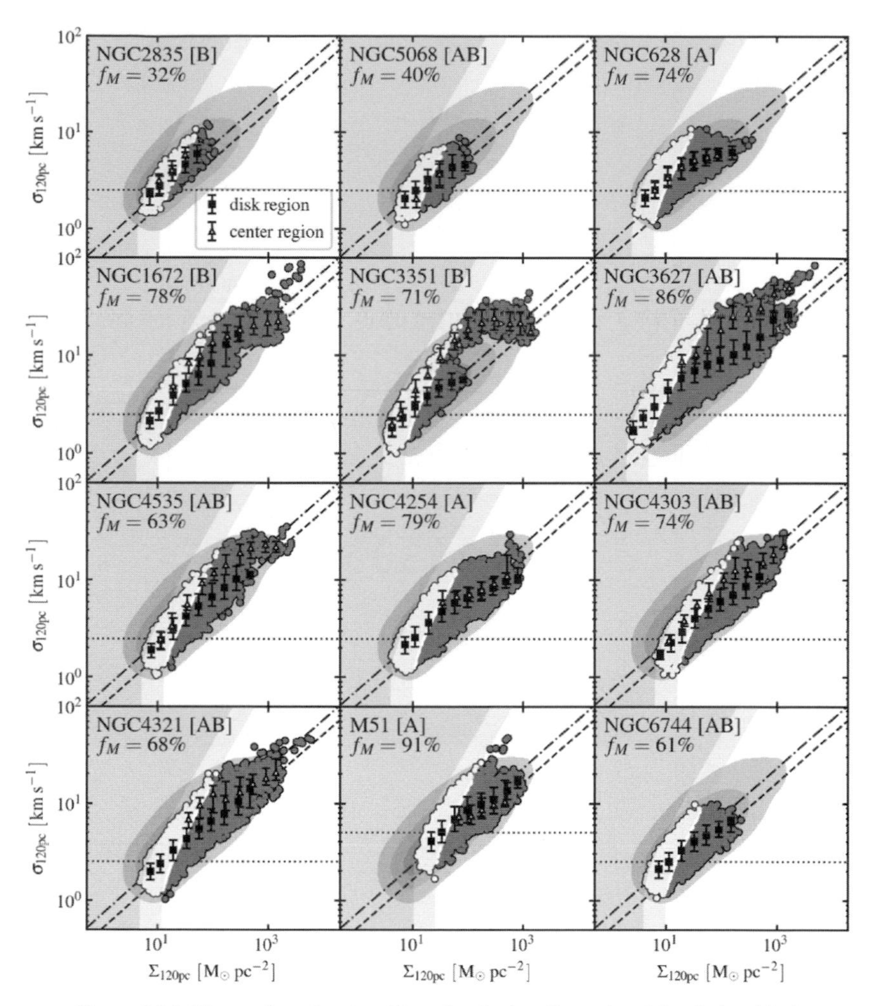

Figure 16.6 The surface density (Σ) and velocity dispersion (σ) relationship in which the diagonals show σ proportional to $\Sigma^{0.5}$ for virial factor $\alpha_{vir} = 1$ (dashed) and 2 (dashed-dotted); f_M is the molecular gas mass fraction. The shading defines a variety of sample limits (Sun et al., 2018).

values not only correlate with one another but also correlate with, for example, the kinetic energy (K) to self-gravitational potential energy (U_g) ratio. For the simple spherical approximation, this is the virial parameter in which

$$a_{vir} \equiv 2K/U_g = (5\sigma^2 R)/\varphi GM,$$

where M, R, and σ refer to the cloud mass, radius, and one-dimensional velocity dispersion, respectively. φ is a geometrical factor that quantifies the density structure inside the cloud. For spherical clouds with a radial density profile $\rho(r)$ proportional to $r^{-\gamma}$:

$$\varphi = (1-\gamma/3)/(1-2\gamma/5) \text{ (Bertoldi \& McKee, 1992).}$$

The other parameter derived from CO flux-deduced Σ and σ values is the internal turbulent pressure, P_{turb}. For a line-of-sight depth ~2 R, P_{turb} is $\sim\rho\sigma^2$ and $\sim\frac{1}{2}R\Sigma\sigma^2$. In the sample survey P_{turb} ~10^3–10^5 Kcm^{-3} characterizes the low-mass galaxy disks, while P_{turb} ~10^5–10^7 Kcm^{-3} characterizes that of the high-mass ones (Sun et al., 2021). Figure 16.7 summarises the Σ and σ relationship for the disks of the sample galaxies, as well as α_{vir} and P_{turb} values for disk and central locations. Line width is shown as a function of surface density (*upper plot*), as

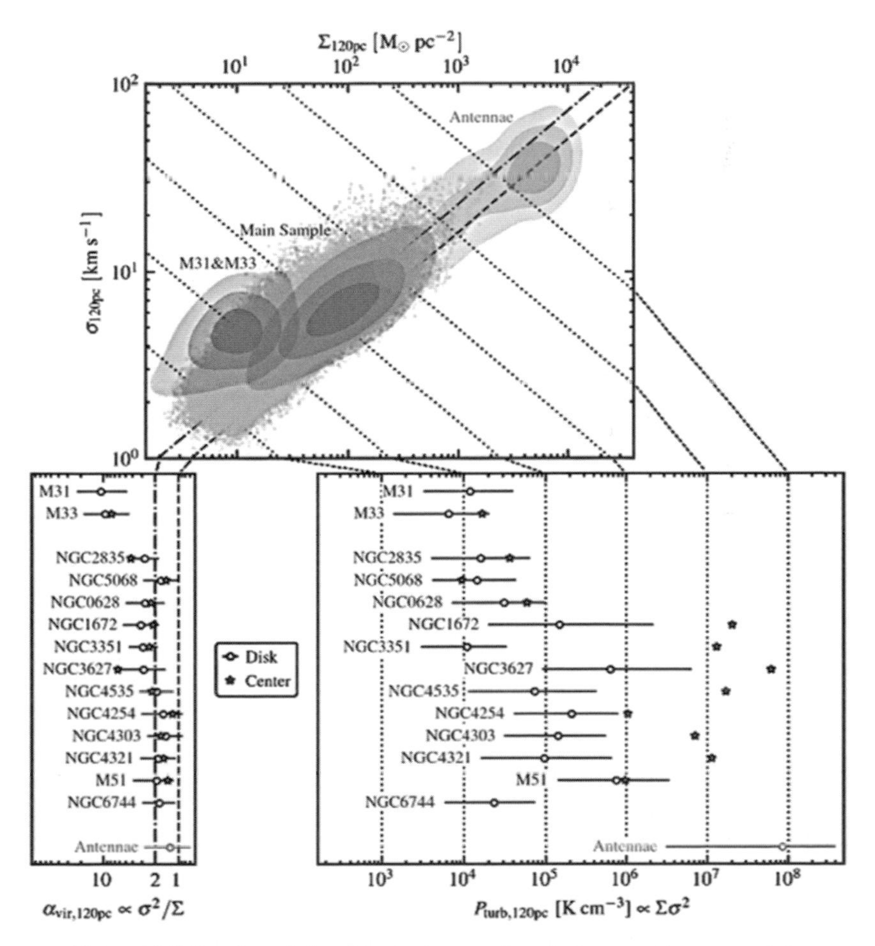

Figure 16.7 Σ and σ relationships in the sample galaxies as well as a_{vir} and P_{turb} values (Sun et al., 2018).

well as the mass-weighted distribution of virial and turbulent pressure parameters (*lower left and lower right plots, respectively*), for all targets at 120 pc resolution.

At this spatial scale, galaxies show a wide range of surface density and velocity dispersion, so a wide range of internal turbulent pressure, alongside a narrow virial parameter range. In general, the internal turbulent pressure varies dramatically within individual galaxies as we would expect for the range of expected environments, and it varies considerably but systematically between galaxies (again as we would expect in this sample with diverse star-formation activity).

16.5 Summary

The dynamical states of GMCs are important for both galactic-scale theories and simulations of star formation. High kinetic energies will reduce star-formation rates for a given mass and density of gas, the evolution of cloud collapse depending primarily on the balance between kinetic and gravitational potential energies. The CANON survey, identifying nearly 200 GMCs within a small sample of external galaxies, confirms some basic molecular cloud correspondences (such as virial mass and velocity dispersion) with their Milky Way equivalents. Higher-resolution observations of nearly 100 galaxies, and out to over double the CANON distance, in the PHANGS-ALMA survey confirm the correlation between gas surface density and velocity distribution.

17

ST16 and N113 in the Large Magellanic Cloud

17.1 Introduction

The larger of two satellite galaxies to our own, the Large Magellanic Cloud (LMC) is a hundred times less massive and a tenth the diameter of the Milky Way. It has a central stellar bar and distorted spiral arms arising from tidal interactions with both the Small Magellanic Cloud and the Milky Way. The LMC is expected to merge with our Galaxy in just a few billion years' time. One of the many interesting comparisons to be made between the two galaxies is the lower metallicity of the LMC and the influence of that on star-formation chemistry traced through molecular emission. By metallicity we mean the abundance of heavier elements arising from nucleosynthesis within the more massive stars or during the cataclysm that is a supernova event. It used to be thought that the lower metallicity value is the result of a galaxy having either fewer high-mass stars or fewer generations of recycled elements. More recent research has shown a more significant impact from star formation that actually ejects metals from low-mass galaxies, aided by gravitational interactions with a neighbouring galaxy such as our own. The mean metallicity of the LMC for some elements can be as much as 50 per cent below that of our own Solar neighbourhood values, although the overall LMC range about the mean is considerable with the older metal-poor star clusters tending to the galactic periphery and the younger metal-richer clusters towards the centre.

While the metal content of the LMC has been studied in the optical, UV, and near-IR, the chemical composition of the LMC's interstellar medium has been extensively studied in the radio frequency band. Single-dish surveys have looked at molecular cloud chemistries ($\leq$ 10 pc) (Tang et al., 2017); interferometry at millimetre wavelengths has looked at clump-scale dense molecular gas (~1–3 pc) (Anderson et al., 2014); early ALMA millimetre and submillimetre observations have already focused on dense molecular gas around YSOs

(<1 pc) (Nayak et al., 2018); and there have also been extensive infrared probes of ice mantles in absorption towards embedded YSOs (Seale et al., 2011). Additionally, several hot cores in the LMC have now been observed with ALMA, which at LMC distances typically offer spatial resolutions of just a few arcseconds (~0.1 pc). Of the small sample of hot cores studied to date, some have shown that organic species such as CH_3OH, H_2CO, and HNCO are underabundant by up to three orders of magnitude compared to typical Galactic hot cores, and we can study these to quantify how low metallicity is impacting on the chemical evolution of star-forming molecular clouds. In this chapter we will look at two particular sites in the LMC, first that of ST16 and then N113.

17.2 ST16

Firstly, consider a newly discovered low-metallicity hot core identified towards an embedded high mass YSO, the infrared source IRAS 05195-6911, otherwise known as ST16. This is located close to the star-forming region N119 (Shimonishi et al., 2020), a massive spiral-shaped HII region extending over a 23 k pc^2 area and harbouring the open star cluster NGC 1910, shown in Figure 17.1. The N119 region consists of several large-scale expanding filaments forming bubble-shaped nebulae, shown in Figure 17.2. One of the bubbles, L132a, is home to NGC 1910 and includes the luminous blue variable S-Doradus, identifiable in Figure 17.3. S-Doradus is just under 50 kpc from Earth and a million times brighter than the Sun, one of the brightest known stars in the LMC.

N119 extends over a $10' \times 12'$ region of sky, its centre located at RA $5^h\ 18^m\ 45^s$, DEC -69° 14′ 03″. The embedded YSO ST16 lies a few arc minutes (~60 pc) from N119's centre. The left-hand panel of Figure 17.4 shows the continuum flux distribution at 870 μm centred on the ST16 embedded target, while the right-hand panel shows the higher-temperature distribution of CH_3OH as representative of the hot core emission at the 0.1 pc scale.

A representative spectrum from the recent ALMA survey that includes CH_3OH lines is shown in Figure 17.5, and a rotation diagram analysis with results for CH_3OH at two distinct transitions is shown in Figure 17.6.

The rotation diagram analysis, as mentioned in earlier chapters, is appropriate where multiple lines with different excitation energies are detected, and where optically thin conditions and local thermodynamic equilibrium (LTE) can reasonably be assumed. The implication of the two transitions is clearly that there are two gas components along the line of sight, that, when averaged

Figure 17.1 The box marks the open cluster NGC 1910 located in the N119 HII region (ESO/Robert Gendler).

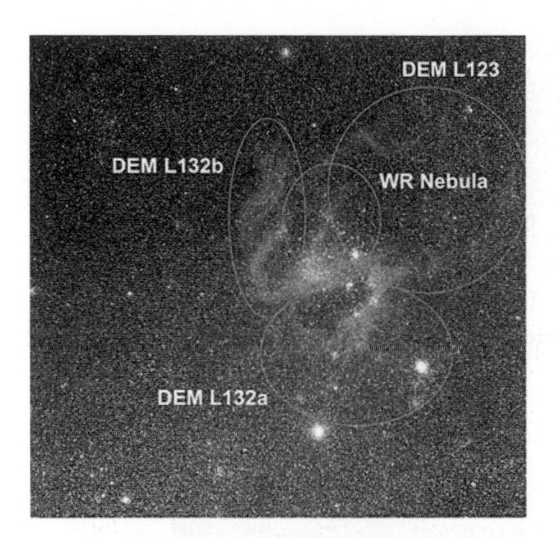

Figure 17.2 The N119 'bubbles' and the location of S-Doradus in the upper centre of L132a (ESO/Robert Gendler).

against results for other species, suggest a hot inner gas at T_{rot} ~150 K and a warm ambient gas at T_{rot} ~50 K.

17.3 Column Densities

While the derivation of column density for unobserved H_2 can follow a standard treatment assuming optically thin dust emission, the differences between LMC conditions and those in our own Galaxy are instructive. For example, using the dust continuum data mapped in Figure 17.4, we can adopt:

$$N_{H2} = (F_v / \Omega) / 2K_v B_v(T_d)\, Zmm_H,$$

where (F_v / Ω) is the continuum flux density (per beam angle), K_v is the mass absorption coefficient of dust grains coated with thin ice mantles appropriate to 870 μm, T_d is the dust temperature, $B_v(T_d)$ the Planck function, *Z* the dust-to-gas mass ratio, *m* the mean atomic mass per hydrogen atom, and m_H the hydrogen mass. The dust-to-gas mass ratio in our own galaxy is about 0.008, whereas in the LMC, scaling by metallicity to ~0.33 $Z_\odot$ gives a ratio value around 0.0027. The dust temperature is obviously a key assumption as is the observation that the submillimetre continuum emission is almost entirely thermal emission from dust grains.

Alternative methods of estimation use their own assumptions. For example, one approach adopts a total visual extinction value (A_V) based on the spectral energy distribution (SED). Figure 17.7 shows that spectral energy distribution

Figure 17.3 The spiral-shaped massive HII region N119 observed in Hα. S-Doradus sits in the crook of the lower spiral 'elbow' (ESO/Robert Gendler).

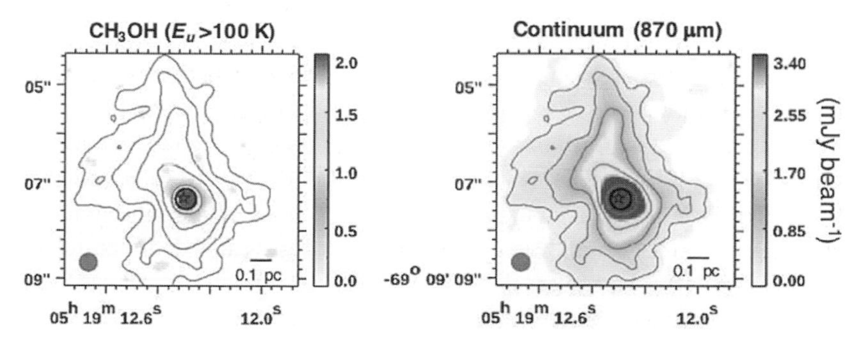

Figure 17.4 Continuum flux distribution for ST16 at 870 mm and the higher-temperature emission of CH_3OH (ALMA/ Shimonishi et al., 2020).

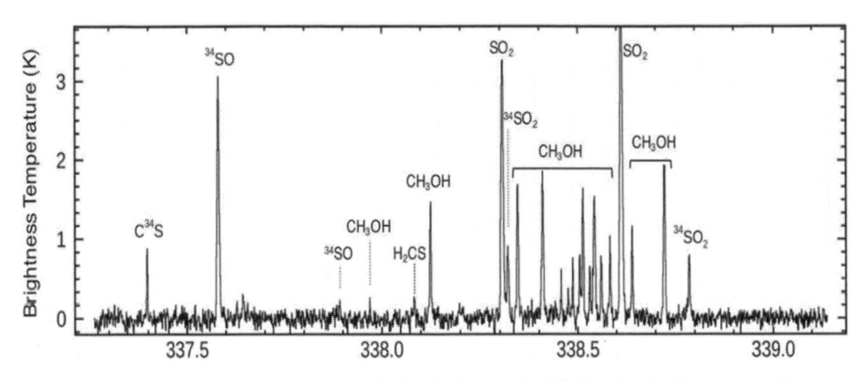

Figure 17.5 An ST16 spectrum obtained from a 0.42″ (0.10 pc) diameter region centred on the continuum and molecular emission peak, with selected lines identified (ALMA/ Shimonishi et al., 2020).

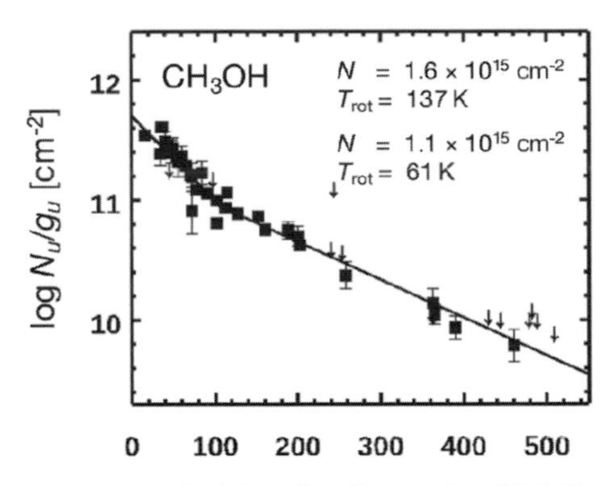

Figure 17.6 Rotation diagram for CH_3OH with the derived column densities and temperatures for two distinct transitions (Shimonishi et al., 2020).

based on a range of observations with a variety of instruments (the bolometric luminosity of the source estimated to be ~3×10^5 $L_\odot$ by integrating the SED from 1 μm to 1,200 μm). Various approximations then bring us to an N_{H2}/A_V value that assumes all hydrogen to be in molecular form. A third model derives A_V from the mid-infrared silicate dust absorption spectrum. In the case of the ST16 ALMA observations, different methods result in fairly consistent results, averaging $N_{H2} = (5.6 \pm 0.6) \times 10^{23}$ cm^{-2} (Shimonishi et al., 2020). This enables us to quantify relative abundances for all the observed species. If we also assume, as a first approximation, a uniform spherical distribution of gas around the protostar, we can estimate an average gas density $n_{H2} = 3 \times 10^6$ cm^{-3} and a total gas mass ~100 $M_\odot$ as benchmarks from which to develop more sophisticated models.

One proviso: the derived column densities are based on total line-of-sight detections because there have yet to be the selective high-temperature hot core

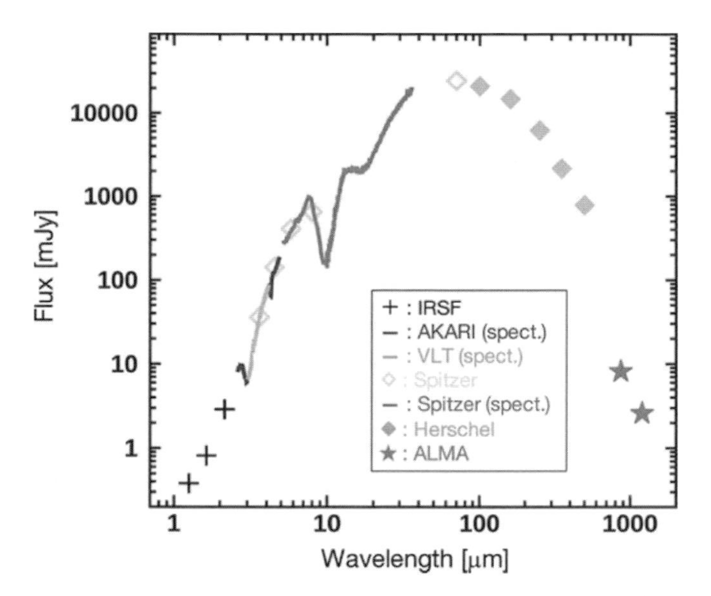

Figure 17.7 The spectral energy distribution of the ST16 high-mass YSO based on several databases (Shimonishi et al., 2020 and references therein).

observations towards ST16 (in, say, the high-excitation lines of CH_3CN discussed in Chapter 15) typically utilized towards Galactic hot cores. Also, in ST16 an average dust temperature of 60 K reflects the mass-weighted average along a line of sight in which the low-temperature component certainly makes a non-negligible contribution. Nonetheless, we can be confident that for those gas-phase species observed in the higher-excitation lines available (including CH_3OH, for example), column density totals almost exclusively originate from the hot core region rather than that of the surrounding core and clump, hence relative fractional abundances do have real meaning.

17.4 Comparative Abundances

In general, the fractional molecular abundances of ST16 are smaller than those of Galactic hot cores. However, different species do show different degrees of decrease. Familiar hot core species CH_3OH and H_2CO, with fractional abundances of $(4.8 \pm 0.9) \times 10^{-9}$ and $(3.8 \pm 0.6) \times 10^{10}$ respectively, are at least an order of magnitude lower than their Galactic counterparts. C_2H, which we have discussed as both a cold dense-cloud radical and an indicator of warm carbon-chain chemistry (WCCC) in earlier chapters, is less abundant but less significantly so. The HCO^+ abundance is essentially the same in both galaxies.

Of nitrogen species, apart from NO whose abundance is comparable, all are less abundant by up to a factor of 10. Sulphur- and silicon-bearing species (CS, H_2CS, and SiO) are all less abundant by over an order of magnitude. SO_2 abundance is down by a factor of four, while SO is not significantly different. OCS, while perhaps comparable or slightly down, has the greatest uncertainty associated with its detections. Figure 17.8 therefore offers a schematic of the overall structure of the ST16 hot core in terms of molecular distribution and temperature profile as deduced from the observations and analysis.

Galactic hot cores show chemical variations of their own, and the small sample of hot cores observed to date in the LMC, not unexpectedly, also show variations. In particular, organic molecules show significant abundance variations in the low-metallicity LMC cores. This does not appear to be due to any site-specific lack of carbon or oxygen, for COMs abundance in the LMC generally scales with the metallicity. Nonetheless, with large chemical diversity among organic species, we must look for other factors than merely the low-metallicity environment. One interesting idea is the 'warm ice chemistry' hypothesis (Shimonishi et al., 2016a, 2016b). This suggests that the low dust extinction values generally in the LMC and specifically in some locations is likely to result in warmer dust than is typical of our own Galaxy and this inhibits the hydrogenation of CO in ice-forming dense clouds. We have previously noted the fundamental nature of this reaction process in initiating the usual dense-cloud gas and grain-surface COMs reaction networks. Low abundances of the usual Galactic products, CH_3OH, HNCO, and H_2CO (although the latter docs also have quite efficient gas-phase production routes) would therefore

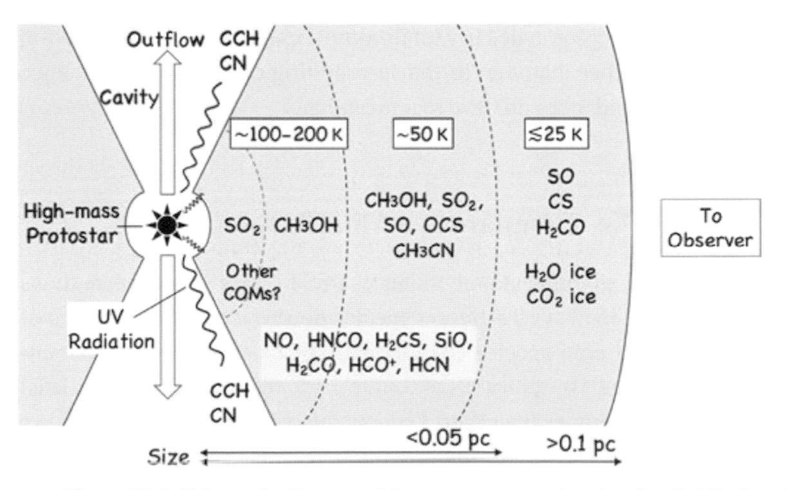

Figure 17.8 Schematic diagram of the temperature and molecular distributions in the ST16 hot core (Shimonishi et al., 2020).

arise from different chemical histories during the ice-formation stage of cloud collapse. Variations in hot core gas-phase make-up would inevitably arise, independent of low metallicity considerations. However, we must not assume comparable timescales for the observed hot cores. Species such as CH_3OH simply diminish in abundance with time as a result of chemical processing. Perhaps we are simply observing this across the hot core sample. In contrast to COMs, abundance of the dominant hot core sulphur species, SO_2, seems invariably to scale with metallicity across the cores observed to date. Of note is the fact that the isotope abundance ratios for sulphur towards ST16 are estimated as $^{32}S/^{34}S = 17$ and $^{32}S/^{33}S = 53$ (based on SO, SO_2, and CS isotopologues), suggesting that both ^{34}S and ^{33}S are overabundant in the LMC relative to Galactic values.

17.5 A Rotating Envelope

Observations of ST16 show evidence additionally for both a rotating protostellar envelope and an outflow cavity. While the hot core emission corresponds precisely with the continuum centre of ST16, there are no near- to mid-infrared detections (hydrogen recombination lines or fine-structure lines from ionized elements), so the protostar though massive (>100 $R_\odot$) is at an early evolutionary stage well before the emergence of a detectable HII bubble. The infrared spectral characteristics of ST16 strongly suggest at least 10 per cent of elemental oxygen remains trapped in ices along the line of sight. Given hot core temperatures high enough to sublimate ices, this still-depleted oxygen must lie in the cooler foreground envelope. How this compares with the abundances of the dominant ices in our Galaxy (H_2O, CO_2, and CO) and how it contributes to subsequent hot core gas-phase mixes remains to be seen. What is seen, however, in the velocity maps of ^{34}SO and SO_2 is evidence of a rotating protostellar envelope, the first such extragalactic observation.

The SO_2 velocity distribution map shown in Figure 17.9 has a greyscale-indicated offset velocity relative to the systemic velocity. With the high-mass YSO location marked with a central star symbol, the central light grey indicates an equivalence to the systemic velocity, the slightly darker grey to the west (on the right) indicates a negative relative velocity (blue-shifted), and the darker grey to the east (on the left) indicates a positive relative velocity (red-shifted). Given the clearly apparent rotation, the diagonal dotted line indicates the likely outflow axis also inferred from other molecular emission distributions, including C_2H and CN (Shimonishi et al., 2020).

The beam width shown as an oval in the lower left corner of Figure 17.9 corresponds to major and minor axes of 0.090 pc and 0.076 pc respectively, so an

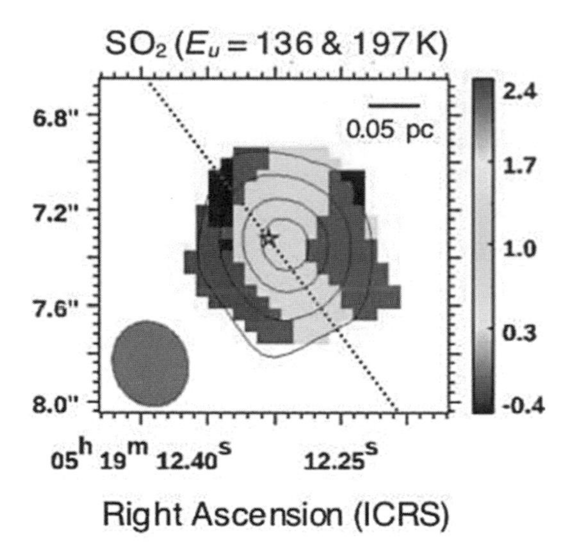

Figure 17.9 An SO_2 velocity map for ST16, red-shifted gas as dark greyscale to the left, blue-shifted dark greyscale to the right. The central YSO is marked as a star and the diagonal dotted line indicates the likely outflow axis (Shimonishi et al., 2020).

area of 0.02 pc^2 in the plane of the sky. Across the roughly 0.2 pc spread of SO_2 emission, there is a velocity separation of 2–3 km s^{-1} between the red-shifted east side and the blue-shifted west side of the core. Since the direction of velocity separation is almost perpendicular to the apparent outflow axis, SO_2 is much more likely to be tracing rotation than outflow motion. The C_2H and CN evidence for a bipolar outflow axis comes from emissions peaking at the north-east and south-west perimeters of the hot core, particularly well collimated on the north side. Since both of these molecules are known to be bright in photo-dissociation regions (PDRs) they may well be sensitive tracers of early stellar UV-irradiation. This would suggest an irradiated PDR-like outflow cavity and, since the dust extinction from the outer edge of the dust clump to the C_2H-CN emission region is in excess of A_V ~20 mag, the UV source is likely to be the evolving high-mass protostar rather than any external radiation field.

17.6 N113

A second source is N113, another of the most prominent star-formation regions in the LMC, containing a massive (~10^5 $M_\odot$) and molecular emission–rich GMC. Its clumpy structure is directly evident in high-density tracer emission of HCO^+ and HCN (e.g. Seale et al., 2012) located within ~6 pc of the peak CO(1-0) emission (Green et al., 2008; Ellingsen et al., 2010). Point-like mid-infrared emission, masers, and compact HII regions superimposed on extended emission and aligned north-west to south-east, show star formation concentrated in the central region of N113. The gas and dust at this location appear compressed by a complex structure of ionized gas bubbles (prominent in Hα detections) engendered by massive stars in several young clusters (Oliveira et al., 2006). SST and HST data show stage 0-II YSOs at masses down to ~3 $M_\odot$ (e.g. Carlson et al., 2012) as well as three massive YSOs (30–40 $M_\odot$) labelled YSO 1, 3, and 4 in Figure 17.10.

Overall, N113 is a complex environment with multiple clumps on a scale ~1 pc in the plane of the sky and cores between 0.1 and 0.2 pc, sites strongly influenced by stellar winds in the west and more-quiescent conditions in the east, showing multiple star-formation locations at a range of evolutionary stages.

17.7 Low-Metallicity COMs

In galaxies with sub-solar metallicities, questions remain as to the formation efficiency of COMs (e.g. Shimonishi et al., 2016a). Low metallicity correlates positively with low dust abundance in a feedback loop with star-formation

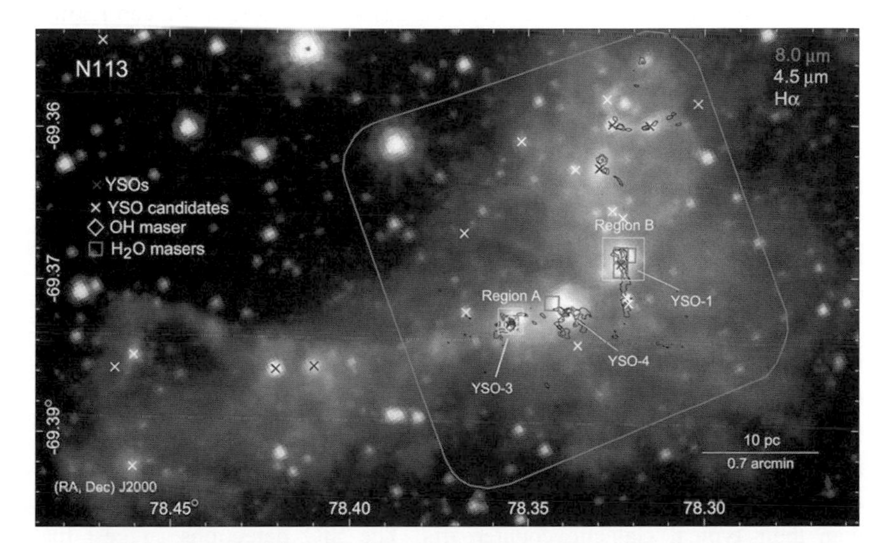

Figure 17.10 The central region of N113 (*inside the rectangle*) targeted by ALMA observations, showing YSOs 1, 3, and 4. The contours represent ^{13}CO (2-1) integrated emission, and the smaller boxes indicate the locations of COMs emission (Sewilo et al., 2018).

rates and UV radiation field intensities. Reduced dust shielding and a lower molecular fraction (essentially less dust, so less atomic H to molecular H_2 conversion) impact on dense-cloud chemistry. The typical gas-to-dust ratio in Galactic dense clouds is ~100 (although significantly lower towards the Galactic Centre and higher in the outer Galaxy). In comparable LMC regions the ratio is almost double that (Acharyya & Herbst, 2015). Apart from the lower elemental abundances of gas-phase heavier atoms (including C, N, O, and S, that contribute so significantly to observable COMs), the greater penetration of UV photons into dense gas warms both gas and dust grains that are present. All of these factors potentially inhibit the formation and survival of COMs which, we reiterate, evolve most efficiently through gas–grain interactions on cold grain surfaces.

Two small, white squares in Figure 17.10 are labelled Regions A and B. These are the two regions within the ALMA observations in which COMs emission was detected (Sewilo et al., 2018). The two images show 1.3 mm continuum distribution, the peaks of which are labelled where concomitant with Hα and molecular line peaks. One site in each region, A1 and B3 respectively, has been identified with COMs emission, each of which are also associated with H_2O masers (and A1 also with OH masers).

Figure 17.11 shows A and B in continuum (contours) and Hα (greyscale) with particular sources A1 and B3 marked. The B3 location is actually brighter in DCN line emission than A1, which suggests it is younger. We can remind ourselves that COMs emission in both A1 and B3 is identifying gaseous products in the envelope within which grain-surface processes have delivered complex products, before protostellar warming has released these into the gas. Earlier chapters have also emphasised that a more deuterated gas is characteristic of lower temperatures, so we also expect the DCN/HCN ratio to drop fairly rapidly with time in warming gas, whether the DCN originates in the gas phase or in ice. Column densities estimated at $\sim 10^{23}$ cm^{-2} translate to particle densities $\sim 10^{6}$ cm^{-3} for both sources (Sewilo et al., 2018).

The three observed COMs are CH_3OH, CH_3OCHO, and CH_3OCH_3, along with SO_2, H_2S, SiO, OCS, ^{13}CS, ^{33}SO, c-C_3H_2, and DCN, shown in the spectra of Figure 17.12. Given we now have two compact (d ~0.17 pc) and hot (T_{rot}~130 K) sources, with high-excitation lines in methanol, high column and number densities, COMs released from ices into the gas phase, and associated masers, we can identify both as hot cores even if they are slightly larger than their typical Galactic counterparts. The SiO detections suggest that shock sputtering of dust and ice may be as important here as thermal evaporation, although the evidence for outflows is unclear. A non-shock origin for SiO through silicon hydride desorption from ices is possible but does require

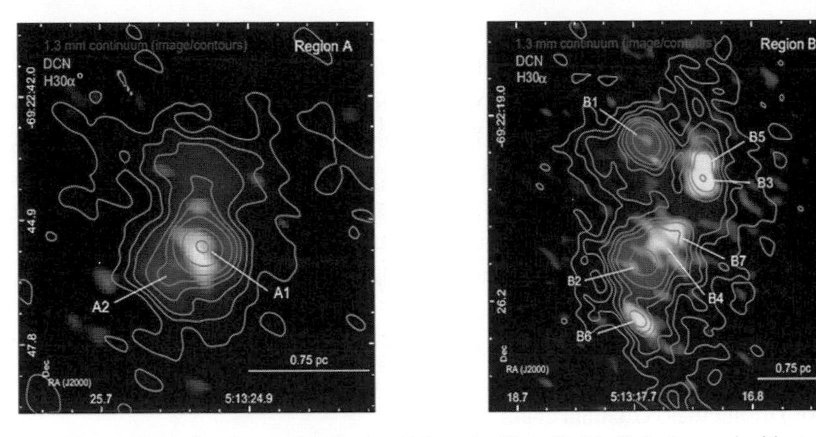

Figure 17.11 Continuum distribution (1.3 mm) with peaks labelled where coincident with Hα and molecular emission peaks (Sewilo et al., 2018).

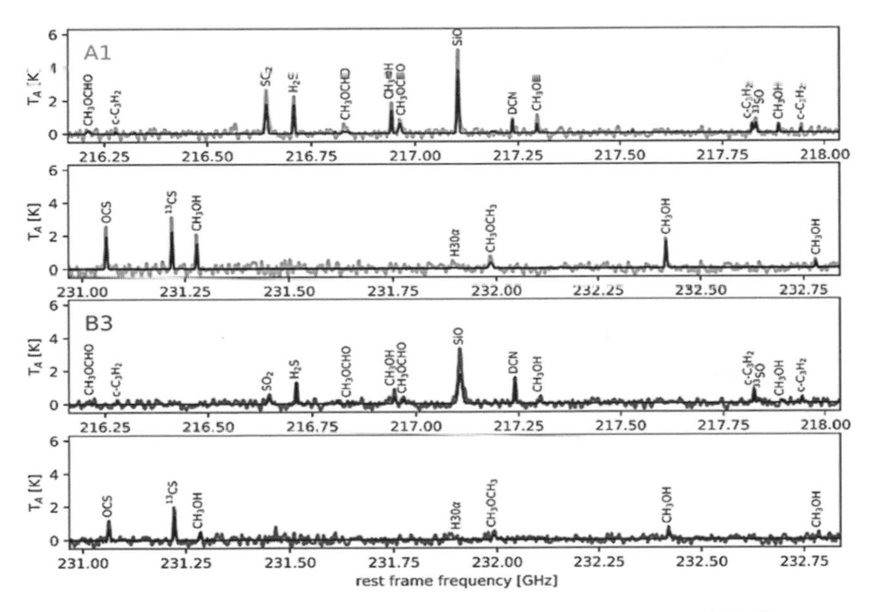

Figure 17.12 Lighter greyscale spectra for A1 (*upper two spectra*) and B3 (*lower two spectra*) with synthetic spectra superimposed in bold (Sewilo et al., 2018).

gas-phase O_2 abundances higher than apparently observed (MacKay, 1996; Sewilo et al., 2018). We have identified CH_3OH repeatedly in preceding chapters as the product of CO hydrogenation on cold dust, and DCN as the product of both cold gas and grain-surface reactions.

As expected for the low-metallicity case, abundances of COMs in N113 are broadly comparable to, but at the lower end of, Galactic values. The general chemistry is similar in both galaxies, and we may expect grain reactions on warm dust (although with reduced efficiency) and post-desorption ion-molecule chemistry to be responsible for CH_3OCH_3 and CH_3OCHO formation (Garrod & Herbst, 2006; Taquet et al., 2016). Both H_2S/CH_3OH and OCS/CH_3OH abundance ratios appear comparable in A1 and B3, encouraging us to think that all three species formed in ices during a shared N113 cold prestellar phase. The sulphur species likely formed through S atom reactions with CO, as discussed in Chapters 14 and 15, with SO and SO_2 commonly formed in post-shock or quiescent, warm, desorbed gas (Hartquist et al., 1980; Charnley, 1997). Finally, c-C_3H_2 is either an ion-molecule cool-gas product or, like C_2H, a PDR indicator produced in the photodestruction of larger refractory organic (PAH) molecules (Guzman et al., 2014).

17.8 Summary

Hot core molecular abundances in the low-metallicity ST16 case are derived and compared with their Galactic equivalents. The central protostar itself is deduced to have a rotating molecular disk evident in SO_2 detections, and an outflow cavity with a UV-irradiated PDR rim detected in C_2H and CN. The lack of evidence for an evolved HII region points to an early stage of HMSF for the ST16 source. The N113 location shows a complex environment with multiple clumps and cores. In both low-metallicity sources, warm dust inhibits formation and survival of COMs, while reaction routes appear broadly comparable with Galactic models.

18

Starburst Galaxy NGC 253

18.1 Introduction

The nuclear starburst galaxy NGC 253 lies at a distance ~3.5 Mpc (about 10,000,000 light years) in the direction of the southern Sculptor constellation. About 70,000 light years across (two-thirds the diameter of our own Galaxy), it is the largest and brightest of the Sculptor Group, the nearest neighbour galaxies to our own Local Group. Intermediate in shape between a spiral and a barred galaxy, it is bright and dusty, the high dust content correlating with rapid star formation. It is a strong X-ray and gamma-ray source, indicating multiple massive central black hole interactions. Caroline Herschel was the first astronomer to identify NGC 253 as a nebula in 1783, using the twenty-foot reflector telescope she shared with her husband William. Figure 18.1 shows a modern ground-based image taken in 2021.

NGC 253 offers an opportunity to study star formation in a relatively extreme environment. Its massive and dense molecular clouds seen in the millimetre and submillimetre (e.g. Sakamoto et al., 2011; Leroy et al., 2015; Meier et al., 2015) are an ideal environment in which super star clusters (SSCs) can form. Such large-scale structures are seen not only at the centres of galaxies such as NGC 253, but in the overlap regions of galaxy mergers such as that of the Antennae. While the extreme conditions in which SSCs form is thought to be rare in the present-day universe, it is believed they were common around the peak of cosmic star-formation rate history (~10.5 billion years ago), which was also the era in which most of today's globular clusters formed. To understand SSCs observable today, is also perhaps to gain insight into the physics of globular cluster formation in the very distant past. High-resolution observations with the Atacama Large Millimetre/Submillimetre Array (ALMA) have identified 14 active, compact, massive SSCs in the central high-density region of NGC 253.

Figure 18.1 NGC 253 (R. Jay GaBany/Cosmotography.com, 2021).

18.2 Super Star Clusters

The 14 SSC sites in NGC 253 are shown in Figure 18.2, detected through 350 GHz dust continuum emission. Previous chapters have illustrated how clusters of high-mass stars feed back energy and momentum into the surrounding ISM, kick-starting fragmentation and core collapse, potentially triggering successive fresh star formation. Outflows from young clusters are particularly influential in clearing away ambient gas and thereby affecting star formation efficiency rate (SFR). We have already referred in the HMSF chapters to the range of physical processes by which gas clearing occurs, including photoionization, radiation pressure, stellar winds, and neighbouring supernovae impacts. Clustered environments promote higher SFRs, and the typically deeply embedded SSCs will have masses $>10^5$ $M_\odot$ within radii ~1 pc. The SFR in the central kiloparsec diameter region of NGC 253 is estimated to be ~2 $M_\odot$ yr^{-1} (Bendo et al., 2015; Leroy et al., 2015) concentrated in dense clumps, knots, and clouds of gas (Turner & Ho, 1985; Ulvestad & Antonucci, 1997; Paglione et al., 2004; Sakamoto et al., 2006, 2011; Ando et al., 2017).

The 14 SSCs are located within these dense gas structures. In their original identification, a resolution of 0.11″ (~1.9 pc) offered minimal resolving power for these sites having smaller radii, between 0.6-1.5 pc (Leroy et al., 2018). Follow-up observation also at 350 GHz with ALMA improved that resolution to 0.028″, which at 3.5 Mpc represents ~0.48 pc (~10^5 AU). Three of the 14

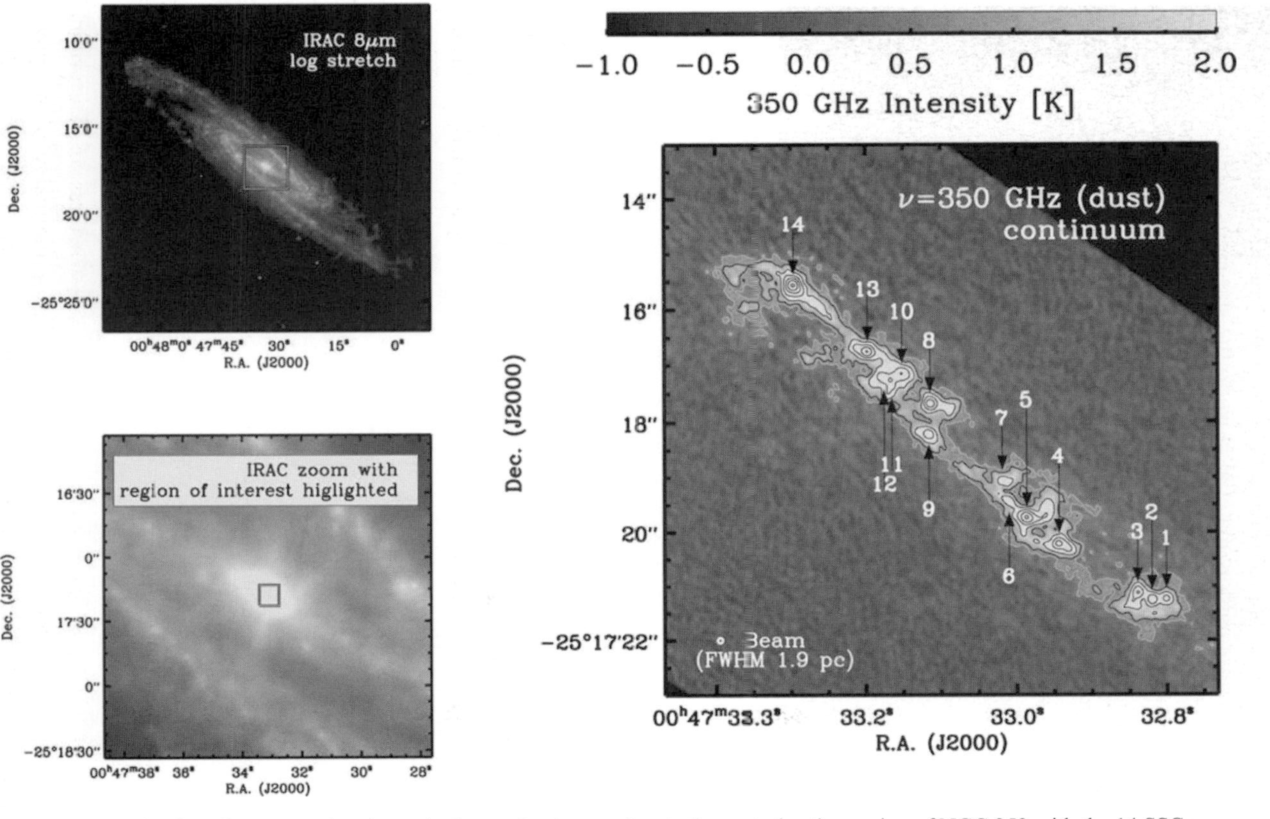

Figure 18.2 Three images tracing the scale from galactic overview to the central active region of NGC 253 with the 14 SSCs identified (Leroy et al., 2018).

show blue-shifted absorption and red-shifted emission line profiles (classic P-Cygni profiles) in several lines. Such a profile arises where sufficiently high-temperature outflow material close to the star emits, while beginning to absorb as it cools in moving further from the central source. Figure 18.3 gives the example of CS (7-6) as a direct signature of massive outflows from the three SSCs (Levy et al., 2021).

Two segments of the full band spectrum (342.5 GHz to 358 GHz) for one of the three SSC locations (SSC 4a) is shown as representative in Figure 18.4 (Levy et al., 2021). Of the many lines detected, apart from the CS and $H^{13}CN$, other lines are also used to determine the systemic velocity so that peak

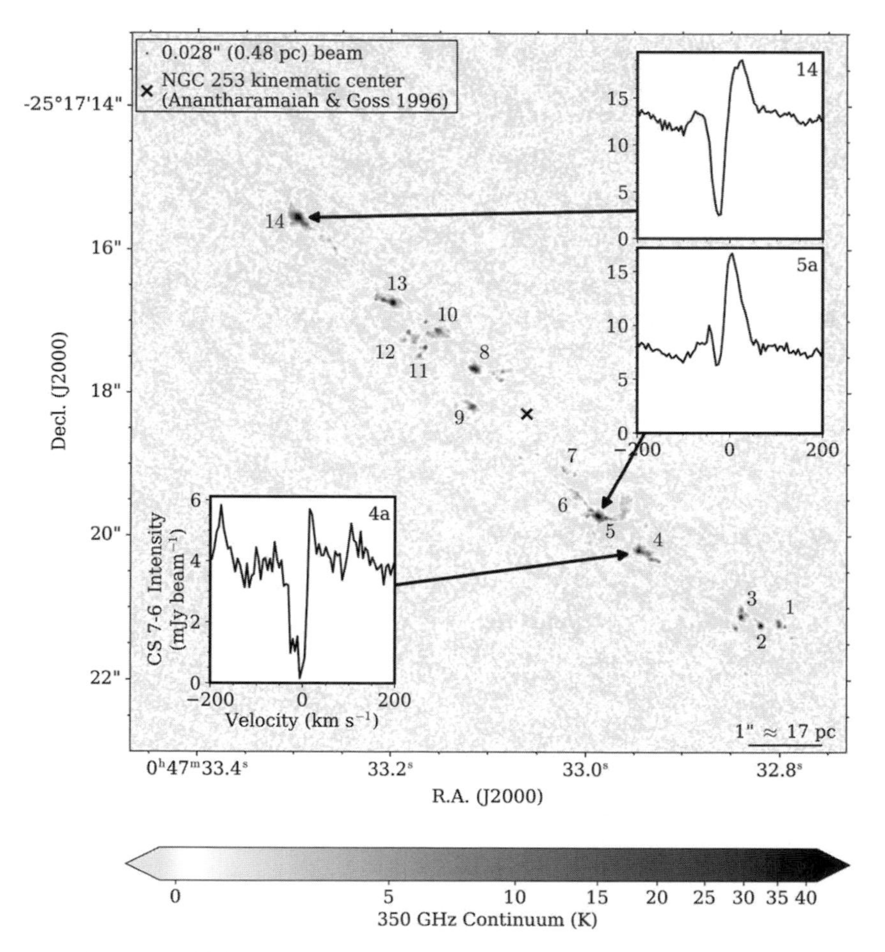

Figure 18.3 CS (7-6) spectra towards the three P-Cygni profile sites in NGC 253, showing long-wavelength emission and shorter-wavelength absorption (Levy et al., 2021).

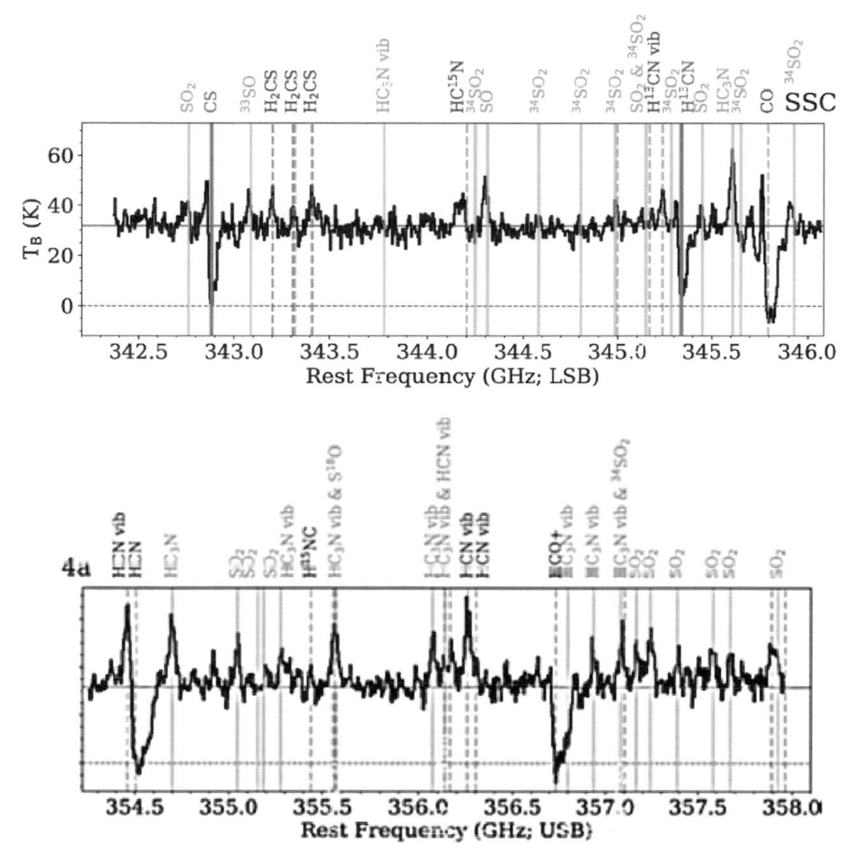

Figure 18.4 A sample full band spectrum for one of the three SSC locations (4a) with P-Cygni profiles evident (Levy et al., 2021).

intensities are mapped within ±200 km s^{-1} of the galaxy's systemic velocity. The strong CO (3–2) emission, while prominent, is less indicative of the SSC itself, since it incorporates large-scale gas motions from clouds along the line of sight, and it is difficult to know how much originates in the cluster location itself. The wider distribution of CO is evident in Figure 18.5.

The most useful emission lines in assessing physical conditions in the outflow are CS (7-6) and H^{13}CN (4-3), both of which show bright lines and, having high critical densities of 3–5 × 10^{6} cm^{-3} and 0.8–2 × 10^{7} cm^{-3} respectively, undoubtedly probe gas that is localized in the clusters themselves rather than in less-dense foreground material. Figure 18.6 shows the peak intensity maps for CS and H^{13}CN. The contours indicate the dust continuum at 350 GHz delineating the SSCs and the square plots zoom in to 1″ (~17 pc) scale, clearly showing the localized and spatially resolved emission.

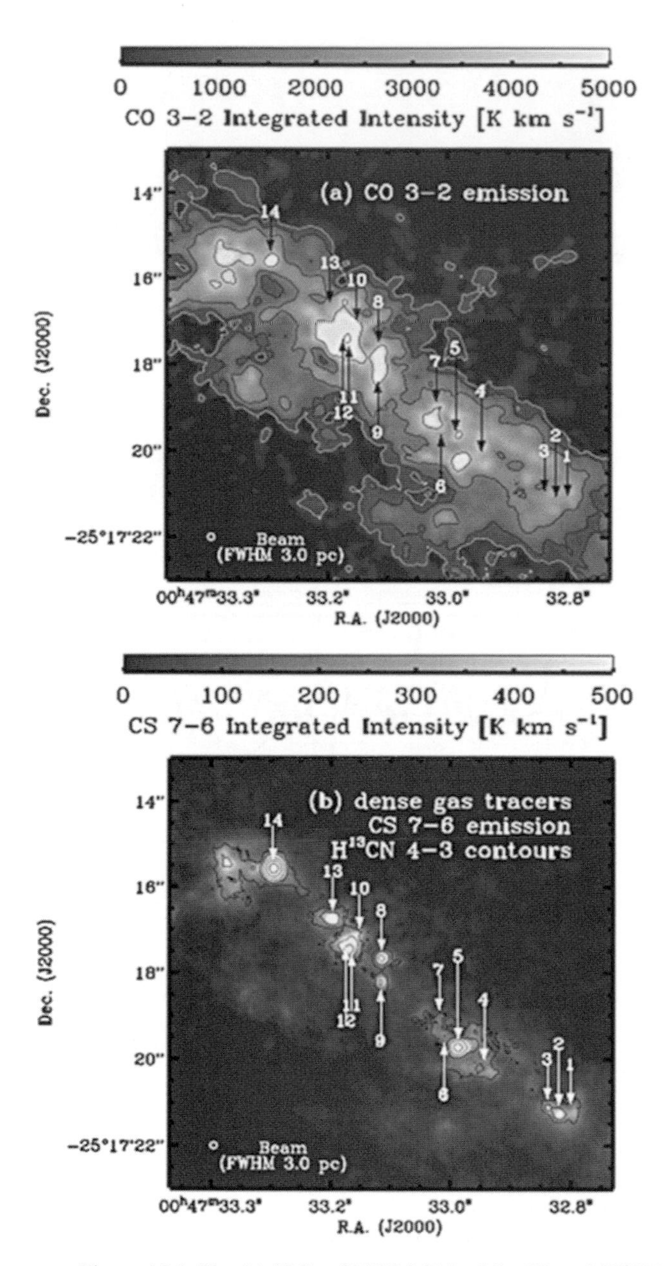

Figure 18.5 The 14 SSCs of NGC 253 in CO, CS, and $H^{13}CN$ emission (Leroy et al., 2018).

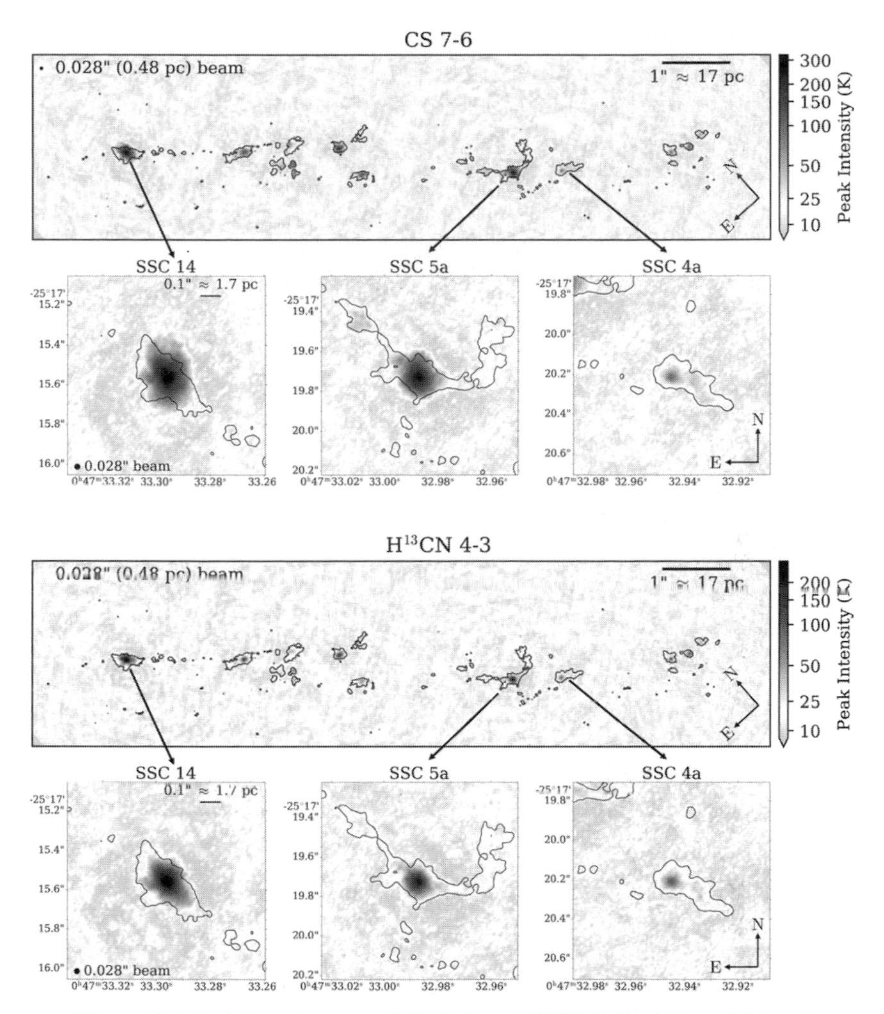

Figure 18.6 Peak intensity maps of CS (7-6) and $H^{13}CN$ (4-3) within ± 200 km s^{-1} of systemic velocity in NGC 253 (Levy et al., 2021)

18.3 Physical Parameters from Absorption Lines

While all the clusters show different velocities compared to the ambient bulk gas motions, at 0.5 pc resolution many of the SSC sources are shown to fragment into a bright, massive primary cluster surrounded by smaller satellite clusters, as is evident in Figure 18.6. SSC 4, for example, appears as six dust clumps, in which the primary location, SSC 4a, shows a peak intensity four times brighter than any of the smaller associated clumps. Focusing on the three locations (4a, 5a, and 14) in which P-Cygni profiles indicate outflow activity, what can be deduced about the physical properties of these outflow phenomena?

Consider Figure 18.7 which shows how absorption line profiles can be fitted (or not) to Gaussian curves and spherical outflow model profiles, CS (7-6) in panel (a), $H^{13}CN$ (4-3) in panel (b). The shaded background regions show where other lines can contaminate the spectrum, but best fits can still give a velocity for the bulk of outflow material at maximum absorption. If the maximum outflow velocity is 2σ from the average, for a true Gaussian 95 per cent of the material traced has an outflow velocity slower than the maximum. Given a mean velocity and a cluster size, the crossing time (the time taken for a parcel of gas to travel from the centre of the cluster to the edge) can be estimated. Equally, from the best-fit absorption to continuum ratio, optical depth and column density can be derived. From column density and projected size, H_2 mass in the outflow can also be estimated, if we assume, for example, a near-spherical outflow. In fact, opening-angle constraints have been modelled against line profiles to show near-spherical does reflect the facts (Levy et al., 2021). Given the mass, crossing time, and mean outflow velocity, we can calculate the mass outflow rate, the gas removal timescale, the radial momentum injected per unit stellar mass in the cluster, and the total kinetic energy of the outflow (e.g. see Appendices A and B in Levy et al., 2021).

For the three P-Cygni primary clusters, their outflows show maximum velocities ~40-50 km s^{-1}. While these maxima are greater than the escape velocity for gas to leave the cluster environment, the bulk of material traced shows lesser values – for example, in SSC 4a only 20 per cent reaches escape velocity. The assumption is therefore that the bulk of material is eventually regathered by the cluster, the gas crossing time being just a few 10^4 years, which is a lower limit for the outflow age. The outflow masses are obviously large, those derived from $H^{13}CN$ (4-3) being larger than those derived from CS (7-6). The difference may be an optical depth effect (CS being optically thicker), or that the estimated relative molecular abundances are significantly inaccurate, the latter particularly having been extrapolated to parsec scale cluster size from observations of the centre of NGC 253 several hundred times larger (Martin et al., 2006). The $H^{13}CN$-derived masses are thought to be more accurate (Levy et al., 2021).

Short gas crossing times, large outflow masses, and apparent outflow mass rates suggest the outflow activity in these objects is relatively short lived. Perhaps observations of the three SSC outflows identified in NGC 253 are catching a small fraction of those actually present in this short-lived phase.

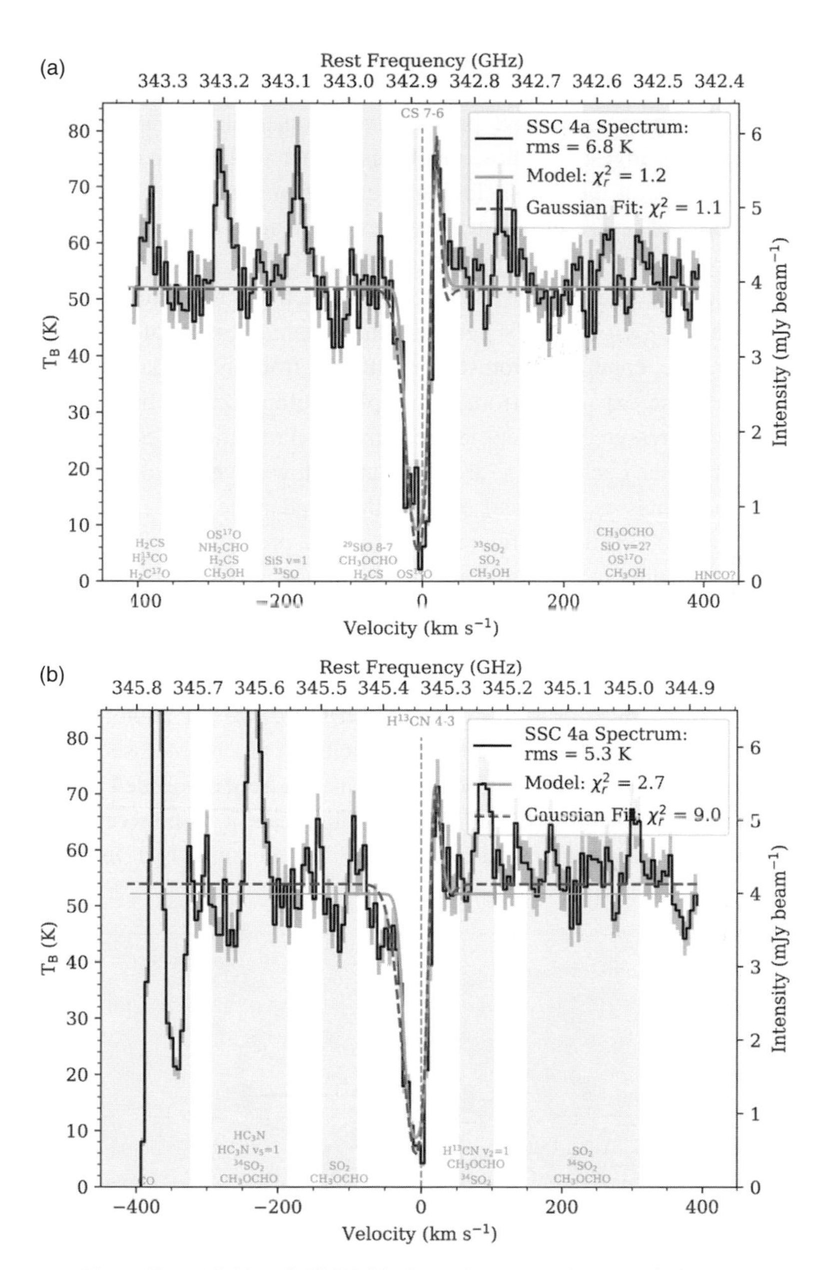

Figure 18.7 CS (a) and $H^{13}CN$ (b) absorption spectra (Levy et al., 2021).

18.4 Outflow Dynamics

Once the physical parameters have been derived from molecular absorption data, it is possible to investigate those mechanisms that are likely to drive outflow activity in each of the SSCs. Having decided that the bulk of observed outflow gas does not escape the cluster confines, these are all feedback mechanisms that determine the star formation efficiency (SFE). Given the stellar masses of these clusters are at least 10^5 $M_{\odot}$, there are likely to be hundreds of O-stars in each SSC (Kroupa, 2001). Among the potential drivers of SSC outflows are, therefore, combined protostellar outflows from individual sources, multiple supernovae expansion fronts, multiple photoionization fronts, UV (direct) radiation pressure, dust-reprocessed (indirect) radiation pressure, and multiple stellar winds. Each of these are efficient outflow drivers across a range of different cluster masses, radii, and ages, as illustrated in the mass-radius diagram of Figure 18.8. While this figure simplifies the predictions of feedback efficiency by excluding momentum carried or duration, it does offer strongly suggestive guidance. In the case of our three SSCs observed with outflows, analysis has shown the most likely combination will be dust-reprocessed radiation pressure and stellar winds (Levy et al., 2021). All three clusters are likely too young for supernovae to have exploded and too dense for photoionization or UV (direct) radiation pressure to be efficient. Dust reprocessed radiation pressure arises when the majority of UV emission from embedded young stars is absorbed by dense surrounding dust. This heats it to between tens and hundreds of degrees Kelvin and re-emission is predominantly at infrared

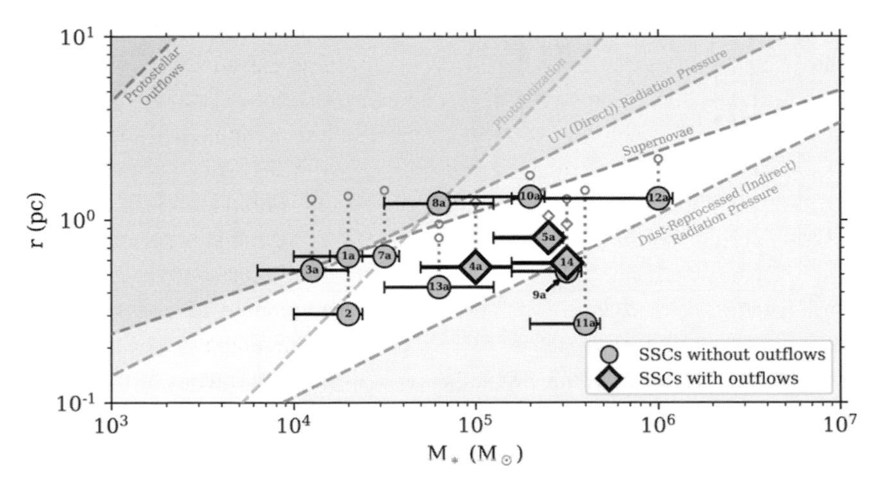

Figure 18.8 A cluster mass-radius diagram (Levy et al., 2021; adapted from Krumholz et al., 2019).

wavelengths. The radiation field produced by the heated dust generates an expansion pressure, the energy density of which can be considerable (Barnes et al., 2020). The continuous ejection of atomic and sub-atomic material from multiple young O-stars (stellar winds), particularly as they cool, has also been modelled to offer sufficient mechanical pressure to at least contribute significantly to cluster outflow (Lancaster et al., 2021).

18.5 Cluster Evolution Chemistry

There are also potential chemical insights into cluster and outflow development suggested directly from likely reaction pathways. SSC evolutionary stages are far from clear cut, especially since cyclic clustering is likely, as we have noted with regard to re-accretion of gas that fails to escape its environment. Nonetheless, typical timescales for cluster formation are taken to be around five times the free-fall time, with complete dispersal around eight times, and within which gas actively collapses to form stars on a timescale between one and two free-fall (Skinner & Ostriker, 2015). Likely lower limits for the overall ages of each cluster have been derived, all close to 10^5 years (Rico-Villas et al., 2020). These values are derived from comparisons with ratios of protostellar luminosities to zero main sequence star (ZAMS) luminosities, so undoubtedly short of the true age.

Alternative timescale estimates have also been considered and these include chemical-clock sequences. Given the multiple uncertainties in physical circumstances and their temporal development, we can recognise that chemical timescales may only give us very approximate possibilities for the general direction of evolutionary timescales of large gas and dust structures. This contrasts with some of the tightly focused locations we have examined in, for example, the chapters on LMSF. Nonetheless, a number of longstanding clock models have still proved useful even on the SSC scale, and the observed CS chemistry we have already mentioned is relevant in the case we are considering. Whatever the sulphur reservoirs in ices that we discussed in detail in Chapters 14 and 15, once desorbed into the more active gas-phase conditions associated with star formation, hydrogenated or polysulphane species will convert efficiently to SO, then SO_2, and eventually to CS, H_2CS, and OCS (e.g. Hatchell et al., 1998). With $[SO]/[SO_2]$ abundance ratios proving a rather direct and reliable potential chemical clock (Charnley, 1997), and SO_2 being widely used as a tracer for low-velocity outflows in stellar cores (e.g. Wright et al., 1996; Liu et al., 2012), we might look similarly to CS abundance.

It has been shown that the relative molecular abundance ratio $[CS]/[H_2]$ varies between 10^8–10^{10} in dense gas conditions as a function of three factors: age, O_2 abundance, and temperature (Charnley, 1997). This modelling shows CS to be most abundant at $\sim 10^4$ years, provided O_2 abundance is low and temperatures are < 100 K. Estimates of the kinetic temperatures of the SSCs in NGC 253 vary from 200 to 300 K (Rico-Villas et al., 2020). At our putative cluster age $\sim 10^5$ years and a temperature of 300 K, while not at maximum CS abundance, a modelled $[CS]/[H_2]$ abundance ratio $\sim 4 \times 10^{-9}$ (Charnley, 1997) is quite close to the $5.0 \pm 0.6 \times 10^{-9}$ derived from rotation diagram analysis of spectral survey results (Martin et al., 2006). We earlier noted the possible optical thickness of CS emission towards SSC 4a against the probably more reliable values derived for $H^{13}CN$. An additional comparison of CS, SO, and SO_2 line ratios across the three clusters has suggested an excitation enhancement by a factor of two or three might be closer to the reality for CS, and which is more in line with the derived value for $H^{13}CN$ (Krieger et al., 2020). The chemical routes to $H^{13}CN$ depend less on nitrogen fractionation than on $^{12}C/^{13}C$ ratios. To reverse our previous deduction and justify a lower $H^{13}CN$ abundance, more closely aligned to that of CS, would require we assume that ratio to be > 300, a value much larger than the highest value measured in NGC 253 to date (Martin et al., 2010) which seems, therefore, unlikely.

18.6 The Wider Molecular Sample

One of the interesting observations across the 14 SSC sources is the variation in chemical complexity of the gas mix (Meier et al., 2015; Kreiger et al., 2020). For example, SSC 14 appears the richest protocluster, with 19 species detected through 55 submillimetre emission lines. In SSCs 2, 3, and 13, almost all those species are also observed, with the exception of $^{34}SO_2$. In contrast, the only bright species in SSC 7 are CO, HCN, HCO^+, and CS. Whether these are all intrinsic chemical differences or the result of physical differences of location, such as the relative excitation environment, is a critical question. SSCs 8 to 12 are all located within ~20 pc of one another in the plane of the sky and are comparable in total brightness in CS, HCN, and HCO^+, while differing significantly in the number of lines and species detected. If the projected spatial proximities reflect the reality, the implication is that the chemical differentiation is likely to be intrinsic.

Multiple components of similar peak intensity are evident in many sources for CO, CS, HCN, and HCO^+. In some cases, the centre of the trough between double peaks in CN, HCN, and HCO^+ coincides closely (< 5 km s^{-1}) with the peak position of other detected lines for the same species. This may be due to temperature gradients in the SSCs, self-absorption within the gas, or absorption against a background source (Krieger et al., 2020). Similar absorption features are seen in HCN and HCO^+ in Galactic molecular clumps associated with inflow and outflow motions (e.g. Wyrowski et al., 2016) and in the Galactic proto-SSC Sgr B2 we met in Chapter 7 is attributed to self-absorption (Mills & Battersby, 2017). However, since none of these features drop below zero in continuum-subtracted spectra, they could arise from layers of surrounding gas at higher or lower velocity. Absorption seems most likely, not least because the other molecules seem almost certainly from comparable central locations, but also because CO, HCN, and HCO^+ have lower critical densities and likely higher opacities relative to the other species. The problem, where absorption is ubiquitous (which is the case for the majority of sources), is that intensities and derived column densities will be difficult-to-define underestimates.

Of note is the application of familiar language to relatively novel situations. For example, in extra-galactic research HCN and HCO^+ are taken to be 'dense gas tracers', with transitions typically considered to trace densities close to their critical densities, which means >10^5 cm^{-3}. In Galactic studies, however, probably arising from subthermal excitation, HCN emission can be seen in gas densities of just a few 100 cm^{-3}, well below critical density. In as-yet ill-defined remote sources such as NGC 253, we always need to look for additional evidence to support our deductions.

In comparing the line ratios of HCN, HCO^+, HNC, and CS with chemical models of photodissociation regions (PDRs) and X-ray dominated regions (XDRs), data suggests that PDR is favoured as an energy source over XDR in the SSCs of NGC 253 (Krieger et al., 2020). This does not imply that UV heating is the dominant energy input, even though expected to be significant for embedded proto-SSCs. Large numbers of young O/B stars could potentially emit enough short-wavelength radiation to create a mild XDR, but mechanical heating from low-velocity shocks associated with accretion or outflows can dominate the energy budget even while UV only contributes. Early observations of NGC 253 suggested exactly that through comparison with Galactic centre observations (Martin et al., 2006). However, with the spatial resolution of early ALMA studies at ~50 pc, an inner nuclear disk (where the SSCs are located) was distinguished in which PDR chemistry and shocks were identifiable. Outside this inner disk it seems evident that an outer nuclear disk surrounds in which shocks do dominate, as has been seen in other extra-galactic bar regions (Meier et al., 2015).

18.7 Summary

Fourteen super star cluster sites are identified in the central bar of NGC 253, and the factors influencing their star formation efficiency are considered. Molecular emission clearly shows red-shifted emission and blue-shifted absorption line profiles (P-Cygni) characteristic of outflows. While separation of large-scale motions in CO along the line of sight is difficult, CS and HCN undoubtedly trace localised and spatially resolved emission within the clusters, rather than the foreground gas. The SSCs are shown to fragment into primary clusters surrounded by smaller satellite clusters. From column density and projected sizes, outflow mass and other physical parameters are estimated, and outflow drivers and feedback mechanisms are discussed. The application of chemical clocks, particularly involving sulphur species, is explored and wider molecular comparisons made.

Appendices

A Galactic and Extragalactic Molecules

As of the end of 2021, over 260 different molecular species had been identified in the interstellar and circumstellar Galactic environment. Just under 70 have been observed in Extragalactic sources. One comprehensive database of these species can be found at the University of Cologne database on https://cdms.astro.uni-koeln.de/classic/molecules.

B Observational and Modelling Databases

The following representative databases offer some of the essential information required in the conversion of molecular line observations into meaningful knowledge about physical conditions associated with star formation.

Firstly, comes the correct identification of line emission for a specific molecular species, for which spectroscopic databases such as the following provide the necessary information:

- Cologne Database for Molecular Spectroscopy (CDMS): www.astro.uni-koeln.de/cdms/

Muller, H. S. P., Schloder, F., Stutzki, J., Winnewisser, G. The Cologne Database for Molecular Spectroscopy, CDMS: A useful tool for astronomers and spectroscopists. *Journal of Molecular Structure*, 742, 215–227 (2005). DOI: https://doi.org/10.1016/j.molstruc.2005.01.027.

Secondly, a collisional excitation database such as the following offers the collisional coefficients necessary to convert emission line data into molecular abundances:

- Leiden Atomic & Molecular Database (LAMDA): http://home.strw.leidenuniv.nl/~moldata

Schoier, F. L., van der Tak, F. F. S., van Dishoeck, E. F., Black, J. H.: An atomic and molecular database for analysis of submillimetre line observations. *A&A*, 432, 369–379 (2005). DOI: https://doi.org/10.1051/0004–6361:20041729.

Thirdly, chemical reaction databases such as the following offer reactions and rate data necessary for modelling interstellar and circumstellar clouds:

- The Kinetic Database for Astrochemistry (KIDA): http://kida.astrochem-tools.org/

Wakelam, V., Herbst, E., Loison, J. C. et al. A KInetic Database for Astrochemistry (KIDA). *ApJS*, 199, 21 (2012). DOI: https://doi.org/10.1088/0067–0049/199/1/21.

- The UMIST Database for Astrochemistry (UDFA): http://udfa.net

Woodall, J., Agundez, M., Markwick-Kemper, A. J., Millar, T. J. The UMIST database for astrochemistry 2006. *A&A*, 466, 1197–1204 (2007). DOI: https://doi.org/10.1051/0004–6361:20064981.

C Wavelength, Frequency, and Energy

Band	Wavelength	Frequency (Hz)	Energy
Radio/millimetre	0.1–1,000 cm	3×10^{7}–3×10^{11}	
Sub-millimetre	300–1,000 mm	3×10^{11}–1×10^{12}	
Far-infrared	10–300 mm	1×10^{12}–3×10^{13}	
Near-infrared	0.8–10 mm	3×10^{13}–4×10^{14}	
Optical	4,000–8,000 Å	4×10^{14}–7×10^{14}	
Ultraviolet	3,000–4,000 Å	7×10^{14}–1×10^{15}	
Far UV	912–3,000 Å	1×10^{15}–3×10^{15}	4–13.6 eV
Extreme UV	100–912 Å	3×10^{15}–2×10^{16}	13.6–100 eV
Soft X-ray	-	2×10^{16}–4×10^{17}	0.1–2 keV
X-ray	-	4×10^{17}–2×10^{20}	2–1,000 keV
Gamma ray	-	2×10^{20}–2×10^{23}	1–1,000 MeV

D Generic Reaction Processes and Rates

	Reaction	Rate	Unit	Note
Photodissociation	$AB + h\nu \rightarrow A + B$	10^{-9}	s^{-1}	(1)
Neutral–neutral	$A + B \rightarrow C + D$	3×10^{-11}	$cm^3\ s^{-1}$	(2)
Ion–molecule	$A^+ + B \rightarrow C^+ + D$	2×10^{-9}	$cm^3\ s^{-1}$	(3)
Charge transfer	$A^+ + B \rightarrow A + B^+$	10^{-9}	$cm^3\ s^{-1}$	(3)

	Reaction	Rate	Unit	Note
Radiative association	A + B → AB + $h\nu$			(4)
Dissociative recombination	$A^+ + e^- \rightarrow C + D$	3×10^{-7}	$cm^3\ s^{-1}$	-
Collisional association	A + B + M → AB + M	10^{-32}	$cm^6\ s^{-1}$	(3)
Associative detachment	$A^- + B \rightarrow AB + e$	10^{-9}	$cm^3\ s^{-1}$	(3)
Cosmic ray ionisation	$A + CR \rightarrow A^+ + e^- + CR$	4×10^{-16}	s^{-1}	-

Notes: (1) Rate in an unshielded radiation field. (2) Assumed radical–radical reaction, exothermic and without activation energy. (3) Rate in exothermic direction. (4) Rates highly reactant specific.

E Interstellar and Circumstellar Physical Parameters

	Density n (cm^{-3})	Temperature T (K)	Scale d (pc)
Diffuse cloud	~100	~80	3
Molecular cloud	~10	~1,000	10
Prestellar core	10^5–10^6		10–30
Protoplanetary disk (outer)	~10^4	10	
Protoplanetary disk (inner)	10^{10}	500	
Photodissociation region	10^3–10^5	100–1000	0.01–0.1
Hot core	10^6–10^{10}	150–1000	10^{-2}
Protostellar outflow	10^6–10^7	200–3000	~10^{-2}
H_2O masers	~10^9	~500	~10^{-5}
HII regions	>100	10,000	
Evolved star envelope	~10^{10}	2,000–3,500	

F Elemental Abundances

Average approximate elemental abundances in the solar neighbourhood relative to the total number of hydrogen nuclei.

H	1
He	9×10^{-2}
O	5×10^{-4}
C	3×10^{-4}
N	7×10^{-5}
Si	3×10^{-5}
Mg	4×10^{-5}
Fe	3×10^{-5}
S	1×10^{-5}
Na, Ca	2×10^{-6}

G Interstellar Ice Composition

Representative interstellar ice abundances relative to H_2O observed through absorption against background sources (Tielens, 2021).

	Quiescent Cloud	Low-mass protostar	High-mass protostar	Comets
H_2O	100	100	100	100
CO (total)	25	5	13	23
CO (in H_2O ice)	3	10	6	-
CO (pure ice)	22	4	3	-
CO_2 (total)	21	19	13	6
CO_2 (in H_2O ice)	18	(10)	9	-
CO_2 (CO mix)	3	(10)	2	-
CH_4	<3	<1.4	1.5	0.6
CH_3OH	10	30	18	2.4
H_2CO	-	-	6	1.1
HCOOH	<1	<1	7	0.09
OCS	<0.2	<0.1	0.2	0.4
NH_3	<8	<11	15	0.7
OCN^-	<0.5	<0.2	3.5	(0.1)

Research Journal Abbreviations

A&A	Astronomy and Astrophysics
A&AS	Astronomy and Astrophysics Supplement
AAS	American Astronomical Society
ACS	American Chemical Society
AJ	Astronomical Journal
ApJ	Astrophysical Journal
ApJS	Astrophysical Journal Supplement
Ap&SS	Astrophysics and Space Science
ARA&A	Annual Review of Astronomy and Astrophysics
ARNPS	Annual Review of Nuclear and Particle Science
ASIC	Advanced Study Institute Conference
ASP	Astronomical Society of the Pacific
ASPC	Astronomical Society of the Pacific Conference Series
ASSP/L	Astrophysics & Space Science Proceedings/Library
Chem Rev	Chemical Reviews
Chem Phys	Chemical Physics
FaDi	Faraday Discussions
IAUS	International Astronomy Union Symposium
JRASC	Journal of the Royal Astronomical Society of Canada
LNP	Lecture Notes in Physics (Springer-Verlag)
MNRAS	Monthly Notices of the Royal Astronomical Society
Msl.conf	'Molecules in Space and Laboratory' ISBN 9782901057581
NatAs	Nature Astronomy
Natur	Nature
PASJ	Publications of the Astronomical Society of Japan

PASP	Publications of the Astronomical Society of the Pacific
Prpl.conf	Protostars & Planets, University of Arizona Press
RMAA	Revista Mexicana de Astronomia y Astrofisica
RSPTA	Philosophical Transactions of the Royal Society A
RvMP	Reviews of Modern Physics

References

Acharyya, K., & Herbst, E., 2015, *ApJ*, 812, 142
Agundez, M., & Wakelam, V., 2013, *Chem Rev*, 113, 8710
Ahmadi, A., Kuiper, R., & Beuther, H., 2019, *A&A*, 632, 50
Akeson, R., & Carlstrom, J., 1996, *ApJ*, 470, 528
ALMA Partnership et al., 2015, *ApJ*, 808, L4, 1
Altenhoff, W., Downes, D., Goad, L. et al., 1970, *A&AS*, 1, 319
Altwegg, K., Balsiger, H., Berthelier, J. et al., 2017, *RSPTA*, 375, 2097, 253
Anderson, A., Meier, D., Ott, J. et al., 2014, *ApJ*, 793, 37
Anderson, L., Sormani, M., Ginsburg, A. et al., 2020, *ApJ*, 901, 51
Ando, R., Nakanishi, K., Kohno, K. et al., 2017, *ApJ*, 849, 81
André, P., Konyves, V., Arzoumanian, D., & Palmeirim, P., 2014, *ASSP*, 36, 225
André, P., Men'shchikov, A., Bontemps, S. et al., 2010, *A&A*, 518, L102
André, P., Ward-Thompson, D., & Barsony, M., 1993, *ApJ*, 406, 122
Andrews S. , Terrell, M., Tripathi, A. et al., 2018, *ApJ*, 865, 157
Andron, I., Gratier, P., Majumdar, L. et al., 2018, *MNRAS*, 481, 5651
Armijos-Abendano, J. , Martin-Pintado, J. , Lopez, E. , 2020a, *ApJ*, 895, 57
Armijos-Abendano, J., Banda-Barragan, W. E., Martin-Pintado, J. et al., 2020b, *MNRAS* 499, 4918
Arthur, S., Kurtz, S., Franco, J., & Albarran, M., 2004, *ApJ*, 608, 282
Avalos, M., Lizano, S., Rodriguez, L. et al., 2006, *ApJ*, 641, 406
Avalos, M., Lizano, S., & Rodriguez, L., 2009, *ApJ*, 690, 1084
Bacmann, A., Caux, E., Hily-Bant, P. et al., 2010, *A&A*, 521, 42
Baganoff, F., Bautz, M., Ricker, G. et al., 2001, *AAS*, 199, 8503
Ballesteros-Paredes, J., & Hartmann, L., 2007, *RMAA*, 43, 123
Bally, J., Cunningham, N., Moeckel, N. et al., 2011, *ApJ*, 727, 113
Bally, J., Walawender, J., Johnstone, D. et al., 2008, *Handbook of Star Forming Regions, Volume I: The Northern Sky ASP Monograph Publications*, Vol. 4. Edited by Bo Reipurth, p. 308
Balucani, N., Ceccarelli, C., & Taquet, V., 2015, *MNRAS*, 339, 16
Barnard, E. E., *Catalogue of 349 dark objects in the sky*, 1927
Barnes, A., Longmore, S., Dale, J. et al., 2020, *MNRAS*, 498, 4906
Beckwith, S., & Sargent, A., 1993, *ApJ*, 402, 280

Belloche, A., Garrod, R., Muller, H. et al., 2009, *A&A*, 499, 215
Belloche, A., Garrod, R., Muller, H. et al., 2019, *A&A*, 628, 10
Belloche, A., Maury, A., Maret, S. et al., 2020, *A&A*, 635, 198
Belloche, A., Menten, K., Comito, C. et al., 2008, *A&A*, 482, 179
Belloche, A., Muller, H., Garrod, R., & Menten, K., 2016, *A&A*, 587, 91
Belloche, A., Muller, H., Menten, K. et al., 2013, *A&A*, 559, 47
Belloche, A., Muller, H., Menten, K. et al., 2014, *A&A*, 561, 1
Beltran, M. T., Cesaroni, R., Codella, C. et al., 2006, *Nature*, *443*, 427
Beltran, M., Cesaroni, R., Moscadelli, L., & Codella, C., 2007, *A&A*, 471, 13
Beltran, M., Cesaroni, R., Neri, R. et al., 2005, *A&A*, 435, 901
Beltran, M., Cesaroni, R., Zapata, L. et al., 2011, *A&A*, 532, 91
Beltran, M., Olmi, L., Cesaroni, R. et al., 2013, *A&A*, 552, 123
Bendo, G., Beswick, R., D'Cruze, M. et al., 2015, *MNRAS*, 450, 80
Benjamin, R., Churchwell, E., Babler, B., Bania, T. et al., 2003, *PASP*, 115, 953
Bergin, E., Langer, W., & Goldsmith, P., 1995, *ApJ*, 441, 22
Bergner, J., Guzman, V., Oberg, K. et al., 2018, *ApJ*, 857, 69
Bergner, J., Martin-Domenech, R., Oberg, K. et al., 2019, *ACS Earth Space Chem*, 3, 8, 1564
Bergner, J., Oberg, K., Garrod, R., & Graninger, D., 2017, *ApJ*, 841, 120
Bertoldi, F., & McKee, C., 1992, *ApJ*, 388, 495
Beuther, H., Mottram, J., Ahmadi, A. et al., 2018, *A&A*, 617, 100
Beuther, H., & Nissen, H., 2008, *ApJ*, 679, 121
Beuther, H., Schilke, P., Sridharan, T. et al., 2002, *ApJ*, 566, 945
Beuther, H., Tackenberg, J., Linz, H. et al., 2012, *ApJ*, 747, 43
Beuther, H., Zhang, Q., Bergin, E. et al., 2007, *A&A*, 468, 1045
Beuther, H., Zhang, Q., Greenhill, L. et al., 2004, *ApJ*, 616, 31
Bisbas, T., Papadopoulos, P., & Viti, S., 2015, *ApJ*, 803, 37
Bjerkeli, P., Jorgensen, J., & Brinch, C., 2016a, *A&A*, 587, 145
Bjerkeli, P., Jorgensen, J., Bergin, E. et al., 2016b, *A&A*, 595, 39
Bjerkeli, P., Liseau, R., Nisini, B. et al., 2013, *A&A*, 552, 8
Black, J., & van Dishoeck, E., 1987, *ApJ*, 322, 412
Blake, G., Sutton, E., Masson, C., & Phillips, T., 1987, *ApJ*, 315, 621
Bockelee-Morvan, D., & Biver, N., 2017, *RSPTA*, 375, 2097
Boehle, A., Ghez, A., Schodel, R. et al., 2016, *ApJ*, 830, 17
Boger, G., & Sternberg, A., 2005, *ApJ*, 632, 302
Bolatto, A., Leroy, A., Rosolowsky, E. et al., 2008, *ApJ*, 686, 948
Bonfand, M., Belloche, A., Garrod, R. et al., 2019, *Proceedings of the Annual meeting of the French Society of Astronomy and Astrophysics*, 69
Bonfand, M., Belloche, A., Menten, K. et al., 2017, *A&A*, 604, 60
Bonnell, I., & Bate, M., 2002, *MNRAS*, 336, 659
Bonnell, I., Bate, M., Clarke, C., & Pringle, J., 2001, *MNRAS*, 323, 785
Bonnell, I., Bate, M., & Zinnecker, H., 1998, *MNRAS*, 298, 93
Bontemps, S., Andrfe, P., Terebey, S., & Cabrit, S., 1996, *A&A*, 311, 858
Boogert, A., Pontoppidan, K., Knez, C. et al., 2008, *ApJ*, 678, 985
Boogert, A., Schutte, W., Helmich, F. et al., 1997, *A&A*, 317, 929
Bottinelli, S., Ceccarelli, C., Lefloch, B. et al., 2004b, *ApJ*, 615, 354
Bottinelli, S., Ceccarelli, C., Neri, R. et al., 2004a, *ApJ*, 617, 69

Bottinelli, S., Wakelam, V., Caux, E. et al., 2014, *MNRAS*, 441, 1964
Breen, S., & Ellingsen, S., 2011, *MNRAS*, 413, 2339
Breen, S., Ellingsen, S., Caswell, J. et al., 2016, *MNRAS*, 459, 4066
Breen, S., Fuller, G., Caswell, J. et al., 2015, *MNRAS*, 450, 4109
Brown, P., Charnley, S., & Millar, T., 1988, *MNRAS*, 231, 409
Burge, C., van Loo, S., Falle, S., & Hartquist, T., 2016, *A&A*, 596, 28
Burkhardt, A., Shingledecker, C., Le Gal, R. et al., 2019, *ApJ*, 881, 32
Calcutt, H., Jorgensen, H., Muller, L. et al., 2018, *A&A*, 616, 90
Calcutt, H., Viti, S., Codella, C. et al., 2014, *MNRAS*, 443, 3157
Calcutt, H., Willis, E., Jorgensen, J. et al., 2019, *A&A*, 631, 137
Campbell, M., Harvey, P., Lester, D., & Clark, D., 2004, *ApJ*, 600, 254
Carey, S., Clark, F., Egan, M. et al., 1998, *ApJ*, 508, 721
Carey, S., Noriega-Crespo, A., Mizuno, D. et al., 2009, *PASP*, 121, 76
Carlhoff, P., Nguyen Luong, Q., Schilke, P. et al., 2013, *A&A*, 560, 24
Carlson, L., Sewilo, M., Meixner, M. et al., 2012, *A&A*, 542, 66
Carlstrom, J., & Vogel, S., 1989, *ApJ*, 337, 408
Carral, P., & Welch, W., 1992, *ApJ*, 385, 244
Caselli, P., Hasegawa, T., & Herbst, E., 1993, *ApJ*, 408, 548
Caselli, P., Walmsley, C., Terzieva, R., & Herbst, E., 1998, *ApJ*, 499, 234
Caswell, J., Green, J., & Phillips, C., 2013, *MNRAS*, 431, 1180
Cazaux S., Caselli, P., & Spaans, M., 2011, *ApJ*, 741, 34
Cazaux, S., Tielens, A., Ceccarelli, C. et al., 2003, *ApJ*, 593, 51
Ceccarelli, C., Bacmann, A., Boogert, A. et al., 2010, *A&A*, 521, 22
Ceccarelli, C., Caselli, P., Fontani, F. et al., 2017, *ApJ*, 850, 176
Ceccarelli, C., Castets, A., Caux, E. et al., 2000, *A&A*, 355, 1129
Cernicharo, J., Marcelino, N., Roueff, E. et al., 2012, *ApJ*, 759, 43
Cesaroni, R., Beltran, M., Moscadelli, L. et al., 2019, *A&A*, 624, 100
Cesaroni, R., Churchwell, E., Hofner, P. et al., 1994, *A&A*, 288, 903
Cesaroni, R., Hofner, P., Walmsley, C., & Churchwell, E., 1998, *A&A*, 331, 709
Cesaroni, R., Pestalozzi, M., Beltran, M. et al., 2015, *A&A*, 579, 71
Cesaroni, R., Walmsley, C., & Churchwell, E., 1992, *A&A*, 256, 618
Chandler, C., Brogan, C., Shirley, Y. et al., 2005, *ApJ*, 632, 371
Chang, Q., & Herbst, E., 2016, *ApJ*, 819, 145
Charnley, S., 1997, *ApJ*, 481, 396
Charnley, S., & Rodgers, S., 2005, *IAUS*, 231, 237
Charnley, S., Tielens, A., & Millar, T. et al., 1992, *ApJ*, 399, 71
Chen, X., Shen, Z-Q. , Li, X-Q. et al., 2017, *MNRAS*, 466, 4364
Chen, K., Woosley, S., & Whalen, D., 2020, *ApJ*, 893, 99
Chuang, K., Fedoseev, G., Ioppolo, S. et al., 2016, *MNRAS*, 455, 1702
Churazov, E., Khabibullin, I., Sunyaev, R., & Ponti, G., 2017, *MNRAS*, 465, 45
Churchwell, E., 1997, *ApJ*, 479, 59
Churchwell, E., 2002, *ASPC*, 267, 3
Churchwell, E., Povich, M., Allen, D. et al., 2006, *ApJ*, 649, 759
Churchwell, E., Walmsley, C., & Wood, D., 1992, *A&A*, 253, 541
Clark, P., Glover, S. et al., 2013, *ApJ*, 768, 34
Claussen, M., Wilking, B., Benson, P. et al., 1996, *ApJS*, 106, 111
Codella, C., Beltran, M., Cesaroni, R. et al., 2013, *A&A*, 550, 81

Codella, C., Ceccarelli, C., Caselli, P. et al., 2017, *A&A*, 605, 3
Codella, C., Testi, L., & Cesaroni, R., 1997, *A&A*, 325, 282
Colombo, D., Hughes, A., Schinnerer, E. et al., 2014, *ApJ*, 784, 3
Contreras, Y., Schuller, F., Urquhart, J. et al., 2013, *A&A*, 549, 45
Cordiner, M. , & Charnley, S. , 2021, *AAS*, 53, 7
Cortes, P., Hull, C., Girart, J. et al., 2019, *ApJ*, 884, 48
Coutens, A., Jorgensen, J., van der Wiel, M. et al., 2016, *A&A*, 590, 6
Coutens, A., Ligterink, N., Loison, J. et al., 2019, *A&A*, 623, 13
Coutens, A., Willis, E., Garrod, R. et al., 2018, *A&A*, 612, 107
Crockett, N., Bergin, E., Neill, J. et al., 2014, *ApJ*, 787, 112
Crovisier, J., 2004, *ASSL*, 305, 179
Cruz-Diaz, G., Martin-Domenech, R., Munoz Caro, G., & Chen, Y.-J., 2016, *A&A*, 592, 68
Cruz-Diaz, G., Munoz Caro, G., Chen, Y.-J., & Yih, T.-S., 2014, *A&A*, 562, 119
Csengeri, T., Leurini, S., Wyrowski, F. et al., 2016, *A&A*, 586, 149
Csengeri, T., Urquhart, J., Schuller, F. et al., 2014, *A&A*, 565, 75
Cuadrado, S., Goicoechea, J., Cernicharo, J. et al., 2017, *A&A*, 603, 124
Cuadrado, S., Goiceochea, J., Pilleri, P. et al., 2015, *A&A*, 575, 82
Cuadrado, S., Goicoechea, J., Roncero, O. et al., 2016, *Natur*, 537, 207
Cuadrado, S., Salas, P., Goicoechea, J. et al., 2019, *A&A*, 625, 3
Dame, T., Hartmann, D., & Thaddeus, P., 2001, *ApJ*, 547, 792
Danger, G., Borget, F., Chomat, M. et al., 2011, *A&A*, 535, 47
DeFrees, D., McLean, A., & Herbst, E., 1985, *ApJ*, 293, 236
De Pree, C., Gaume, R., Goss, W., & Claussen, M., 1996, *ApJ*, 464, 788
De Simone, M., Codella, C., Ceccarelli, C. et al., 2020, *A&A*, 640, 75
Di Francesco, J., Myers, P., Wilner, D. et al., 2001, *ApJ*, 562, 770
Dobashi, K., Uehara, H., Kandori, R. et al., 2005, *PASJ*, 57, 1
Donovan-Meyer, J., Koda.,J., Momose, R. et al., 2012, ApJ, 744, 42
Donovan-Meyer, J., Koda, J., Momose, R. et al., 2013, *ApJ*, 772, 107
Draine, B., 2011, *ApJ*, 732, 100
Drozdovskaya, M., Walsh, C., Visser, R. et al., 2015, *MNRAS*, 451, 3836
Druard, C., Braine, J., Schuster, K. et al., 2014, *A&A*, 567, 118
Druard, C., & Wakelam, V., 2012, *MNRAS*, 426, 354
Dunham, M., Stutz, A. et al., 2014, *Protostars & Planets VI*, University of Arizona Press, 195
Egusa, F., Hirota, A., Baba, J., & Muraoka, K., 2018, *ApJ*, 854, 90
Ellingsen, S., Breen, S., Caswell, J. et al., 2010, *MNRAS*, 404, 779
Elmegreen, B., 2000, *ApJ*, 539, 342
Engargiola, G., Plambeck, R., Rosolowsky, E., & Blitz, L., 2003, *ApJS*, 149, 343
Enoch, M., Evans, N., Sargent, A., & Glenn, J., 2009, *ApJ*, 692, 973
Esplugues, G., Casaux, S., Caselli, P. et al., 2019, *MNRAS*, 486, 1853
Etxaluze, M., Goicoechea, J., Cernicharo, J. et al., 2013, *A&A*, 556, 137
Evans, N., Dunham, M., Jorgensen, J. et al., 2009, *ApJS*, 181, 321
Faesi, C., Lada, C., & Forbrich, J., 2018, *ApJ*, 857, 19
Faure, A., Halvick, P., Stoecklin, T. et al., 2017, *MNRAS*, 469, 612
Favre, C., Fedele, D., Semenov, D. E. et al., 2018, *ApJ*, 862L, 2
Fayolle, E., Oberg, K., Jorgensen, J. et al., 2017, *NatAs*, 1, 703

Fedoseev, G., Chuang, K., van Dishoeck, E. et al., 2016, *MNRAS*, 460, 4297
Fedoseev, G., Ioppolo, S., Zhao, D. et al., 2015, *MNRAS*, 446, 439
Feng, S., Beuther, H., Zhang, Q. et al., 2016, *ApJ*, 828, 100
Flower, D., & Pineau des Forets, G., 1995, *MNRAS*, 275, 1049
Fontani, F. Pascucci, I. , Caselli, P. et al., 2007, *A&A*, 470, 639
Forster, J., & Caswell, J., 1987, *IAUS*, 115, 174
Fuente, A., Goicoechea, J., Pety, J. et al., 2017, *ApJ*, 851, 49
Fuente, A., Martin-Pintado, J., Cernicharo, J., & Bachiller, R., 1993, *A&A*, 276, 473
Fuente, A., Martin-Pintado, J., Neri, R., & Moriarty-Schieven, G., 1996, *A&A*, 310, 286
Fuente, A., Martin-Pintado, J., Rodriguez-Fernandez, N. et al., 2000, *A&A*, 354, 1053
Fuente, A., Navarro, D., Caselli, P. et al., 2019, *A&A*, 624, 105
Fuente, A., Rodriguez-Franco, A., Garcia-Burillo, S. et al., 2003, *A&A*, 406, 899
Fukui, Y., & Kawamura, A., 2010, *ARA&A*, 48, 547
Fukui, Y., Mizuno, N., Yamaguchi, R. et al., 1999, *PASJ*, 51, 745
Furuya, R., Cesaroni, R., Codella, C. et al., 2002, *A&A*, 390, 1
Galli, P., Bouy, H., Olivares, J. et al., 2020, *A&A*, 643, 148
Galvin-Madrid, R., Rodriguez, L., Ho, P., & Keto, E., 2021, ApJ, 674, 33
Garay, G., & Rodriguez, L., 1990, *ApJ*, 362, 191
Garrod, R., 2013, *ApJ*, 765, 60
Garrod, R., 2019, *ApJ*, 884, 69
Garrod, R., Belloche, A., Muller, H., & Menten, K., 2017, *A&A*, 601, 48
Garrod, R., & Herbst, E., 2006, *A&A*, 457, 927
Garrod, R., Wakelam, V., & Herbst, E., 2007, *A&A*, 467, 1103
Garrod, R., Widicus Weaver, S., & Herbst, E., 2007, *A&A*, 467, 1103
Garrod, R., Widicus Weaver, S., & Herbst, E., 2008, *ApJ*, 682, 283
Gasiprong, N., Cohen, R., & Hitawarakorn, B., 2002, *MNRAS*, 336, 47
Gaume, R., & Claussen, M., 1990, *ApJ*, 351, 538
Gaume, R., Claussen, M., de Pree, C. et al., 1995, *ApJ*, 449, 663
Gaume, R., Fey, A., & Claussen, M., 1994, *ApJ*, 432, 648
Geballe, T., Baas, F., Greenberg, J., & Schutte, W., 1985, *A&A*, 146, 6
Genzel, R., & Downes, D., 1977, *A&A*, 61, 117
Geppert, W., Hamberg, M., Thomas, R. et al., 2006, *FaDi*, 133, 177
Gerin, M., Goicoechea, J., Pety, J., & Hily-Blant, P., 2009, *A&A*, 494, 977
Gerin, M., de Luca, M., Black, J. et al., 2010, *A&A*, 518, L110
Gibb, A., Wyrowski, F., & Mundy, L., 2004, *ApJ*, 616, 301
Gieser, C., Beuther, H., Semenov, D. et al., 2021, *A&A*, 648, 66
Ginard, D., Gonzalez-Garcia, M., Fuente, A. et al., 2012, *A&A*, 543, 27
Ginsburg, A., Bally, J., Barnes, A. et al., *ApJ*, 853, 171
Ginsburg, A., & Kruijssen, J., 2018, *ApJ*, 864, 17
Girichidis, P., Federrath, C., Banerjee, R., & Klessen, R., 2011, *MNRAS*, 413, 2741
Glover, S., Federrath, C., Mac Low, M., & Klessen, R., 2010, *MNRAS*, 404, 2
Glover, S., & Mac Low, M., 2011, *MNRAS*, 412, 337
Goicoechea, J., Aguado, A., Cuadrado, S. et al., 2021, *A&A*, 647, 7
Goicoechea, J., Cuadrado, S., Pety, J. et al., 2017, *A&A*, 601, 9
Goicoechea, J., & Le Bourlot, J., 2007, *A&A*, 467, 1
Goicoechea, J., Pety, J., Cuadrado, S. et al., 2016, *Nature*, 537, 207
Goicoechea, J., Pety, J., Gerin, M. et al., 2009, *ASPC*, 420, 43

Goldsmith, P., Snell, R., Hasegawa, T., & Ukita, N., 1987, *ApJ*, 314, 392
Goumans, T., Uppal, M., & Brown, W., 2008, *MNRAS*, 384, 1158
Graedel, T., Langer, W., & Frerking, M., 1982, *ApJS*, 48, 321
Graninger, D., Wilkins, O., & Oberg, K., 2016, *ApJ*, 819, 140.
Gratier, P., Pety, J., Guzman, V. et al., 2013, *A&A*, 557, 101
Green, J., Caswell, J., Fuller, G. et al., 2008, *MNRAS*, 385, 948
Grenier, I., Black, J., & Strong, A., 2015, *ARA&A*, 53, 199
Guelin, M., Langer, W., & Wilson, R., 1982, *A&A*, 107, 107
Guzman, V., Goicoechea, J., Pety, J. et al., 2013, *A&A*, 560, 73
Guzman, V., Pety, J., Gratier, P. et al., 2014, *FaDi*, 168, 103
Habart, E., Abergel, A., Walmsley, C. et al., 2005, *A&A*, 437, 177
Hajigholi, M., Persson, C., Wirstrom, E. et al., 2016, *A&A*, 585, 158
Hamberg, M., Zhaunerchyk, V., Vigren, E. et al., 2010, *A&A*, 522, 90
Hanaoka, M., Kaneda, H., Suzuki, T. et al., 2019, *PASJ*, 71, 6
Hartmann, L., 2002, *ApJ*, 578, 914
Hartquist, T., Dalgarno, A., & Oppenheimer, M., 1980, *ApJ*, 236, 182
Hatchell, J., Fuller, G., & Millar, T., 2001, *A&A*, 372, 281
Hatchell, J., Thompson, M., Millar, T., & Macdonald, G., 1998, *A&AS*, 133, 29
Hattori, Y., Kaneda, H., Ishihara, D. et al., 2016, *PASJ*, 68, 37
Heaton, B., Little, L., & Bishop, I., 1989, *A&A*, 213, 148
Heaton, B., Little, L., Yamashita, T. et al., 1993, *A&A*, 278, 238
Heaton, B., Matthews, N., Little, L., & Dent, W., 1985, *MNRAS*, 217, 485
Herbst, E., & van Dishoeck, E., 2009, *ARA&A*, 47, 427
Herpin, F., Marseille, M., Wakelam, V. et al., 2009, *A&A*, 504, 853
Higuchi, A., Hasegawa, T., Saigo, K. et al., 2015, *ApJ*, 815, 106
Higuchi, A., Sakai, N., Watanabe, Y. et al., 2018, *ApJS*, 236, 52.
Hily-Blant, P., Maret, S., Bacmann, A. et al., 2010, *A&A*, 521, 52
Hindson, L., Thompson, M., Urquhart, J. et al., 2010, *MNRAS*, 408, 1438
Hoare, M., Kurtz, S., Lizano, S. et al., 2007, *Protostars and Planets V*, University of Arizona Press, 181
Hoffman, I., Goss, W., Palmer, P., & Richards, A., 2003, *ApJ*, 598, 1061
Hofner, P., & Churchwell, E., 1996, *A&AS*, 120, 283
Hogerheijde, M., 1998 PhD, Leiden University
Hogerheijde, M., Jansen, D., & van Dishoeck, E., 1995, *ApJ*, 294, 792
Hollenbach, D., Kaufman, M., Bergin, E., & Melnick, G., 2009, *ApJ*, 690, 1497
Hollenbach, D., Kaufman, M., Neufeld, D. et al., 2012, *ApJ*, 754, 105
Hollenbach, D., & Tielens, A., 1999, *RvMP*, 71, 173
Hollis, J., Jewell, P., Lovas, F. et al., 2004a, *ApJ*, 610, 21
Hollis, J., Jewell, P., Lovas, F. et al., 2004b, *ApJ*, 613, 45
Hollis, J., Lovas, F., Remijan, A. et al., 2006b, *ApJ*, 643, 25
Hollis, J., Pedelty, J., Boboltz, D. et al., 2003, *ApJ*, 596, 235
Hollis, J., Remijan, A., Jewell, P. et al., 2006a, *ApJ*, 642, 933
Hoq, S., Jackson, J., Foster, J. et al., 2013, *ApJ*, 777, 157
Horn, A., Mollendal, H., Sekiguchi, O. et al., 2004, *ApJ*, 611, 605
Hosokawa, T., Yorke, H., & Omukai, K., 210, *ApJ*, 721, 478
Hudson, R., & Loeffler, M., 2013, *ApJ*, 773, 109
Hughes, A., Harvey, J., Marble, A., & Pevtsov, A., 2013, *ApJ*, 779, 46

Huttemeister, S., Wilson, T., Mauersberger, R. et al., 1995, *A&A*, 294, 667
Imai, H., Omi, R., Kurayama, T. et al., 2011, *PASJ*, 63, 1293
Ioppolo, S., Cuppen, H., van Dishoeck, E., & Linnartz, H., 2011, *MNRAS*, 410, 1089
Jacobsen, S., Jorgensen, J., van der Wiel, M. et al., 2018, *A&A*, 612, 72
Jansen, D., Spaans, M., Hogerheijde, M., & van Dishoeck, E., 1995, *A&A*, 303, 541
Jimenez-Escobar, A., & Munoz Caro, G., 2011, *A&A*, 536, 91
Jimenez-Serra, I., Vasyunin, A., Caselli, P. et al., 2016, *ApJ*, 830, 6
Jin, M., & Garrod, R., 2020, *ApJS*, 249, 26
Johnston, K., Palmer, P., Wilson, T. et al., 1983, *ApJ*, 271, 89
Johnstone, D., Hendricks, B, Herczeg, G., & Bruderer, S., 2013, *ApJ*, 765, 133
Jones, P., Burton, M., Cunningham, M. et al., 2008, *MNRAS*, 386, 117
Jones, P., Burton, M., Tothill, N., & Cunningham, M., 2011, *MNRAS*, 411, 2293
Jorgensen, J., Favre, C., Bisschop, S. et al., 2012, *ApJ*, 757L, 4
Jorgensen, J., Hogerheijde, M., Blake, G. et al., 2004, *A&A*, 415, 1021
Jorgensen, J., Johnstone, D., Kirk, H., & Myers, P., 2007, *ApJ*, 656, 293
Jorgensen, J., van der Wiel, M., Coutens, A. et al., 2016, *A&A*, 595, 117
Jorgensen, J., Visser, R., Sakai, N. et al., 2013, *ApJ*, 779, 22
Kahane, C., Ceccarelli, C., Faure, A., & Caux, E., 2013, *ASPC*, 476, 323
Kalcheva, I., Hoare, M., Urquhart, J. et al., 2018, *A&A*, 615, 103
Kaplan K., Dinerstein, H., Oh, H. et al., 2017, *ApJ*, 838, 152
Kauffmann, J., Pillai, T., Shetty, R. et al., 2010, *ApJ*, 716, 433
Kenyon, S., Hartmann, L., Strom, K., & Strom, S., 1990, *AJ*, 99, 869
Keto, E., 2007, *ApJ*, 666, 976
Keto, E., Proctor, D., Ball, R. et al., 1992, *ApJ*, 401, 113
Keto, E., Zhang, Q., & Kurtz, S., 2008, *ApJ*, 672, 423
Kim, W., Wyrowski, F., Urquhart, J. et al., 2020, *A&A*, 644, 160
Kirsanova, M., & Wiebe, D., 2019, *MNRAS*, 486, 2525
Klaassen, P., Johnston, K., Urquhart, J. et al., 2018, *A&A*, 611, 99
Knapp, G., & Morris, M., 1976, *ApJ*, 204, 415
Kobayashi, H., Ishiguro, M., Chikada, Y. et al., 1989, *PASJ*, 41, 141
Koda, J., 2009, *AAS*, 41, 456
Kohno, M., Tachihara, K., Torii, K. et al., 2021, *PASJ*, 73, 129
Konig, C., Urquhart, J., Csengeri, T. et al., 2017, *A&A*, 599, 139
Krieger, N., Bolatto, A., Leroy, A. et al., 2020, *ApJ*, 897, 176
Kristensen, L., Klaassen, P., Mottram, J. et al., 2013, *A&A*, 549, 6
Kristensen, L., van Dishoeck, E., Bergin, E. et al., 2012, *A&A*, 542, 8
Kroupa, P. , 2001, *ASPC*, 243, 387
Krumholz, M., McKee, C., & Bland-Hawthorn, J., 2019, *ARA&A*, 57, 227
Kuan, Y.-J., Huang, H.-C., Charnley, S. et al., 2004, *ApJ*, 616, 27
Kuan, Y.-J., Mehringer, D., & Snyder, L., 1996, *ApJ*, 459, 619
Kuan, Y.-J., & Snyder, L., 1994, *ApJS*, 94, 651
Kurtz, S., 2005, *IAUS*, 227, 111
Laas, J., & Caselli, P., 2019, *A&A*, 624, 108
Lada, C., 1987, *IAUS*, 115, 1L
Ladjelate, B., Andre, P., Konyves, V. et al., 2020, *A&A*, 638, 74
Lancaster, L., Ostriker, E., Kim, J.-G., & Kim, C.-G., 2021, *ApJ*, 922, 3
Le Gal, R., Herbst, E., Dufour, G. et al., 2017, *A&A*, 605, 88

Le Petit, F., Ruaud, M., Bron, E. et al., 2016, *A&A*, 585, 105
Leroy, A., Bolatto, A., Ostriker, E. et al., 2015, *ApJ*, 801, 25
Leroy, A., Bolatto, A., Ostriker, E. et al., 2018, *ApJ*, 869, 126
Leroy, A., Schinnerer, E., Hughes, A. et al., 2021, *ApJS*, 257, 43
Leung, C., Herbst, E., & Huebner, W., 1984, *ApJS*, 56, 231
Leurini, S., Rolffs, R., Thorwirth, S. et al., 2006, *A&A*, 454, 47
Levy, R., Bolatto, A., Leroy, A. et al., 2021, *ApJ*, 912, 4
Li, J., Wang, J., Qiao, H. et al., 2020, *MNRAS*, 492, 556
Li, S., Wang, J., Zhang, Z.-Y. et al., 2017, *MNRAS*, 466, 248
Ligterink, N., Coutens, A., Kofman, V. et al., 2017, *MNRAS*, 469, 2219
Lin, Y., Liu, H. et al., 2016, *ApJ*, 828, 32
Lis, D., Gerin, M., Phillips, T., & Motte, F., 2002, *ApJ*, 569, 322
Lis, D., & Goldsmith, P., 1990, *ApJ*, 356, 195
Lis, D., Goldsmith, P., Carlstrom, J., & Scoville, N., 1993, *ApJ*, 402, 238
Liu, H., Jimenez-Serra, I., Ho, P. et al., 2012, *ApJ*, 756, 10
Liu, H.-L, Liu, T., Evans, N. et al., 2021, *MNRAS*, 505, 2801
Liu, S.-Y., & Snyder, L., 1999, *ApJ*, 523, 683
Liu, T, Evans, N., Kee-Tae, K. et al., 2020, *MNRAS*, 496, 2812
Loinard, L., Zapata, L., Rodriguez, L. et al., 2013, *MNRAS*, 430L, 10
Loison, J-C. , Wakelam, V., & Hickson, K., 2014, *MNRAS*, 443, 398
Lombardi, M., Bouy, H., Alves, J., & Lada, C., 2014, *A&A*, 566, 45
Lombardi, M., Lada, C., & Alves, J., 2008, *A&A*, 480, 785
Loomis, R., Cleeves, L., Oberg, K. et al., 2018, *ApJ*, 859, 131
Loomis, R., Zaleski, D., Steber, A. et al., 2013, *AAS*, 22135210
Looney, L., Mundy, L., & Welch, W., 2000, *ApJ*, 529, 477
Lumsden, S., Hoare, M., Urquhart, J. et al., 2013, *ApJS*, 208, 11
Lykke, J., Coutens, A., Jorgensen, J. et al., 2017, *A&A*, 597, 53
Macdonald, G., Gibb, A., Habing, R., & Millar, T., 1996, *A&AS*, 119, 333
Macdonald, G., & Habing, R., 1994, *LNP*, 459, 291
Macdonald, G., Habing, R., & Millar, T., 1995, *Ap&SS*, 224, 177
MacKay, D., 1996, *MNRAS*, 278, 62
MacKay, D., 1999, *MNRAS*, 304, 61
Maity, S., Kaiser, R., & Jones, B., 2014, *ApJ*, 789, 36
Manigand, S., Coutens, A., Loison, J. et al., 2021, *A&A*, 645, 53
Manigand, S., Jorgensen, J., Calcutt, H. et al., 2020, *A&A*, 635, 48
Maret, S., & Bergin, E., 2007, *ApJ*, 664, 956
Marrone, D., Baganoff, F., Morris, M. et al., 2008, *ApJ*, 682, 373
Martin, S., Aladro, R., Martin-Pintado, J., Mauersberger, R., 2010, *A&A*, 522, 62
Martin, S., Mauersberger, R., Martin-Pintado, J. et al., 2006, *ApJS*, 164, 450
Matthews, N., Little, L., Macdonald, G. et al., 1987, *A&A*, 184, 284
Maxia, C., Testi, L., Cesaroni, R., & Walmsley, C., 2001, *A&A*, 371, 287
McKee, C., & Tan, J., 2002, *Natur*, 416, 59
McKee, C., & Tan, J., 2003, *ApJ*, 585, 850
McLaughlin, D., & Pudritz, R., 1996, *ApJ*, 469, 194
McLaughlin, D., & Pudritz, R., 1997, *ApJ*, 476, 750
Medina, S., Urquhart, J., Dzib, S. et al., 2019, *A&A*, 627, 175
Meier, D., Walter, F., Bolatto, A. et al., 2015, *ApJ*, 801, 63

Melton, E., 2020, *AJ*, 159, 200
Menten, K., Reid, M., Forbrich, J., & Brunthaler, A., 2007, *A&A*, 474, 515
Meyer, D., Vorobyov, E., Kuiper, R., & Kley, W., 2017, *MNRAS*, 464, 90
Migenes V., Johnston, K., Pauls, T., & Wilson, T., 1989, *ApJ*, 347, 294
Millar, T., & Herbst, E., 1990, *A&A*, 231, 466
Millar, T., Herbst, E., & Charnley, S., 1991, *ApJ*, 369, 147
Millar, T., Macdonald, G., & Gibb, A., 1997, *A&A*, 325, 1163
Millar, T., Macdonald, G., & Habing, R., 1995, *MNRAS*, 273, 25
Mills, E., & Battersby, C., 2017, *ApJ*, 835, 76
Minier, V., & Booth, R., 2002, *A&A*, 387, 179
Minier, V., Booth, R., & Conway, J., 2000, *A&A*, 362, 1093
Minissale, M., Dulieu, F., Cazaux, S., & Hocuk, S., 2016, *A&A*, 585, 24
Momose, R., Kopda, J., Kennicutt, R. et al., 2013, *ApJ*, 772, 13
Mookerjea, B., Casper, E., Mundy, L., & Looney, L., 2007, *ApJ*, 659, 447
Mookerjea, B., Ossenkopf, V., Ricken, O. et al., 2012, *A&A*, 542, 17
Moscadelli, L., Cesaroni, R., Beltran, M., & Rivilla, V., 2021, *A&A*, 650, 142
Moscadelli, L., Goddi, C., Cesaroni, R., & Beltran, M., 2007, *IAUS*, 242, 135
Moscadelli, L., Rivilla, V., Cesaroni, R. et al., 2018, *A&A*, 616, 66
Moscadelli, L., Sanna, A., Goddi, C. et al., 2019, *A&A*, 631, 74
Motte, F., Bontemps, S., Schilke, P. et al., 2007, *A&A*, 476, 1243
Motte, F., Bontemps, S., & Louvet, F., 2018b, *ARA&A*, 56, 41
Motte, F., Schilke, P., & Lis, D., 2003, *ApJ*, 582, 277
Mowat, C., Hatchell, J., Rumble, D. et al., 2017, *MNRAS*, 467, 812
Mumma, M., & Charnley, S., 2011, *ARA&A*, 49, 471
Mundy, L., Wootten, A., Wilking, B. et al., 1992, *ApJ*, 385, 306
Murga, M., Kirsanova, M., Vasyunin, A. et al., 2020, *MNRAS*, 497, 2327
Murillo, N., van Dishoeck, E., van der Wiel, M. et al., 2018, *A&A*, 617, 120
Murphy, T., Cohen, M., Ekers, R. et al., 2010, *MNRAS*, 405, 1560
Murray, N., & Chang, P., 2012, *ApJ*, 746, 75
Nagy, Z., Van der Tak, F., Ossenkopf, V. et al., 2013, *A&A*, 550, 96
Narayanan, G., Walker, C., & Buckley, H., 1998, *ApJ*, 496, 292
Natta, A., Walmsley, C., & Tielens, A., 1994, *ApJ*, 428, 209
Navarro-Almaida, D., Le Gal, R., Fuente, A. et al., 2020, *A&A*, 637, 39
Nayak, P., Subramaniam, A., Choudhury, S., & Sagar, R., 2018, *A&A*, *616*, 187
Neill, J., Muckle, M., Zaleski, D. et al., 2012, *ApJ*, 755, 153
Neufeld, D., Godard, B., Gerin, M. et al., 2015, *A&A*, 577, 49
Neufeld, D., Goicoechea, J., Sonnentrucker, P. et al., 2010, *A&A*, 521, L10
Nguyen-Luong, Q., Motte, F., Carlhoff, P. et al., 2013, *ApJ*, 775, 88
Nguyen-Luong, Q., Anderson, L., Motte, F. et al., 2017, *ApJ*, 844, 25
Nobukawa, M., Ryu, S., Tsuru, T., & Koyama, K., 2011, *ApJ*, 739, 52
Oba, Y., Tomaru, T., Lamberts, T. et al., 2018, *NatAs*, 2, 2280
Oberg, K., 2016, *Chem Rev*, arXiv:1609.03112
Oberg, K., Boogert, A., Pontoppidan, K. et al., 2011a, *ApJ*, 740, 109
Oberg, K., Boogert, A., Pontoppidan, K. et al., 2011b, *IAUS*, 280, 65
Oberg, K., Garrod, R., van Dishoeck, E., & Linnartz, H., 2009a, *A&A*, 504, 891
Oberg, K., Guzman, V., Furuya, K. et al., 2015, *Nature*, 520, 198
Oberg, K., Linnartz, H., Visser, R., & van Dishoeck, E., 2009b, *ApJ*, 693, 1209

Oberg, K., van Dishoeck, E., Linnartz, H. et al., 2010, *ApJ*, 718, 832
Occhiogrosso, A., Vasyunin, A., Herbst, E. et al., 2014, *A&A*, 564, 123
Offner, S., Bisbas, T., Viti, S., & Bell, T., 2013, *ApJ*, 770, 490
Okoda, Y., Oya, Y., Sakai, N. et al., 2018, *ApJ.*, 864, 250
Okoda, Y., Oya, Y., Sakai, N. et al., 2020, *ApJ*, 900, 400
Oliveira, J., van Loon, J., Stanimirovic, S., & Zijlstra, A., 2006, *MNRAS*, 372, 1509
Olmi, L., Cesaroni, R., Hofner, P. et al., 2003, *A&A*, 407, 225
Olmi, L., Cesaroni, R., & Walmsley, C., 1993, *A&A*, 276, 489
Ossenkopf, V., & Henning, Th. , 1994, *A&A*, 291, 943
Osterbrock, D., & Ferland, G., 2006, *Astrophysics of Gaseous Nebulae & Active Galactic Nuclei*, University Science Books, Sausalito, CA
Oya, Y., Sakai, N., Lopez-Sepulcre, A. et al., 2016, *ApJ*, 824, 880
Oya, Y., Sakai, N., Sakai, T. et al., 2014, *ApJ*, 795, 152
Oya, Y., Sakai, N., Watanabe, Y. et al., 2017, *ApJ*, 837, 174
Paglione, T., Yam, O., Tosaki, T., & Jackson, J., 2004, *ApJ*, 611, 835
Palau, A., Ballesteros-Paredes, J., Vazquez-Semadeni, E. et al., 2015, *MNRAS*, 453, 3785
Palumbo, M., Geballe, T., & Tielens, A., 1997, *ApJ*, 479, 839
Palumbo, M., Tielens, A., & Tokunaga, A., 1995, *ApJ*, 449, 674
Papadopoulos, P., Bisbas, T., & Zhang, Z-Y. , 2018, *MNRAS*, 478, 1716
Papadopoulos, P., Thi, W.-F., & Viti, S., 2004, *MNRAS*, 351, 147
Parise, B., Du., F., Liu, F.-C. et al., 2012, *A&A*, 542, 5
Paron, S., Areal, M., & Ortega, M., 2018, *A&A*, 617, 14
Pattle, K., Ward-Thompson, D., Kirk, J. et al., 2015, *MNRAS*, 450, 1094
Peng, T., Despois, D., Brouillet, N. et al., 2013, *A&A*, 554, 78
Perez-Beaupuits, J., Wiesemeyer, H., Ossenkopf, V. et al., 2012, *A&A*, 542, 13
Peters, T. , MacLow, M.-M. , Banerjee, R. et al., 2010a, *ApJ*, 719, 831
Peters, T. , Klessen, R. , MacLow, M.-M. et al., 2010b, *ApJ*, 725, 134
Pety, J., Schinnerer, E., Leroy, A. et al., 2013, *ApJ*, 779, 43
Pety, J., Teyssier, D., Fosse, D. et al., 2005, *A&A*, 435, 885
Pillai, T., Kauffmann, J., Wyrowski, F. et al., 2011, *A&A*, 530, 118
Pilleri, P., Fuente, A., Cernicharo, J. et al., 2012, *A&A*, 544, 110
Pilleri, P., Fuente, A., Gerin, M. et al., 2014, *A&A*, 561, 69
Pilleri, P., Trevino-Morales, S., Fuente, A. et al., 2013, *A&A*, 554, 87
Plambeck, R., Wright, M., Friedel, D. et al., 2009, *ApJ*, 704, 25
Plessis, S., Carrasco, N. et al., 2012, *Icarus*, 219, 254
Podio, L., Lefloch, B., Ceccarelli, C. et al., 2014, *A&A*, 565, 64
Pols, S., Schworer, A., Schilke, P. et al., 2018, *A&A*, 614, 123
Pratap, P., Megeath, S., & Bergin, E., 1999, *ApJ*, 517, 799
Pratap, P., Menten, K., & Snyder, L., 1994, *ApJ*, 430, 129
Purser, S., Lumsden, S., Hoare, M. et al., 2016, *MNRAS*, 460, 1039
Qin, S., Schilke, P., Rolffs, R. et al., 2011, *A&A*, 530, 9
Raga, A., Canto, J., Binette, L., & Calvet, N., 1990, *ApJ*, 364, 601
Rea, N., Esposito, P., Pons, J. et al., 2013, *ApJ*, 775, 34
Reid, M., & Ho, P., 1985, *ApJ*, 288, 17
Reipurth, B., Rodriguez, L., Anglada, G., & Bally, J., 2002, *AJ*, 124, 1045
Remijan, A., Hollis, J., Lovas, F. et al., 2008, *ApJ*, 675, 85

Requena-Torres, M., Martin-Pintado, J., Rodriguez-Franco, A. et al., 2006, *A&A*, 455, 971
Rico-Villas, F., Martin-Pintado, J., Gonzalez-Alfonso, E. et al., 2020, *MNRAS*, 491, 4573
Ridge, N., Di Francesco, J. et al., 2006, *AJ*, 131, 2921
Rivera-Ortiz, L., Ho, P., Fernandez-Lopez, M. et al., 2020, *ApJ*, 902, 47
Rivera-Soto, R., Galvan-Madrid, R., Ginsburg, A., & Kurtz, S., 2020, *ApJ*, 899, 94
Riviere-Marichalar, P., Fuente, A., Goicoechea, J. et al., 2019, *A&A*, 628, 16
Rizzo, J., Fuente, A., & Garcia-Burillo, S., 2005, *ApJ*, 634, 1133
Roberge, W., Jones, D., Lepp, S., & Dalgarno, A., 1991, *ApJS*, 77, 287
Roberts, H., & Millar, T. J., 2006, Phil. Trans. R. Soc. A, 364, 3063
Robitaille, J., Scaife, A., Carretti, E. et al., 2018, *A&A*, 617, 101
Rodgers, S., & Charnley, S., 2001, ApJ, 553, 613
Rodgers, S., & Charnley, S., 2003, *ApJ*, 585, 355
Rodriguez, L., Zapata, L., & Ho, P., 2007, *ApJ*, 654, 143
Rodriguez-Franco, A., Martin-Pintado, J., & Fuente, A., 1998, *A&A*, 329, 1097
Rollig, T., Houck, J., Sloan, G. et al., 2007, *A&A*, 467, 187
Roshi, D. A., Churchwell, E., & Anderson, L. D., 2017, *ApJ*, 838, 144
Rosolowsky, E., Keto, E., Matsushita, S., & Willner, S., 2007, *ApJ*, 661, 830
Ruaud, M., Loison, J., Hickson, K. et al., 2015, *MNRAS*, 447, 4004
Sadavoy, S., Di Francesco, J., Andre, P. et al., 2014, *ApJ*, 787, 18
Sadavoy, S., Di Francesco, J., Bontemps, S. et al., 2010, *ApJ*, 70, 1247
Sakai, N., Oya, Y., Sakai, T. et al., 2014, *ApJ*, L38, 791
Sakai, N., Sakai, T., & Yamamoto, S., 2008, *ApJ*, 673, 71
Sakai, N., Sakai, T., Hirota, T. et al., 2009, *ApJ.*, 697, 769
Sakai, N., & Yamamoto, S., 2013, *Chem Rev*,113, 12, 8981
Sakamoto, K., Ho, P., Iono, D. et al., 2006, *ApJ*, 636, 685
Sakamoto, K., Mao, R., Matsushita, S. et al., 2011, *ApJ*, 735, 19
Sanchez-Monge, A., ALMA 2019: https://doi.org/10.5281/zenodo.3585445
Sanchez-Monge, A., Kurtz, S., Palau, A. et al., 2013, *ApJ*, 766, 114
Sanchez-Monge, A., Schilke, P., Schmiedeke, A. et al., 2017, *A&A*, 604, 6
Sanhueza, P., Contreras, Y., Wu, B. et al., 2019, *ApJ*, 886, 102
Schenewerk, M., Snyder, L., Hillis, J. et al., 1988, *ApJ*, 328, 785
Schilke, P., Groesbeck, T., Blake, G. et al., 1997, *ApJS*, 108, 301
Schilke, P., Pineau des Forets, G., Walmsley, C., & Martin-Pintado, J., 2001, *A&A*, 372, 291
Schinnerer, E., Meidt, S., Pety, J. et al., 2013, *ApJ*, 779, 42
Schmiedeke, A., Schilke, P., Moller, Th. et al., 2016, *A&A*, 588, 143
Schoier, F., Jorgensen, J., van Dishoeck, E., & Blake, G., 2002, *A&A*, 390, 1001
Schruba, A., Leroy, A., Kruijssen, J. et al., 2017, *ApJ*, 835, 278
Schuller, F., Csengeri, T., Urquart, J. et al., 2017, *A&A*, 601, 124
Schuller, F., Menten, K., Contreras, Y. et al., 2009, *A&A*, 504, 415
Seale, J., Looney, L., Chen, C. et al., 2011, *ApJ*, 727, 36
Seale, J., Looney, L., Wong, T. et al., 2012, *ApJ*, 751, 42
Semenov, D., Hersant, F., Wakelam, V. et al., 2010, *A&A*, 522, 42
Semenov, D., & Wiebe, D., 2011, *ApJS*, 196, 25
Sewilo, M., Churchwell, E, Kurtz, S. et al., 2004, *AAS*, 20513909

Sewilo, M., Churchwell, E., Kurtz, S. et al., 2008, *ApJ*, 681, 350
Sewilo, M., Churchwell, E., & Kurtz, S., 2011, *ApJS*, 194, 44
Sewilo, M., Indebetouw, R., Charnley, S. et al., 2018, *ApJ*, 853, 19
Shimonishi, T., Dartois, E., Onaka, T., & Boulanger, F., 2016a, *A&A*, 585, 107
Shimonishi, T., Das, A., Sakai, N. et al., 2020, *ApJ*, 891, 164
Shimonishi, T., Onaka, T., Kawamura, A., & Aikawa, Y., 2016b, *ApJ*, 827, 72
Shingledecker, C., Tennis, J., Le Gal, R., & Herbst, E., 2018, *ApJ*, 861, 20
Shu, F., 1977, *ApJ*, 214, 488
Shu, F., Adams, F., & Lizano, S., 1987, *ARA&A*, 25, 23
Simons, M., Lamberts, T., & Cuppen, H., 2020, *A&A*, 634, 52
Skinner, M., & Ostriker, E., 2015, *ApJ*, 809, 187
Smith, R., Longmore, S., & Bonnell, I., 2009, MNRAS, 400, 1775
Sollins, P., Hunter, T., Battat, J. et al., 2005, *ApJ*, 616, 35
Spitzer, L., 1978, *JRASC*, 72,349
Sternberg, A., & Dalgarno, A., 1995, *APJS*, 99, 565
Strong, A., Moskalenko, I., & Ptuskin, V., 2007, *ARNPS*, 57, 285
Sun, J., Leroy, A., Schruba, A. et al., 2018, *ApJ*, 860, 172
Surcis, G., Vlemmings, W., van Langevelde, H. et al., 2019, *A&A*, 623, 130
Tan, J., Beltran, M., Caselli, P. et al., 2014, *prpl.conf.*, 149
Tan, J., Kong, S., Zhang, Y. et al., 2016, *ApJ*, 821, L3
Tanaka, K., Tan, J., & Zhang, Y., 2016, *ApJ*, 818, 52
Tang, X., Henkel, C., Chen, C. et al., 2017, *A&A*, 600, 16
Taquet, V., Ceccarelli, C., & Kahane, C., 2012, *ApJ*, 748, 3
Taquet, V., Lopez-Sepulcre, A., Ceccarelli, C. et al., 2015, *ApJ*, 804, 81.
Taquet, V., Wirstrom, E., & Charnley, S., 2016, *ApJ*, 821, 46
Tauber, J., Lis, D., Keene, J. et al., 1995, *A&A*, 297, 567
Tercero, B., Cuadrado, S., Lopez, A. et al., 2018 *A&A*, 620, 6
Terrier, R., Ponti, G., Belanger, G. et al., 2010, *ApJ*, 719, 143
Teyssier, D., Fosse, D., Gerin, M. et al., 2004, *A&A*, 417, 135
Thiel, V., Belloche, A., Menten, K. et al., 2019, *A&A*, 623, 68
Thompson, M., Macdonald, G., & Millar, T., 1999, *A&A*, 342, 809
Tielens, A., 1993, *Dust and Chemistry in Astronomy*, Institute of Physics Publishing, Philadelphia
Tielens, A., 2006, *The Physics and Chemistry of the Interstellar Medium*, Cambridge University Press
Tielens, A., 2021, *Molecular Astrophysics*, Cambridge University Press
Tielens, A., & Hagen, W., 1982, *A&A*, 114, 245
Tielens, A., & Hollenbach, D., 1985, *ApJ*, 291, 747
Tiwari M., Menten, K., Wyrowski, F. et al., 2019, *A&A*, *626*, 28
Tobin, J., Looney, L., Li, Z. et al., 2016, *ApJ*, 818, 73
Tothill, N., Lohr, A., Parshley, S. et al., 2009, *ApJS*, 185, 98
Townsley, L., Broos, P., Garmire, G. et al., 2014, *ApJS*, 213, 1
Trevino-Morales, S., Fuente, A., Sanchez-Monge, A. et al., 2016, *A&A*, 593, 12
Trevino-Morales, S., Pilleri, P., Fuente, A. et al., 2014, *A&A*, 569, 19
Turner, B., 1970, *PASP*, 82, 996
Turner, B., & Ho, P., 1985, *ApJ*, 299, 77
Tychoniec, L., Hull, C., Tobin, J., & van Dishoeck, E., 2018, *ApJS*, 238, 19

Ulvestad, J., & Antonucci, R., 1997, *ApJ*, 488, 621
Urquhart, J., Csengeri, T., Wyrowski, F. et al., 2014a, *A&A*, 568, 41
Urquhart, J., Figura, C., Wyrowski, F. et al., 2019, *MNRA*S, 484, 4444
Urquhart, J., Konig, C., Giannetti, A. et al., 2018, *MNRAS*, 473, 1059
Urquhart, J., Moore, T., Csengeri, T. et al., 2014b, *MNRAS*, 443 1555
Urquhart, J., Moore, T., Schuller, F. et al., 2013a, MNRAS 431, 1752
Urquhart, J., Thompson, M., Moore, T. et al., 2013b, *MNRAS*, 435, 400
Useli-Bacchitta, F., Bonnamy, A., Mulas, G. et al., 2010, *CP*, 371, 16
Van Buren, D., Mac Low, M., Wood, D., & Churchwell, E., 1990, *ApJ*, 353, 570
Van der Tak, F., Boonman, A., Braakman, R., & van Dishoeck, E., 2003, *A&A*, 412, 133
Van der Wiel, M., Jacobsen, S. et al., 2019, *A&A*, 626, 93
Van Dishoeck, E., & Black, J., 1987, *ASIC*, 210, 241
Van Dishoeck, E., & Black, J., 1988, *ApJ*, 334, 771
Van Dishoeck, E., Blake, G., Draine, B., & Lunine, J., 1993, *Protostars & Planets III*, University of Arizona Press, 163
Van Kempen, T., van Dishoeck, E., Hogerheijde, M., & Gusten, R., 2009, *A&A*, 508, 259
Vastel, C., Ceccarelli, C., Lefloch, B., & Bachiller, R., 2014, *ApJ*, 795, 2
Vasyunin, A., & Herbst, E., 2013, *ApJ*, 769, 34
Vidal, M., Dickinson, C., Davies, R., & Leahy, J., 2015, *MNRAS*, 452, 656
Vidal, T., Loison, J-C, Jaziri, A. et al., 2017, *MNRAS*, 469, 435
Vigren, E., Zhaunerchyk, V., Hamberg, M. et al., 2012, *ApJ*, 757, 34
Visser, R., van Dishoeck, E., & Doty, S., 2011, *IAUS*, 280, 138.
Visser, R., van Dishoeck, E., Doty, S., & Doullemond, C., 2009, *A&A*, 495, 881
Viti, S., Collings, M. , Dever, J. et al., 2004, *MNRAS*, 354, 1141
Vogel, S., Genzel, R., & Palmer, P., 1987, *ApJ*, 316, 243
Wakelam, V., Ceccarelli, C., Castets, A. et al., 2005, *A&A*, 437, 149
Walawender, J., Bally, J., Francesco, J. et al., 2008, *Handbook of Star Forming Regions I*, ASP Monograph 4, ed. B. Reipurth, 346
Walsh, A., Bourke, T., & Myers, P., 2006, *ApJ*, 637, 860
Walsh, A., Burton, M., Hyland, A., & Robinson, G., 1998, MNRAS, 301, 640
Walsh, C., & Ilee, J. D., 2020, *IAU*, 15, 463
Wang, K.-S., Kuan, Y-J, Liu, S.-Y., & Charnley, S., 2010, *ApJ*, 713, 1192
Wang, Y., Du, F., Semenov, D. et al., 2021, *A&A*, 648, 72
Watson, A., & Hanson, M., 1997, *ApJ*, 490, 165
Watt, G., 1983, *MNRAS*, 205, 321
Watt, S., & Mundy, L., 1999, *ApJS*, 125, 143
Weilbacher, P., Monreal-Ibero, A., Kollatschny, W. et al., 2015, *A&A*, 582, 114
Willacy, K., & Williams, D., 1993, MNRAS, 260, 635
Willis, E., Garrod, R., Belloche, A. et al., 2020, *A&A*, 636, 29
Wilson, C., & Scoville, N., 1990, *ApJ*, 363, 435
Wilson, T., & Rood, R., 1994, *ARA&A*, 32, 191
Wong, K., Blanton, M., Burles, S. et al., 2011, *ApJ*, 728, 119
Wood, D., & Churchwell, E., 1989, *ApJS*, 69, 831
Wootten, A., 1989, *ApJ*, 337, 858.
Wright, M., Plambeck, R., & Wilner, D., 1996, *ApJ*, 469, 216
Wu, Y., Liu, X., Chen, X. et al., 2019, *MNRAS*, 488, 495.

Wyrowski, F., Gusten, R., Menten, K. et al., 2012, *A&A*, 542, 15
Wyrowski, F., Gusten, R., Menten, K. et al., 2016, *A&A*, 585, 149
Wyrowski, F., Schilke, P., Hofner, P., & Walmsley, C., 1997, *ApJ*, 487, 171
Yang, A., Thompson, M., Tian, W. et al., 2019, *MNRAS*, 482, 2681
Yang, A., Thompson, M., Urquhart, J., & Tian, W., 2018, *ApJS*, 235, 3
Yang, A., Urquhart, J., Thompson, M. et al., 2021, *A&A*, 645, 110
Zaleski, D., Seifert, N., Steber, A. et al., 2013, *ApJ*, 765, 10
Zapata, L., Ho, P., Fernandez-Lopez, M. et al., 2020, *ApJ*, 902, 47
Zapata, L., Schmid-Burgk, J., Ho, P. et al., 2009, *ApJ*, 704, 45
Zhang, C., Yuan, J., Li, G. et al., 2017, *A&A*, 598, 76
Zhang, Y., Higuchi, A., Sakai, N. et al., 2018, *ApJ*, 864, 76
Zhou, S., 1995, *ApJ*, 442, 685

Chemical Index

Subject Index